Lecture Notes of the Institute for Computer Sciences, Social Informatics and Telecommunications Engineering 537

The LNICST series publishes ICST's conferences, symposia and workshops.
LNICST reports state-of-the-art results in areas related to the scope of the Institute.
The type of material published includes

- Proceedings (published in time for the respective event)
- Other edited monographs (such as project reports or invited volumes)

LNICST topics span the following areas:

- General Computer Science
- E-Economy
- E-Medicine
- Knowledge Management
- Multimedia
- Operations, Management and Policy
- Social Informatics
- Systems

Prakash Pareek · Nishu Gupta · M. J. C. S. Reis
Editors

Cognitive Computing and Cyber Physical Systems

4th EAI International Conference, IC4S 2023
Bhimavaram, Andhra Pradesh, India, August 4–6, 2023
Proceedings, Part II

Springer

Editors
Prakash Pareek ⓘ
Vishnu Institute of Technology
Bhimavaram, Andhra Pradesh, India

Nishu Gupta ⓘ
Norwegian University of Science
and Technology
Gjøvik, Norway

M. J. C. S. Reis ⓘ
University of Trás-os-Montes e Alto Douro
Vila Real, Portugal

ISSN 1867-8211 ISSN 1867-822X (electronic)
Lecture Notes of the Institute for Computer Sciences, Social Informatics
and Telecommunications Engineering
ISBN 978-3-031-48890-0 ISBN 978-3-031-48891-7 (eBook)
https://doi.org/10.1007/978-3-031-48891-7

This Springer imprint is published by the registered company Springer Nature Switzerland AG
The registered company address is: Gewerbestrasse 11, 6330 Cham, Switzerland

Paper in this product is recyclable.

Preface

We are delighted to introduce the proceedings of the fourth edition of the European Alliance for Innovation (EAI) International Conference on Cognitive Computing and Cyber Physical Systems (EAI IC4S 2023), hosted by Vishnu Institute of Technology, Bhimavaram, Andhra Pradesh, India during 4–6 August 2023 in hybrid mode. This conference has together researchers, developers and practitioners around the world who are leveraging and developing intelligent computing systems and cyber physical systems so that communication becomes smarter, quicker, less expensive and accessible in bundles. The theme of EAI IC4S 2023 was "Cognitive computing approaches with machine learning techniques and advanced communications".

The technical program of EAI IC4S 2023 consisted of 70 full papers, which were presented in online mode, i.e. on a web platform, and also in offline mode. The above papers were presented by the registered authors in fourteen technical sessions under five different tracks. The conference tracks were: Track 1 – Machine Learning and its Applications; Track 2 – Cyber Security and Signal Processing; Track 3 – Image Processing; Track 4 – Smart Power Systems; and Track 5 – Smart City Eco-system and Communications. Apart from the high-quality technical paper presentations, the technical program also featured two keynote speeches and one plenary talk. The two keynote speakers were Nishu Gupta from the Department of Electronic Systems, Faculty of Information Technology and Electrical Engineering, Norwegian University of Science and Technology (NTNU) in Gjøvik, Norway and Manuel J. Cabral S. Reis from UTAD University Engineering Department, Portugal. The plenary talk was presented by Anil Gupta, Associate Director, C-DAC, Pune, India on the role of cyber physical systems in assistive technology.

Coordination with the steering chair, Imrich Chlamtac, was essential for the success of the conference. We sincerely appreciate his constant support and guidance. Manuel J. Cabral S. Reis successfully served as general chair for this edition and helped the conference to proceed smoothly. It was also a great pleasure to work with such an excellent organizing committee team for their hard work in organizing and supporting the conference. In particular, the Organizing Committee chaired by Nishu Gupta, and the Technical Program Committee, chaired by Prakash Pareek coordinated and completed the peer-review process of technical papers and made a high-quality technical program. We are also grateful to Conference Managers Sara Csicsayova and Kristina Havlickova for their support and to all the authors who submitted their papers to the EAI IC4S 2023 conference.

We sincerely appreciate the management and administration of Vishnu Institute of Technology, Bhimavaram (VITB), Andhra Pradesh, India and especially Chairman of Sri Vishnu Educational Society (SVES), Shri K. V. Vishnu Raju, Vice Chairman of SVES, Sri Ravichandran Rajagopal, Secretary of SVES, Shri K. Aditya Vissam and D. Suryanarayana, Director and Principal of VITB, K. Srinivas, Vice Principal, VITB and

N. Padmavathy, Dean R&D, VITB for giving their support to us as the main host Institute of EAI IC4S 2023.

We strongly believe that EAI IC4S 2023 conference provided a good forum for all researchers, developers and practitioners to discuss scientific and technological aspects relevant to cognitive computing and cyber physical systems. We also expect that future EAI IC4S conferences will be as successful and stimulating as indicated by the contributions presented in this volume.

December 2023

Prakash Pareek
Nishu Gupta
M. J. C. S. Reis

Conference Organization

Steering Committee

Imrich Chlamtac — University of Trento, Italy

Organizing Committee

General Chair

Manuel J. Cabral S. Reis — UTAD University, Portugal

General Co-chairs

Mohammad Derawi — Norwegian University of Science and Technology, Norway

Ahmad Hoirul Basori — King Abdulaziz University, Saudi Arabia

Organizing Chair

Nishu Gupta — Norwegian University of Science and Technology, Norway

Local Chair

D. Suryanarayana — Vishnu Institute of Technology, India

Organizing Secretaries

Pravesh Kumar — Vaagdevi College of Engineering, India

Prakash Pareek — Vishnu Institute of Technology, India

Technical Program Committee Chair

Prakash Pareek — Vishnu Institute of Technology, India

Technical Program Committee Co-chair

Zhihan Lv   Uppsala University, Sweden

Web Chair

Ariel Soales Teres   Federal Institute of Maranhão, Brazil

Sponsorship and Exhibit Chair

Nishit Malviya   IIIT Ranchi, India

Publications Chair

Luy Nguyen   Vietnam National University Ho Chi Minh
                                   City-University of Technology, Vietnam

Publicity and Social Media Chair

Sumita Mishra   Amity University, India

Publicity and Social Media Co-chair

S. Mahaboob Hussain   Vishnu Institute of Technology, India

Technical Program Committee

Thanos Kakarountas	University of Thessaly, Greece
Anil Gupta	C-DAC, India
D. Suryanarayana	Vishnu Institute of Technology, India
D. J. Nagendra Kumar	Vishnu Institute of Technology, India
William Hurst	Wageningen University and Research, The Netherlands
Felix Härer	University of Fribourg, Switzerland
Ahmad Hoirul Basori	King Abdulaziz University, Kingdom of Saudi Arabia
Mukesh Sharma	NTU, Singapore
Anuj Abraham	A*STAR, Singapore
Sumathi Lakshmiranganatha	Los Alamos National Laboratory, USA
Jitendra Kumar Mishra	IIIT Ranchi, India

Lakhindar Murmu	IIIT Naya Raipur, India
Nishit Malviya	IIIT Ranchi, India
Daniele Riboni	University of Cagliari, Italy
Ashok Kumar	NIT Srinagar, India
Dipen Bepari	NIT Raipur, India
Shubhankar Majumdar	NIT Meghalaya, India
N. Padmavathy	Vishnu Institute of Technology, India
R. V. D. Ramarao	Vishnu Institute of Technology, India
Amit Kumar Dubey	Technology Innovation Institute, UAE
Dharmendra Kumar	Madan Mohan Malaviya University of Technology, India
R. Srinivasa Raju	Vishnu Institute of Technology, India
Venkata Naga Rani Bandaru	Vishnu Institute of Technology, India
G. K. Mohan Devarakonda	Vishnu Institute of Technology, India
Atul Kumar	IIITDM Jabalpur, India
Somen Bhattacharjee	IIIT Dharwad, India
Sumit Saha	NIT Rourkela, India
K. Srinivas	Vishnu Institute of Technology, India
Amit Bage	NIT Hamirpur, India
Sridevi Bonthu	Vishnu Institute of Technology, India
P. Sita Rama Murty	Vishnu Institute of Technology, India
Idamakanti Kasireddy	Vishnu Institute of Technology, India
Pragaspathy S.	Vishnu Institute of Technology, India
Abhinav Kumar	NIT Allahabad, India
Sadhana Kumari	BMS College of Engineering, India
Ashish Singh	Kalinga Institute of Industrial Technology, India
Vegesna S. M. Srinivasavarma	SSN Institutions, India
B. V. V. Satyanarayana	Vishnu Institute of Technology, India
G. Prasanna Kumar	Vishnu Institute of Technology, India
V. S. N. Narasimha Raju	Vishnu Institute of Technology, India
Sunil Saumya	IIIT Dharwad, India
Lokendra Singh	KL University, Vijaywada, India
A. Prabhakara Rao	KL University, Hyderabad, India
Ravi Ranjan	Government Engineering College, Vaishali, India
S. Sugumaran	Vishnu Institute of Technology, India
Akash Kumar Pradhan	MVGR College of Engineering, India
Abhishek Pahuja	KL University, Vijaywada, India
Ajay Kumar Kushwaha	Bharati Vidyapeeth (Deemed to be University), India
D. M. Dhane	Bharati Vidyapeeth (Deemed to be University), India
Naveen Kumar Maurya	Vishnu Institute of Technology, India

N. P. Nethravathi	Reva University, India
Sasidhar Babu S.	Reva University, India
Gopal Krishna	Presidency University, India
Mayur Shukla	Oriental College of Technology, India
Bhanu Pratap Singh	LNCT, India
Priyanka Bharti	Reva University, India
Laxmi B. Rannavare	Reva University, India
Vikram Palodiya	Sreenidhi Institute of Science and Technology, India
Madhumita Mishra	Reva University, India
D. R. Kumar Raja	Reva University, India
Ashwin Kumar U. M.	Reva University, India
Syed Jahangir Badashah	Sreenidhi Institute of Science and Technology, India
Vipul Agarwal	KL University, Vijaywada, India
Veeramni S.	Amrita University, India
Gireesh Gaurav Soni	SGSITS, India
Lalit Purohit	SGSITS, India
S. Mahaboob Hussain	Vishnu Institute of Technology, India
Megha Kuliha	SGSITS, India
R. C. Gurjar	SGSITS, India
Sumit Kumar Jindal	Vellore Institute of Technology, India
P. Ramani	SRM Institute of Science and Technology, India
T. J. Nagalakshmi	Saveetha University, India
Rajkishor Kumar	Vellore Institute of Technology, India
Avinash Chandra	Vellore Institute of Technology, India
Naveen Mishra	Vellore Institute of Technology, India
Abhishek Tripathi	Kalasalingam Academy of Research and Education, India
Tripta	Government Engineering College, Vaishali, India

Contents – Part II

Smart City Eco-System and Communications

Contents – Part I

Cyber Security and Signal Processing

Image Processing

Smart Power Systems

Performance Evaluation of Off Grid Hybrid Distribution System with ANFIS Controller

Naga Venkata Karthik Gandham and Pragaspathy Subramani(✉)

Electrical and Electronics Engineering, Vishnu Institute of Technology, Bhimavaram, Andhra Pradesh 534202, India
pathyeee@yahoo.co.in

Abstract. Off-grid hybrid distribution systems are innovative and environmentally friendly ways to address the energy needs of remote or isolated communities or facilities by combining a number of power generation and energy storage sources. These systems are made to function independently from the primary electrical grid, which makes them perfect for areas with little or no access to centralized electrical infrastructure. In this work, designed an effective microgrid system with integration of distributed power sources such as wind turbine and Photovoltaic system (PV). Moreover, adaptive neuro fuzzy inference system (ANFIS) is developed in this study to track the maximum power from PV panels under various uncertain conditions. Further, due to intermittent nature of the wind speed and irradiance levels, battery system is designed in this work for stable and reliable operation of microgrid system. Moreover, the battery performance and ANFIS-based maximum power point tracking (MPPT) are investigated. As per simulation findings, ANFIS based MPPT approach provides an effective outcome in terms of settling time, efficiency and accuracy in suggested microgrid systems.

Keywords: Microgrid · ANFIS · PMSG · MPPT · Battery Bank

1 Introduction

Smart city is the state of art for the modern day living and perhaps it influences the unbounded technology to roll over the culture and certainly enhances the functional outcomes via vital information and communication. Internet of Things (IoT) on the other hand, sustains the aforementioned concept of smart city and invites wider adoption in the sectors like transportation, shadowing, disaster monitoring, natural calamities, electrifications, public and private analytics etc. Due to a worldwide shortage of traditional energy sources, the DG sources incorporation has expanded dramatically in recent years. The usage of these resources is affected by a number of problems, such as their high upfront cost and the inability to manage environmental conditions like variations in wind and solar irradiation. Power should be shifted from one source to another using an intelligent energy control system that monitors the demand, the environment and the energy storage devices.

© ICST Institute for Computer Sciences, Social Informatics and Telecommunications Engineering 2024
Published by Springer Nature Switzerland AG 2024. All Rights Reserved
P. Pareek et al. (Eds.): IC4S 2023, LNICST 537, pp. 3–18, 2024.
https://doi.org/10.1007/978-3-031-48891-7_1

Altering the parameters of a hybrid renewable energy system (HRES), which draws power from many renewable resources, is best accomplished through the usage of a Microgrid. Controller settings, power management techniques, and system dimensions that maximize efficiency for HRES are the primary topic of a number of studies in the literature [1, 2]. In [2], hybrid Microgrid architecture was built out of photovoltaic (PV) and wind turbine (WT) generators. A mathematical steady-state model of the interconnected wind and solar Microgrid system was developed, and two different control mechanisms were implemented. This hybrid Microgrid model was simulated with the MATLAB/SIMULINK software. The best method for controlling the load was used. Many case studies including the relocation of sources and loads were carried out on the test system. The findings demonstrate that the system may be maintained in a steady-state condition by applying the proposed control methods while the network is changed from one operating condition to the next. In [3], the EMS is presented in the form of a Markov decision process (MDP). If you're dealing with an unpredictable situation, reinforcement learning is the way to go how effective data-driven method has also been enhanced by the creation of a comprehensive reward function that reduces the investigation of infeasible activities.

The findings demonstrate that the system may be maintained in a steady-state condition by applying the proposed control methods while the network is changed from one operating condition to the next. In [3], a Markov decision process model of the EMS is presented. Additional publications in the literature [4–6] typically concentrate on performance evaluation, converter design, and integrating renewable and conventional resources. Another alternative presented by the authors of [7] is the Smart Hybrid Energy System (SHES), which is a micro generation system meant to replace the existing diesel generator. To maximize the system's effectiveness, care must be taken with each component to guarantee a constant flow of power to the load. Mixed integer linear programming (MGEM) is used to operate a Microgrid, which is a type of hybrid energy system (HES) that combines sustainable sources like wind and solar with conventional power sources like photovoltaic cells, fuel cells, micro turbines, diesel, and energy storage. The HES's inability to effectively avoid pollution is exacerbated by its use of inefficient and environmentally damaging energy sources like diesel. Intelligent and smart controllers have an impact on HRES performance, although this impact is minimally defined or explained in the study. The topic of this research can be broken down further by the type of connected hybrid microgrid. Droop and power flow regulation are the primary areas of attention for the AC/DC microgrid's power management. For AC/DC grids can apply droop control [8–10]. If the latter is chosen, the interconnection between the two sub-grids is planned largely to deal with the case where a sub-grid fails due to overstress, while keeping adjustments to individual sources to a minimum [9]. By developing a sophisticated controller for a targeted hybrid DG, this study intends to close this gap. This work's key contribution may be summed up as follows: the study presented in this publication is the first to examine the feasibility of deploying a hybrid wind and photovoltaic system in off-grid settings.

This off-microgrid device uses a smart controller to programme in the desired characteristics [11]. When the amount of PV power is greater than the amount of power need to operate the load, the surplus required for alteration of batteries. Here energy produced by the PV is less, the stored energy is supports the requirements. The smart controller also prevents damage to the battery banks from rated, under/over-charging states. The controller must react appropriately in all cases. Whenever the SOC < 81%, the wind turbine will automatically switch down before the PV panels. Before the SOC exceeds the controller-defined "safe margin," in this case 75%, the DG has disconnected. Additional commands to off loads and inverter are sent out the SOC < 21%. After the battery is charged to 75% of its capacity, the inverter's power is cut. The hybrid energy system is mathematically described in portion 2 of the present work. The development of the proposed hybrid controller technique is discussed in portion 3. Portion 4 presents and discusses the results. Portion 5 lays out the major takeaways and directions for future research.

2 Suggested Microgrid System

The hybrid DG and the power electronic components that connect the both sides of AC/DC are the three main components of the system under investigation. Multiple controllers are required for high-powered electrical machinery. The Microgrid system configuration and diagram are depicted in Fig. 1. In the next portion, we'll go to the specifics of how each component is laid out. Figure 2 represents the analogous circuit of a PV cell.

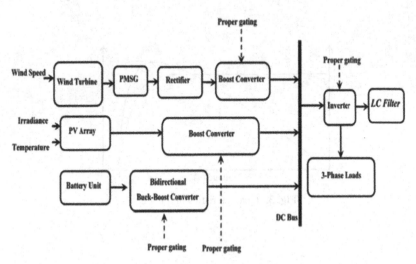

Fig. 1. Setup for the testing environment

From PV Curves

$$I_{pv} = I_{ph} - I_s e^{\left(\frac{q(V_{pv}+R_s I_{pv})}{nkT}-1\right)} - \frac{V_{pv} + R_s I_{pv}}{R_{sh}} \qquad (1)$$

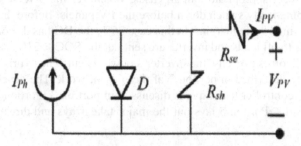

Fig. 2. Analogous circuit of a PV Cell

Where n represents the P-N junction's ideality factor, T represents the temperature in Kelvin. The combined PV system consists of a PV array with three 3×6 strings, capable of producing 3.91 kW at full irradiation of 1001 W/m^2. Figure 3 shows that the PV panel's open circuit voltage is very close to its maximum power output. This panel is rated at 215 W under standard test conditions. The open circuit voltage is 36.6, and the maximum power voltage is 29. Maximum point current is 7.35 A, while the short circuit current is 7.84 A. You can learn more about this panel and its features by consulting [12, 13].

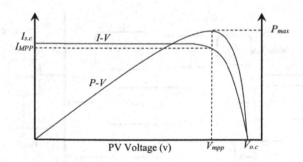

Fig. 3. Characteristics of a PV cell

2.1 WES with PMSG

All wind power systems require wind turbines. They act as the driving force for the electric generators that are mounted on their shafts. Wind turbines that rotate on different axes. The operating speed also categorizes devices into two groups: constant and variable. Figure 4 portrays the extracted mechanical power from the turbines. The P_{wind} is given by

$$P = 0.5\, p_a A\, C_p v^3 \qquad (2)$$

The aerodynamic power coefficient of a wind turbine is affected by the tip-speed ratio and the blade pitch angle, both of which are indicators of performance. Depending on the features of the turbine, a general equation for C_p is provided in (2) [14, 15].

$$C_\rho(\lambda,\ \beta) = C_1\left(\frac{C_3}{\lambda_i} - C_3\beta - C_4\right)e^{\frac{-C_5}{\lambda_i}} + C_6\lambda \qquad (3)$$

where C1 = 0.5176, C2 = 116, C3 = 0.4, C4 = 5, C5 = 21, and C6 = 0.0068, and

$$\lambda = \frac{\omega R}{V_w}$$
$$\frac{1}{\lambda_i} = \frac{1}{\lambda + 0.08\beta} - \frac{0.035}{\beta^3 + 1} \qquad (4)$$

Turbine radius (R) is measured in meters; wind velocity (V_W) is measured in m/s; and angular velocity (ω) is measured in rad/s.

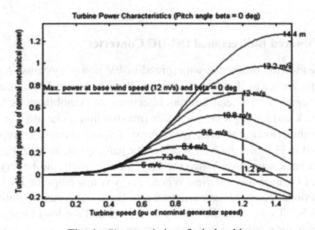

Fig. 4. Characteristics of wind turbine

2.2   Permanent Magnet Synchronous Generators

Various pole numbers are used in the construction of PMSG in WECS. To conserve energy, it operates at the same low speed as the wind turbine. Therefore, PMSG can be linked directly to the wind turbines rotor shaft. Direct-drive operation is an example of this sort of operation because it does not require the installation of a gearbox [16]. The system offers a benefit over DFIG-based systems, which call for a gearbox, in terms of both installation and maintenance costs due to the elimination of gears. The PMSG is flexible in its ability to accommodate varying rotor speeds. The frequency and amplitude of the voltage delivered to the PMSG's stator terminals vary with the Vw. The power converters being used are rated at full capacity, allowing for the most efficient conversion of wind energy possible throughout a broad range of wind speeds in the absence of any extra rotor control. Full-rated power converters not only help in conforming to a variety of grid requirements, but also eliminate the need for extra hardware in fault ride-through situations [17–19]. Figure 5 shows the basic wind energy systems.

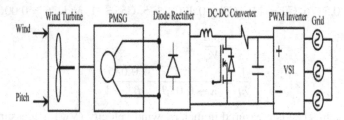

Fig. 5. Standard wind energy conversion using PMSGs

2.3   Battery-Powered Bidirectional DC-DC Converter

The usage of lead-acid batteries is widespread in PV power systems. A bidirectional DC-DC converter is used to charge and discharge these batteries. The authors of [15, 16] explore bidirectional DC-DC converter architectures and combinations for use in PV systems. The buck and boost processes in this investigation make use of a bidirectional chopper. The bidirectional converter in its most frequent design, as reported in this study, is depicted in Fig. 6. Switch S1 receives the gate signal, turning the bidirectional converter into a buck converter. When the PV output is high, the battery system will enter this mode to charge the batteries. When energy is low from the PV system or the grid, the bidirectional converter can function as a boost converter by delivering a gate signal to switch S2. This causes the battery to deplete and the load to be supplied with power. Parameters for a bidirectional converter are developed using expressions.

$$C_H = \frac{D}{R_H f_s \left(\Delta \frac{V_H}{V_H} \right)}$$

$$L_{b,min} \geq \frac{D(1-D)^2 R_H}{2 f_s}$$

(5)

Here RL& RH are the load resistances of bidirectional chopper, respectively. Boost side capacitance is denoted by CH. The minimum value of the inductance is Lb, min, and the switching frequency is FS. Microgrids have various features, which are detailed in Table 1.

Buck Mode:

$$L_{b,min} \geq \frac{(1 - D)R_L}{2f_s} \tag{6}$$

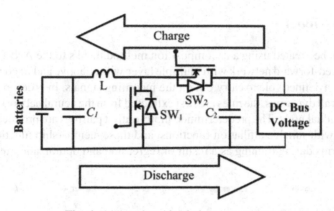

Fig. 6. Bidirectional buck–boost converter

Table 1. Sizing and specifications of the microgrid system.

Load Sizing	DC Bus Voltage	700 Vdc
	Load Power Required	2 kW
Battery Sizing	Batteries capacity	102 Ah
	Battery Voltage	96 Vdc
	Batteries capacity	9.8 kWh
	Batteries strings (parallel)	1
	Batteries per string (series)	4
PV Array Sizing	PV Module	Soltech 1STH-215-P
	Max Power Per Module	213 W
	Max Current	7.35 A
	Max Voltage	29 V
	Parallel Strings	6
	Series Modules per string	3
PMSG	Rated Power	3 kW
	Rated Speed	360 RPM

Table 2. Comparision for praposed and conventional system Performance

	ANFIS based IFO-PID	IFO-PID
Wind Power (W)	9950	9800
PV Power (W)	3300	3000
SSCs Power (W)	13500	13000
Battery Stored Power (W)	2600	2500
Battery Supplied Power (W)	4700	4500

3 ANFIS Model

The rules can be created using a decomposition method thanks to the ANFIS structure. ANFIS is a feed-forward network with multiple layer with adaptive and fixed nodes, each of which has a defined role to carry out on the incoming signals. In order to capture the system's overall dynamics, the rules are first extracted from the neural network at the level of each individual node. The predetermined input-output pairings are produced by fuzzy if-then rules with suitable affiliation functions, and these membership functions take on their final forms during training as a result of regression and optimization techniques.

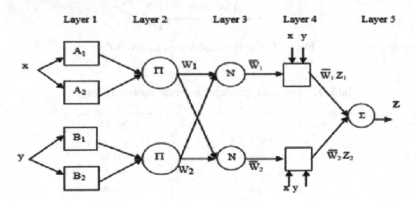

Fig. 7. ANFIS model

A gauge of how effectively the system is simulating the given training data set for a particular set of parameters is provided by the gradient vector (Fig. 7). After obtaining the gradient vector, optimization techniques are used to modify the parameters in order to lower a specified error criterion. When the training and checking errors are under a certain threshold, the system converges. Since the blind control system in this situation is thought of as a first order system, a first order Sugeno fuzzy model will be used to describe it. Due to its transparency and effectiveness, we will use the first order Sugeno fuzzy model for the dynamic application. Sugeno models of higher order complicate the system without adding much value. So, a first order system is used to create the blind control system.

4 Simulation Outcomes

Microgrid system was developed and two different control mechanisms were implemented. This hybrid Microgrid model was simulated with the MATLAB/SIMULINK software and the respective results are shown in Figs. 8, 9, 10, 11, 12, 13, 14, 15, 16, 17, 18, 19, 20. The suggested system's simulation results can be found in below references, along with the parameters that were employed. DC-link fixed-value reference to 240 V. The results of the energy management unit focus of the simulation test. Firstly, at an initial state of charge (SOC) of 80%, the BSS connects of DC load of 8000 watts to the DC-link through two load-side converters. The wind prole among 8 as well as 13 m/s is seen in Fig. 8. Figure 14 depicts the range of wind power (in watts) produced, that is 4000 to 10000 (based on wind speed). At 25°C and a brightness of 600 W per square meter, a PV array can provide 3000 W of power, as depicted in Fig. 10. The combined photovoltaic and wind power generation Pdg is shown in Fig. 11. Current responses place the range of Pdg produced power at among 7,000 as well as 13,000.

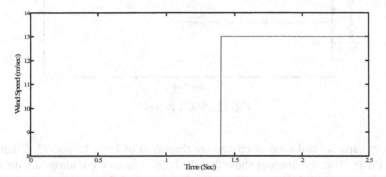

Fig. 8. Wind Speed

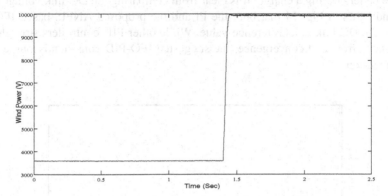

Fig. 9. Wind Power

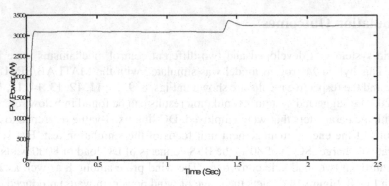

Fig. 10. PV Power

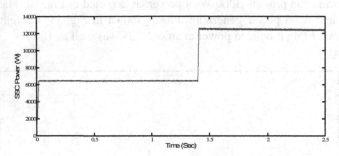

Fig. 11. SSC's Power

Battery capacity and state of charge are displayed in Figs. 12 and 13. When SOC is greater than 20%, the findings show that the battery feeds the microgrid with around 2300 W during the [0–1.4] s period, while the generated Pdg is greater than the load power during the [1.4–2.3] s interval. That's why the microgrid supplies around 4500 W to the battery during a charge. It is clear from comparing the DC-link voltage of the SSCs and LSCs in Fig. 14 that both the PI and the proposed ANFIS based IFO-PID maintain the DC-link at its reference value. While other PID controllers struggle with steady-state errors and convergence, the suggested IFO-PID consistently outperforms the competition.

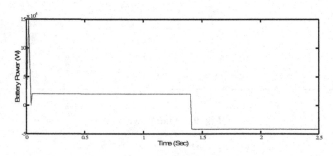

Fig. 12. Battery Power

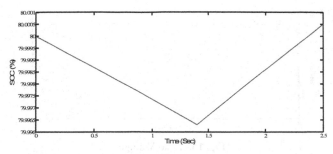

Fig. 13. SOC

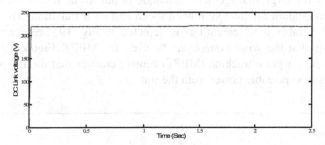

Fig. 14. DC Link Voltage

BSS power, seen in Fig. 12 If you look at Fig. 15, you can see that the proposed energy management system always sends the same 8300 W to the loads. As can be seen in Fig. 16, the suggested ANFIS based IFO-PID is able to maintain a constant output voltage of 220 V. In this subsection, the benefits of the proposed ANFIS based IFO-PID are highlighted by comparing them to those of prior works. Table 2 displays the results of this comparison, ANFIS based IFO-PID and IFO-PID revealing that the suggested method generates more power and shows good performance when compared to the compared control schemes. The proposed control technique effectively handled the hybrid energy, and the goals were met. Current work's load voltage compares favorably to prior approaches and demonstrates improved performance.

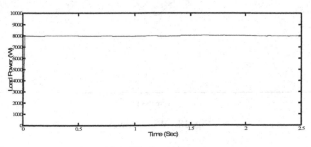

Fig. 15. Load Power

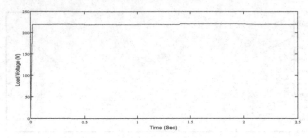

Fig. 16. Load Voltage

The proposed energy management technique is put to the test in Figs. 17 and 18 by simulating a random fluctuation in wind speed and solar radiation. The wind power output during variable wind conditions is depicted in Fig. 19. From what has been stated, it seems that the wind system can be used for MPPT. Figure 20 shows how the maximum power point tracking (MPPT) control ensures that the PV panel always extracts the greatest possible power from the sun.

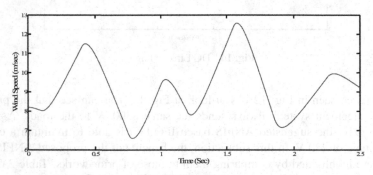

Fig. 17. Wind Speed

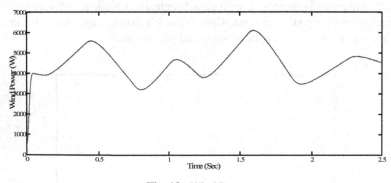

Fig. 18. Wind Power

The combined output of PV and wind is displayed in Fig. 20. As can be observed in Fig. 11, the results show that the output power remains constant at among 5000 as well as

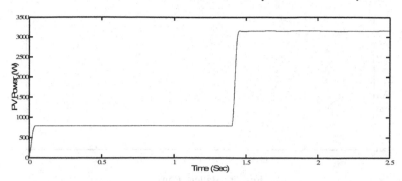

Fig. 19. PV Power

13000 W. The BSS is completely functional in energizing and using mode, and Fig. 21 depicts the power output when subject to random fluctuations. The recommended energy management control to transfer a uniform power to the load of around 8300 W. At last, we have the DC-link voltage response and the load power is plotted out in Fig. 22–23. The reported response demonstrates that the suggested method successfully controls the DC voltage at the specified reference. Consequently, the suggested energy management method assures continuous service and a steady supply of electricity, despite the inevitable fluctuations that are inevitable in any system.

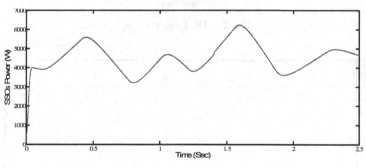

Fig. 20. SSC's Power

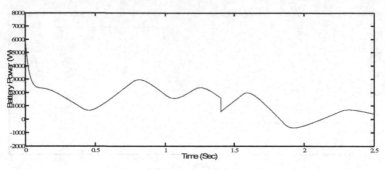

Fig. 21. Battery Power

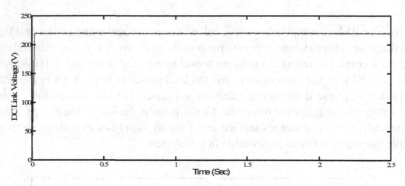

Fig. 22. DC Link Voltage

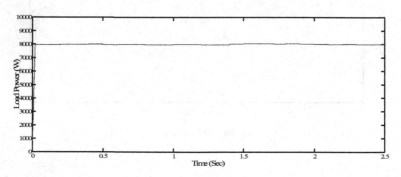

Fig. 23. Load Power

5   Conclusion

A PV system and WECS are suggested for use in the design and control of an independent micro-grid in this paper. The boost converter's control and optimization of power harvesting of PPV was achieved using MPPT with FLC. Here 2 controllers are employed to maintain a stable voltage from the PV to the battery storage system. The goal of controllers is continuously connect to the load. When the solar panels produce more power than is needed, any excess is stored in batteries. The controllers are manages the battery and PV power under dynamic conditions. The controller's actions must be appropriate in every scenario. The controller will not depend on SOC and operates as normal while typical working conditions (21% SOC 81%). When the SOC drops to 81%, a precise instruction is issued to turn off the wind turbine and solar panels. Connecting the wind mill energy and PV modules energy is not possible until the SOC in this controller falls below 75%. When the SOC > 21%, other hand it controls load and inverter. Once the batteries are fully charged, power is restored to the inverter. The problem was resolved by the built independent Microgrid system and associated controllers, which brought electricity to previously unconnected locations. Input penetrations in grid-connected mode of operation, as well as the functionality of the proposed system during fault conditions, will be the focus of future research.

References

1. Pareek, P., Maurya, N.K., Singh, L., Gupta, N., Reis, M.J.C.S.: Study of smart city compatible monolithic quantum well photodetector. In: International Conference on Cognitive Computing and Cyber Physical Systems, pp. 215–224 (2022)
2. Mangu, B., Akshatha, S., Suryanarayana, D., Fernandes, B.G.: Grid-connected PV-wind-battery-based multi-input transformer-coupled bidirectional DC-DC converter for household applications. IEEE J. Emerg. Select. Top. Power Electron. 4(3), 1086–1095 (2016)
3. Roselyn, J.P., et al.: Design and implementation of fuzzy logic based modified real-reactive power control of inverter for low voltage ride through enhancement in grid connected solar PV system. Control. Eng. Pract. 101, 104494 (2020)
4. Divyasharon, R., Banu, R.N., Devaraj, D.: Artificial neural network based MPPT with CUK converter topology for PV systems under varying climatic conditions. In: 2019 IEEE International Conference on Intelligent Techniques in Control, Optimization and Signal Processing (INCOS), pp. 1–6 (2019)
5. Zou, H., Du, H., Ren, J., Sovacool, B.K., Zhang, Y., Mao, G.: Market dynamics, innovation, and transition in China's solar photovoltaic (PV) industry: a critical review. Renew. Sustain. Energy Rev. 69, 197–206 (2017)
6. Wang, P., Wang, W., Meng, N., Xu, D.: Multi-objective energy management system for DC microgrids based on the maximum membership degree principle. J. Modern Power Syst. Clean Energy 6(4), 668–678 (2018)
7. Garmabdari, R., Moghimi, M., Yang, F., Gray, E., Lu, J.: Multi-objective energy storage capacity optimisation considering microgrid generation uncertainties. Int. J. Electr. Power Energy Syst. 119, 105908 (2020)
8. NajiAlhasnawi, B., Jasim, B.H., Esteban, M.D.: A new robust energy management and control strategy for a hybrid microgrid system based on green energy. Sustainability 12(14), 57249 (2020)
9. Zhang, Y., Liu, B., Zhang, T., Guo, B.: An intelligent control strategy of battery energy storage system for microgrid energy management under forecast uncertainties. Int. J. Electrochem. Sci. 9(8), 4190–4204 (2014)
10. Shadmand, M.B., Balog, R.S.: Multi-objective optimization and design of photovoltaic-wind hybrid system for community smart DC microgrid. IEEE Trans. Smart Grid 5(5), 2635–2643 (2014)
11. Pragaspathy, S., Karthikeyan, V., Kannan, R., Korlepara, N.D.P., Krishna, B.: Photovoltaic-based hybrid integration of DC microgrid into public ported electric vehicle. In: Wind and Solar Energy Applications, CRC Press, Tailor and Francis, pp. 287–303 (2023)

12. Bheemraj, T.S., Kumar, Y.A., Karthikeyan, V., Pragaspathy, S.: A hybrid structured high step-up DC-DC converter for integration of energy storage systems in military applications. IEEE Trans. Circuits Syst. II Express Briefs **70**(4), 1545–1549 (2022)
13. Nalli, P.K., Kadali, K.S., Bhukya, R., Palleswari, Y.T.R., Siva, A., Pragaspathy, S.: Design of exponentially weighted median filter cascaded with adaptive median filter. J. Phys. Conf. Ser. **2089**(1), 012020 (2021)
14. Chakravarthi, B.N.C.V., Hari Prasad, L., Chavakula, R.L., VijethaInti, V.V.: Solar energy conversion techniques and practical approaches to design solar PV power station. In: Sustainable and Clean Energy Production Technologies, Springer Nature Singapore, pp. 179–201 (2022)
15. Kannan, R., Karthikkumar, S., Suseendhar, P., Pragaspathy, S., Chakravarthi, B.C.V., Swamy, B.: Hybrid renewable energy fed battery electric vehicle charging station. In: 2021 Second International Conference on Electronics and Sustainable Communication Systems (ICESC), pp. 151–156 (2021)
16. Samy, M.M., Mosaad, M.I., Barakat, S.: Optimal economic study of hybrid PV-wind-fuel cell system integrated to unreliable electric utility using hybrid search optimization technique. Int. J. Hydrogen Energy **46**(20), 11217–11231 (2021)
17. Babatunde, O.M., Munda, J.L., Hamam, Y.: Hybridized off-grid fuel cell/wind/solar PV/battery for energy generation in a small household: a multi-criteria perspective. Int. J. Hydrogen Energy **47**(10), 6437–6452 (2022)
18. Kasireddy, I., Nasir, A.W., Singh, A.K.: Application of FOPID-FOF controller based on IMC theory for automatic generation control of power system. IETE J. Res. **68**(3), 2204–2219 (2019)
19. Gupta, N.K., Kasireddy, I., Singh, A.K.: Design of PID controller using strawberry algorithm for load frequency control of multi-area interconnected power system with and without non-linearity. In: Control Applications in Modern Power Systems: Select Proceedings of EPREC, vol. 870, pp. 153–162 (2022)

Full Swing Logic Based Full Adder for Low Power Applications

D. Durga Prasad[1] ⓘ, B. V. V. Satyanarayana[1](✉) ⓘ, Vijaya Aruputharaj J[2] ⓘ,
Abdul Rahaman Shaik[1] ⓘ, K. Indirapriyadarsini[3] ⓘ, K. V. Subba Rami Reddy[1] ⓘ,
and M. Hemalatha[4] ⓘ

[1] Department of ECE, Vishnu Institute of Technology, Bhimavaram, Andhra Pradesh, India
vvsatya.b@gmail.com
[2] Department of Computer Science, CHRIST (Deemed to be University), Bangalore, India
[3] Department of ECE, DNR College of Engineering and Technology, Bhimavaram, India
[4] Department of ECE, Shri Vishnu Engineering College for Women, Bhimavaram, India

Abstract. During the design of Application-Specific Integrated Circuits, a whole
adder logic circuit plays a significant role. The full adder is a fundamental part of
the majority of VLSI and DSP applications. Power consumption in full adders is
one of the key factors; hence it is necessary to build full adders with low power consumption. Full adders are developed in this work employing full swing AND, OR,
and XOR gates and compared with pass transistor logic (PTL) based AND, OR,
and XOR gates, and complementary metal oxide semiconductor logic (CMOS)
based AND gate, OR gate, and XOR gate. The Mentor Graphics Tool is used to
construct and simulate every planned circuit. After receiving simulation data, we
compared the power consumption, delay and PDP of several complete adder-based
logic designs. In the proposed full swing XOR, the power dissipation and delay
is decreased by 10.5% and 9.8% respectively and hence the full swing full adder
PDP is decreased by 0.6%. As compared to alternative full adder designs based
on logic, full swing by using gates like AND gate, by using the OR gate, and with
the help XOR gate, full adder design consumes less power and hence suitable for
low power applications.

Keywords: Full Adder · TGL · PTL · Mentor Graphics Tool

1 Introduction

The concept of smart cities has gained tremendous momentum in recent years as urban
areas around the world strive to harness the power of technology to enhance the quality of
life for their residents. Smart cities leverage various Internet of Things (IoT) applications
and interconnected systems to efficiently manage resources, improve infrastructure, and
deliver better services. One crucial aspect that underpins the success of smart city initiatives is the implementation of low-power solutions [1]. Low power plays a pivotal
role in enabling the sustainable growth and long-term viability of smart cities, ensuring
that these technological advancements are energy-efficient, environmentally friendly,
and economically feasible.

© ICST Institute for Computer Sciences, Social Informatics and Telecommunications Engineering 2024
Published by Springer Nature Switzerland AG 2024. All Rights Reserved
P. Pareek et al. (Eds.): IC4S 2023, LNICST 537, pp. 19–36, 2024.
https://doi.org/10.1007/978-3-031-48891-7_2

Adders are widely utilized as circuit components in Very Large Scale Integration (VLSI) systems, including microprocessors and Digital Signal Processing (DSP) processors. It serves as the hub for several other processes, including address computation, multiplication, and division. Adders are crucial to the overall system performance in most digital systems [2]. As a result, improving adder's performance is a key objective. The study into low power microelectronics has increased due to the tremendous rise in portable systems like laptops. The cause of this is because battery technology develops more slowly than microelectronics technology [3]. The mobile systems have a finite quantity of electricity at their disposal. Low power design has thus grown to be a significant design restriction [4]. The most computationally intensive applications, such as multimedia processing and DSP, may now be realized in hardware to increase speed of operation. However, as the market for portable electronics grows, so do the researchers' motivations to pursue smaller silicon areas, faster processing, longer battery lives, and improved dependability [5].

The complete adder design is crucial to digital computing. The complete adder design requirements are often several in natures. One of the properties, transistor count, affects the system complexity of arithmetic circuits like multipliers and Arithmetic Logic Units (ALU), among others [6]. The other two key factors in the design of complete adders would be power consumption and speed [7, 8]. Yet, their interactions with one another are paradoxical. To achieve the best design tradeoffs, energy usage per operations and power delayed products have been defined. The right choice of logic types can improve the performance of digital circuits. Various logical approaches frequently emphasize achieving one facet of performance at the price of another. While executing the identical function, the logic styles differ in how they calculate intermediary nodes and transistor counts [9].

Numerous comprehensive adders are designed so far, including those that use static CMOS, with the help of dynamic circuits, and having the large number of transmission gates, with the help of GDI logic, and having the pass transistor logic (PTL) [10, 11]. For the purpose of producing sum and carry outputs, the well-established static CMOS which are adders along with help of complementary pull up PMOS and having the pulling down NMOS networks that need 28 no. of transistors. PTL is a substitute for CMOS that offers implementations of the majority of functionalities with fewer transistors. This may lower total capacitances, increasing speed and reducing power dissipation in the process. The output voltage, however, varies in the PTL-based design because of a certain threshold voltage reduction occurs throughout the circuit input and the circuit output. The Complementary Pass Logic (CPL) and along with the differential restoring circuit that is Swing Restored PTL is a adaptations can be used to remedy this issue (SRPL). However, these logics lead to higher shorts of current, a greater transistors count, and more intricate wiring connections since they require complementary input signals [12].

Another strategy to reduce complexity in logic construction is to use transmission gates. Reference [13] discusses the entire adder implementation that makes use of a transmission gate. It takes 20 transistors; a transmission function adder, which only needs 16, is another option for additional transistor count reduction. Reference [14] discusses this option. As an alternative to CMOS logic, GDI logic is offered. A logic function may be implemented using this low-power design method with fewer transistors. A threshold value V_t separates the output the result that comes maximum or minimum in the voltages from the V_{DD} that is the voltage which results the circuit to work or ground in GDI gates, resulting in a smaller voltage swing at the outputs [15, 16]. It is advantageous for power usage when the voltage swing is reduced. Nevertheless, if cascaded operation is used, this could result in sluggish switching. The reduced output may possibly result in circuit malfunction while operating at low V_{DD} [17].

In order to attain full swing functioning, additional consideration must be required. This work implements an effective technique for digital circuits, comprised of full swing by using gates like AND gate, by using the OR gate, and with the help XOR gate, and using the transistors like pass transistor logic, and by using the CMOS logic.

2 Ultra-Low Power Design Requirement

As smart cities evolve and integrate a multitude of IoT devices and systems, the demand for energy increases exponentially. These devices, such as sensors, actuators, and communication modules, form the backbone of a smart city infrastructure. They collect and exchange vast amounts of data, enabling city authorities to monitor various aspects of urban life, including traffic flow, waste management, energy consumption, and public safety. However, this increased reliance on technology and connectivity poses significant challenges, particularly in terms of power consumption.

1. Energy Efficiency: Smart cities need to optimize energy consumption to ensure sustainable development. By implementing low-power devices and systems, cities can minimize their overall energy footprint. Low-power solutions can significantly extend battery life, reducing the frequency of battery replacements or recharging [18]. This, in turn, reduces the operational costs associated with maintaining and powering the vast network of IoT devices deployed throughout the city.
2. Scalability and Expansion: A crucial characteristic of smart cities is their ability to scale and accommodate future growth. Low-power solutions enable the deployment of a larger number of devices without overburdening the existing power infrastructure [19]. By adopting energy-efficient technologies, smart cities can expand their IoT networks without straining resources or compromising system performance.
3. Environmental Sustainability: The environmental impact of urbanization is a growing concern worldwide. By incorporating low-power devices and systems, smart cities can contribute to reducing greenhouse gas emissions and minimizing their ecological footprint. Energy-efficient IoT solutions decrease the demand for non-renewable energy sources, leading to a more sustainable and eco-friendly urban environment [20].
4. Reliability and Resilience: In smart cities, maintaining the reliability and resilience of IoT systems is crucial for uninterrupted service delivery. Low-power devices can operate for extended periods without needing frequent maintenance, ensuring the continuous functioning of critical applications. Additionally, energy-efficient technologies are less susceptible to power outages, allowing smart cities to remain operational even during unforeseen events or emergencies [21, 22].

In the pursuit of building efficient and sustainable smart cities, low-power solutions have emerged as a vital enabler. By embracing energy-efficient technologies, smart cities can minimize their energy consumption, optimize resource allocation, and reduce their environmental impact [23]. Low-power devices and systems pave the way for scalable and resilient IoT networks, ensuring the long-term viability and success of smart city initiatives. As cities continue to evolve and adapt to the ever-changing technological landscape, prioritizing low power will remain essential in creating truly intelligent, energy-conscious, and livable urban environments [24].

3 Traditional Approach

3.1 XOR Gate using CMOS Logic

A two-input XOR gate is a fundamental digital logic gate that outputs true (1) when the number of true inputs is odd. The operation of a two-input XOR gate using CMOS logic involves the use of both NMOS and PMOS transistors. Dynamic power consumption occurs during switching when the output transitions between logic 0 and logic 1. During this period, both NMOS and PMOS transistors momentarily conduct, causing a short power surge. This causes a brief surge in power consumption, which can be significant in high-frequency operation or large-scale designs. The XOR gate's delay is primarily influenced by the cascaded structure of the inverters within the circuit. This can become a limitation in high-performance applications where ultra-fast operation is essential.

The power-delay product (PDP) represents the trade-off between power consumption and circuit speed. In high-performance systems, optimizing the PDP is crucial, and CMOS XOR gates may not be the most efficient choice. CMOS XOR gates require a relatively higher number of transistors compared to simpler gates like NAND and NOR gates. Each XOR gate needs 14 transistors, which can lead to increased chip complexity and area consumption [25]. In large-scale designs, the higher transistor count can contribute to increased manufacturing costs and may limit the overall chip integration. Due to the complex structure and cascaded nature of the XOR gate, its ability to drive multiple loads can be limited.

The CMOS XOR gates have proven to be versatile and widely used in many digital applications, they do have certain disadvantages, particularly concerning power consumption during switching, delay compared to simpler gates, and the number of transistors required. In some cases, a different gate implementation might be more suitable for achieving the desired balance between power efficiency, speed, and transistor count. The CMOS XOR gate circuit diagram is as shown in Fig. 1.

3.2 XOR Gate Using PTL

PTL XOR gates typically require fewer transistors compared to CMOS XOR gates, which can lead to a smaller silicon area and lower manufacturing costs. Due to the reduced transistor count, PTL XOR gates generally consume less power than CMOS XOR gates, making them more energy-efficient. PTL XOR gates have simpler transistor paths, leading to shorter signal propagation delays and faster switching times compared

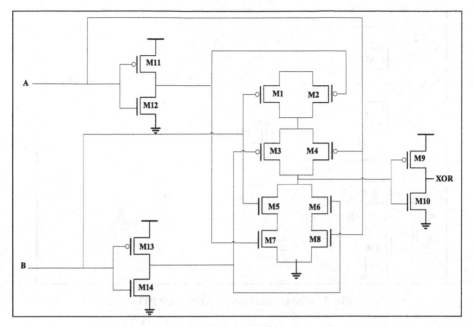

Fig. 1. Circuit schematic of XOR using CMOS logic

to CMOS XOR gates [26]. PTL gates can perform well at lower supply voltages, making them suitable for low-power and energy-constrained applications.

PTL XOR gates suffer from some signal degradation issues due to the presence of parasitic capacitance and resistance in the pass transistors, which can affect the circuit's overall performance. The noise margin (the tolerance for signal fluctuations) of PTL XOR gates is generally lower than that of CMOS XOR gates, making them more susceptible to noise and manufacturing variations. PTL XOR gates have a limited ability to drive multiple loads (fan-out), which can restrict their use in complex circuits with high fan-out requirements. The circuit schematic of XOR using PTL is represented in Fig. 2.

3.3 XOR Gate Using Full Swing Logic

The power consumption of the FA cell can be decreased by designing the XOR gate efficiently, as it is a major consumer of power. Additionally, there are various circuit proposals available for implementing an efficient XOR gate in digital circuits design. The full swing based XOR gate is shown in Fig. 3.

In Fig. 3, due to two NOT gates on the critical route, the double pass transistor XOR gate circuit, which comprises eight transistors, consumes a lot of power. The power dissipated in short-circuits and overall increases with the size of these transistors since doing so results in an intermediate node with a high capacitance. There is a minor increase in critical route latency under ideal PDP circumstances as well [27].

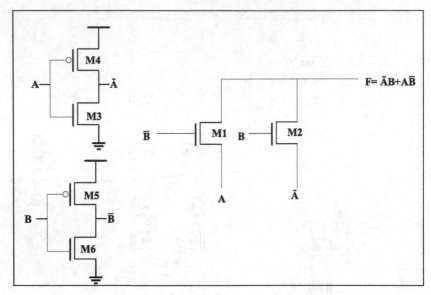

Fig. 2. Circuit schematic of XOR using PTL

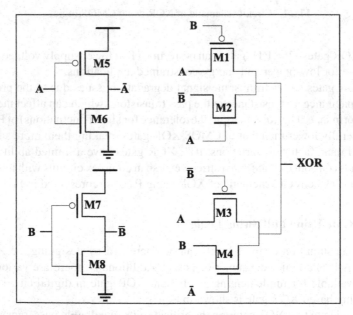

Fig. 3. Circuit schematic of XOR using FSL

Figure 4 shows The PTL logic-based XOR gate circuit, which consists of six transistors, has a better latency and power consumption than an alternative circuit. However, it employs NOT gates on the critical route, which might be problematic. Because an NMOS transistor was used in the XOR circuit, has less latency Because PMOS transistors are slower than NMOS transistors, not gates should be larger in order to maximize XOR speed [28].

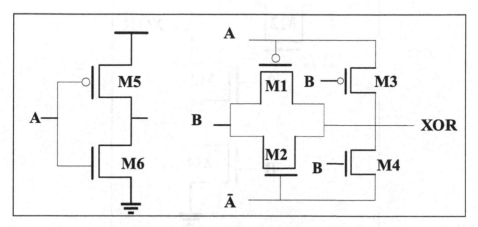

Fig. 4. Circuit schematic of XOR using full swing PTL

4 Proposed Methodology

The non-full-swing XOR circuit of Fig. 5 [29] is efficient in terms of the power and delay. Furthermore, this structure has an output voltage drop problem for only one input logical value.

To solve this problem and give the XOR gates an ideal shape; we propose the circuit shown in Fig. 6. This arrangement creates a perfect swing for all possible input combinations. In the critical route of the circuit, the proposed XOR gate does not contain a NOT gate. Compared to all other circuits, the delay is shorter and the drive strength this better. The proposed XOR gate has one more transistor than the structure shown in Fig. 4, but it operates faster and consumes less power.

The input capacitances in A and B of the XOR circuit depicted in Fig. 6 are not balanced because one of them has to be connected to the input of the NOT gate and the other to the diffuser of the NMOS transistor. As a result, input A, which is likewise linked to the NOT gate, has to be connected to a transistor with a lower input capacitance. Full-swing output, less wiring, good drivable capacitance and simple circuit structure of the proposed XOR circuits are their advantages.

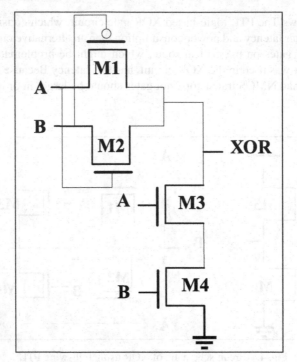

Fig. 5. Circuit schematic of non-full swing XOR

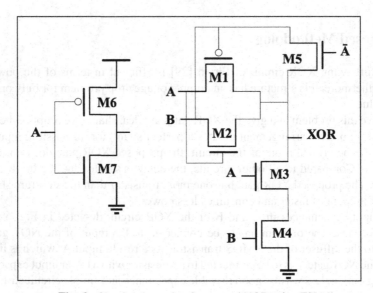

Fig. 6. Circuit schematic of proposed XOR using FSL

5   Design, Implementation and Simulation Results

The PTL XOR gate is designed using pass transistors as the main building blocks. It consists of transmission gates or pass gates that pass or block signals based on the control signals. The CMOS XOR gate is designed using Complementary Metal-Oxide-Semiconductor (CMOS) technology. It employs a combination of NMOS and PMOS transistors to create the XOR function. Full Swing Logic is a design style that enhances noise immunity and signal integrity. The XOR gate designed using Full Swing Logic combines the benefits of both PTL and CMOS techniques to achieve high performance.

Mentor Graphics provides a suite of electronic design automation tools. The design of the XOR gates (PTL, CMOS, and Full Swing Logic) is implemented using the appropriate tool from the Mentor Graphics suite. This involves drawing the circuit schematics, setting up the transistor models, specifying the technology parameters, and defining the logic function of the XOR gate.

After implementation, the XOR gates are simulated using the Mentor Graphics tool to verify their functionality and performance. The outputs waveform of every circuitry may be seen when all the circuitry have been constructed and modeled to use the mentor graphical tool.

5.1   Implementation of XOR Gate Using CMOS Logic

For CMOS XOR gate, the simulation will demonstrate its higher noise immunity, better signal integrity, and high fan-out capability. But it might exhibit increased power consumption and longer switching delays. Figures 7 and 8 depict the simulated design and waveforms of an XOR gate employing CMOS respectively.

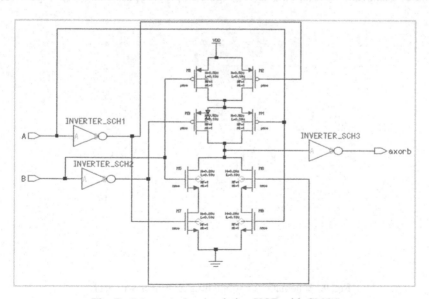

Fig. 7. Schematic for simulating XOR with CMOS

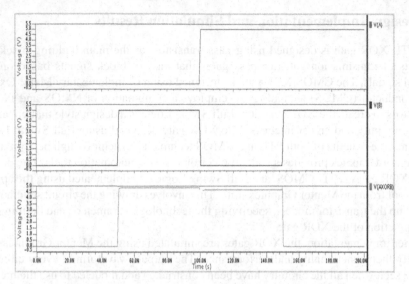

Fig. 8. Waveforms for simulating XOR gate with CMOS

5.2   Implementation of XOR Gate Using PTL

For Pass Transistor Logic (PTL) XOR gate, the simulation will validate its reduced transistor count, low power consumption, and faster switching speed. However, it may show limitations in noise margin and signal degradation. Figures 9 and 10 depict the simulated design and waveforms of an XOR gate employing pass transistors respectively.

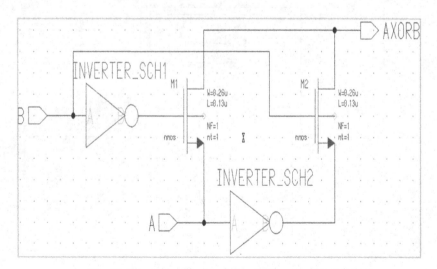

Fig. 9. Schematic for stimulating XOR gate with PTL

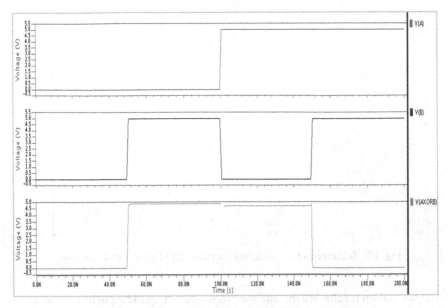

Fig. 10. Waveforms for stimulating XOR gate with PTL

5.3 Implementation of Proposed XOR Gate Using FSL

For full swing logic XOR gate, the simulation results will aim to show the advantages of combining PTL and CMOS techniques, such as improved noise immunity, signal integrity, and moderate power consumption. Figures 11 and 12 demonstrate the simulation design and waveforms for a XOR gate utilizing a full-swing XOR gate.

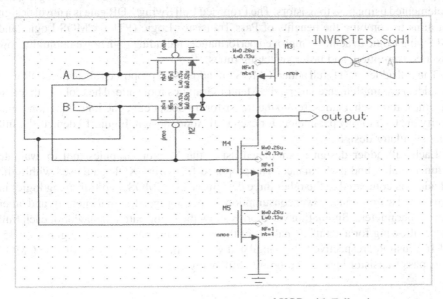

Fig. 11. Schematic for simulating proposed XOR with Full swing

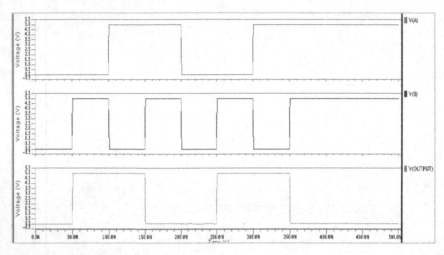

Fig. 12. Waveforms for simulating proposed XOR gate with Fullswing

The simulation results will include waveforms depicting the input and output behavior of each XOR gate, and they will be analyzed to evaluate the design's performance and adherence to specifications. Any design improvements or optimizations can be made based on these simulation results before proceeding to the physical fabrication process.

5.4 Implementation of Full Adder Using Proposed XOR Gate

The CMOS Full Adder is designed using CMOS technology, utilizing a combination of NMOS and PMOS transistors to perform addition. The PTL full adder is designed using pass transistors, and it performs addition using XOR gates and other logic gates implemented using pass transistors. The Proposed Full Swing XOR gate is a novel design that aims to combine the benefits of Pass Transistor Logic (PTL), CMOS Logic, and Full Swing Logic to achieve improved performance, including high noise immunity, low power consumption, and faster switching times.

The full adders (PTL full adder, CMOS full adder, and proposed full swing XOR gate full adder) are implemented using the Mentor Graphics tool suite. The tool is utilized to draw the circuit schematics, set up the transistor models, define the logic functions, and specify the technology parameters for each Full Adder design. Figure 13 depicts the full adder simulator design.

Each full adder design is simulated using the Mentor Graphics tool to evaluate its functionality and performance. The proposed full swing XOR gate used within the full adder is compared to the other three designs (PTL, CMOS, and full swing logic) in terms of noise immunity, power consumption, switching speed, signal integrity, and other relevant parameters. Simulation waveforms show the input-output behavior of each Full Adder, allowing for a detailed analysis of their respective performance. Figures 14, 15, and 16 illustrate, respectively, the simulated waveforms for full adders utilizing CMOS logic, pass transistor logic and proposed full swing circuitry.

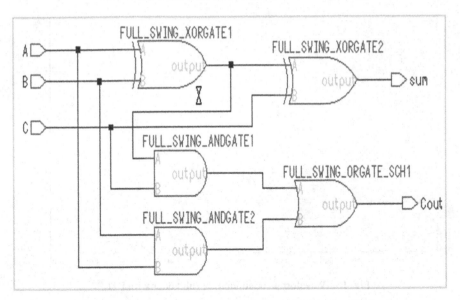

Fig. 13. Schematic of full adder using proposed full swing XOR gate

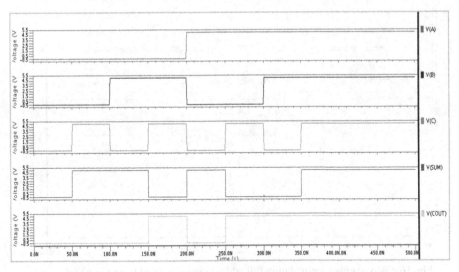

Fig. 14. Waveforms for simulating full adder with CMOS logic

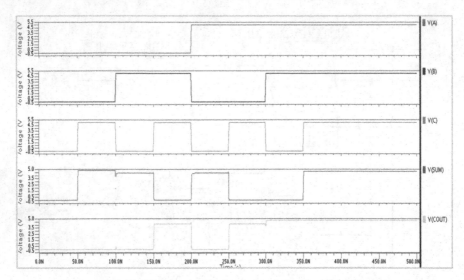

Fig. 15. Waveforms for simulating full adder with PTL.

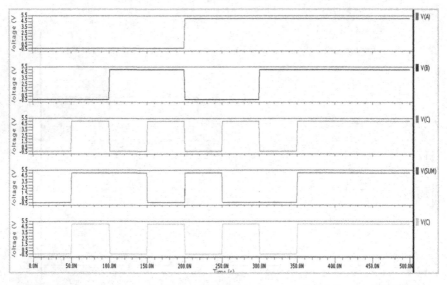

Fig. 16. Waveforms for simulating full adder using proposed full swing XOR gate

6 Performance Evaluation

The Proposed Full Swing XOR gate combines various logic techniques to achieve better noise immunity, making it more robust against signal disturbances and manufacturing variations. The proposed Full Swing XOR gate strikes a balance between the low-power advantages of PTL XOR gates and the power efficiency of CMOS XOR gates, offering moderate power consumption. While Full Swing XOR gates may not be as fast as PTL gates, they generally provide faster switching times compared to CMOS XOR gates. In regards to power consumption,delay and PDP, the following Table 1 compares several XOR gate designs.

Table 1. Performance evaluation of XOR gates in various logics

XOR Designs	No. of Transistors	Power Dissipation (μW)	Delay (pS)	PDP (aJ)
Ref [24]	14	2.48	26.1	64.7
Ref [11]	6	2.14	23.6	50.5
Ref [27]	4	2.46	21.5	52.9
Ref [12]	8	2.50	25.8	64.5
Ref [30]	8	2.48	24.3	60.2
Proposed	**7**	**2.22**	**21.9**	**48.6**

The simulation results are compared to identify the strengths and weaknesses of each Full Adder design. The Proposed Full Swing XOR gate's impact on the overall performance of the Full Adder is evaluated, and it is compared against the traditional PTL, CMOS, and Full Swing Logic implementations. The comparison aims to demonstrate how the proposed Full Swing XOR gate enhances the Full Adder's performance, potentially providing advantages in various aspects over the other logic designs. Comparing several full adder designs in regards to power dispersion, delay and PDP is shown in Table 2 below.

Table 2. Performance evaluation of Full adder using XOR gates in various logics

XOR Designs	Power Dissipation (μW)	Delay (pS)	PDP (aJ)
CMOS [7]	3.61	212	765.3
DPL [16]	4.89	98.8	483.1
CMOS [11]	3.98	119.2	474.4
CPL [8]	6.88	63.7	438.3
HFA [12]	3.71	116.8	433.3
Proposed	**3.92**	**109.9**	**430.8**

Based on the simulation results and comparison, designers can make informed decisions about which Full Adder implementation is most suitable for their specific application. Any design improvements or optimizations can also be made based on the findings before moving on to the physical fabrication process.

7 Conclusion

In this work, the use of a full swing logic-based full adder that incorporates the proposed full swing XOR gate offers significant advantages for low-power applications. This design aims to strike a balance between power efficiency, speed, and performance, making it well-suited for energy-constrained systems. The proposed full swing XOR gate demonstrates notable improvements in power consumption compared to conventional adders. It showcases competitive switching speeds in comparison to PTL and CMOS-based full adders. Its design principles emphasize fast signal propagation and reduced signal degradation, contributing to improved delay performance. It also excels in achieving a favorable Power-Delay Product. It balances power consumption and delay characteristics, offering an optimal compromise between energy efficiency and speed for low-power applications. These adders can be used for efficient and high-performance arithmetic circuits suitable for battery-powered devices, IoT applications, and other low-power electronic systems.

References

1. Pareek, P., Maurya, N.K., Singh, L., Gupta, N., Reis, M.J.C.S.: Study of smart city compatible monolithic quantum well photodetector. In: Gupta, N., Pareek, P., Reis, M. (eds.) Cognitive Computing and Cyber Physical Systems. IC4S 2022. Lecture Notes of the Institute for Computer Sciences, Social Informatics and Telecommunications Engineering, vol. 472, pp. 215–224. Springer, Cham (2022). https://doi.org/10.1007/978-3-031-28975-0_18
2. Purohit, S., Margala, M.: Investigating the impact of logic and circuit implementation on full adder performance. IEEE Trans. Very Large Scale Integr. (VLSI) Syst. 20(7), 1327–1331 (2012)
3. Singh, R., Akashe, S.: Modeling and analysis of low power 10 T full adder with reduced ground bounce noise. J. Circ. Syst. Comput. 23(14), 1–14 (2014)
4. Satyanarayana, B.V.V., Durga Prakash, M.: Device and circuit level design, characterization and implementation of low power 7T SRAM cell using heterojunction tunneling transistors with oxide overlap. Microprocess. Microsyst. 77 (2020)
5. Ykuntam, Y.D., Penumutchi, B., Gubbala, S.: Design of speed and area efficient non linear carry select adder (NLCSLA) architecture using XOR less adder module. In: Chakravarthy, V., Bhateja, V., Flores Fuentes, W., Anguera, J., Vasavi, K.P. (eds.) Advances in Signal Processing, Embedded Systems and IoT . Lecture Notes in Electrical Engineering, vol. 992, pp. 81–91. Springer, Singapore (2023). https://doi.org/10.1007/978-981-19-8865-3_7
6. Patel, R., Parashar, H., Wajid, M.: Faster arithmetic and logical unit CMOS design with reduced number of transistors. In: Computer Networks and Information Technologies: Second International Conference on Advances in Communication. Network, and Computing, CNC 2011, pp. 519–522. Springer, Berlin Heidelberg (2011)
7. Lee, P.M., Hsu, C.H., Hung, Y.H.: Novel 10-T full adders realized by GDI structure. In: 2007 International Symposium on Integrated Circuits, pp. 115–118. IEEE (2007)
8. Chaddha, K.K., Chandel, R.: Design and analysis of a modified low power CMOS full adder using gate-diffusion input technique. J. Low Power Electron. 6(4), 482–490 (2010)
9. Hassoune, I., Flandre, D., O'Connor, I., Legat, J.D.: ULPFA: a new efficient design of a power-aware full adder. IEEE Trans. Circ. Syst. I 57(8), 2066–2074 (2008)
10. Murthy, G.R., Senthilpari, C., Velrajkumar, P.: A new 6-T multiplexer based full-adder for low power and leakage current optimization. IEICE Electron. Express 9(17), 1434–1441 (2012)

11. Morgenshtein, A., Fish, A., Wagner, I.A.: Gate-diffusion input (GDI): a power-efficient method for digital combinatorial circuits. IEEE Trans. Very Large Scale Integr. (VLSI) Syst. **10**(5), 566–581 (2002)
12. Misra, A., Birla, S., Singh, N., Dargar, S.K.: High-performance 10-transistor adder cell for low-power applications. IETE J. Res. 1–19 (2022)
13. Durga Prasad, D., Dileep M., Rama Krishna, C.H.: Design and implementation of full adder using different XOR gates. Int. J. Innov. Technol. Explor. Eng. **9**(4), 1422–1426 (2020)
14. Parvathi, M.: Machine learning based interconnect parasitic R, C, and power estimation analysis for adder family circuits. In: 2022 Sixth International Conference on I-SMAC (IoT in Social, Mobile, Analytics and Cloud, pp. 957–962. IEEE, Dharan, Nepal (2022)
15. Ramesh, A.P.: Implementation of low power high speed adder's using GDI logic. Int. J. Innov. Technol. Explor. Eng. **8**(11), 1291–1298 (2019)
16. Ramesh, A.P.: Implementation of low power carry skip adder using reversible logic. Int. J. Recent Technol. Eng. **8**(3) (2019)
17. Shoba, M., Nakkeeran, R.: GDI based full adders for energy efficient arithmetic applications. Eng. Sci. Technol. Int. J. **19**(1), 485–496 (2016)
18. Satyanarayana, B.V.V., Durga Prakash M.: Lower subthreshold swing and improved miller capacitance heterojunction tunneling transistor with overlapping gate. In: Materials Today Proceedings. Vol. 45, pp.1997–2001, Springer (2021)
19. Godi, P.K., Krishna, B.T., Kotipalli, P.: Design optimisation of multiplier-free parallel pipelined FFT on field programmable gate array. IET Circ. Dev. Syst. **14**(7), 995–1000 (2020)
20. Meriga, C., Ponnuri, R.T., Satyanarayana, B.V.V., Gudivada, A.A.K., Panigrahy, A.K., Prakash, M.D.: A novel teeth junction less gate all around FET for improving electrical characteristics. SILICON **14**, 1979–1984 (2022)
21. Bhuvaneswary, N., Prabu, S., Karthikeyan, S., Kathirvel, R., Saraswathi, T.: Low power reversible parallel and serial binary adder/subtractor. In: Balas, V.E., Solanki, V.K., Kumar, R. (eds.) Further Advances in Internet of Things in Biomedical and Cyber Physical Systems. Intelligent Systems Reference Library, vol. 193, pp. 151–159. Springer, Cham (2021). https://doi.org/10.1007/978-3-030-57835-0_12
22. Satyanarayana, B.V.V., Prakash, M D.: Design analysis of GOS-HEFET on lower subthreshold swing SOI. Analog Integr. Circ. Sign. Process. **109**, 683–694 (2021)
23. Ramesh, G., Manikandan, P., Naveen, P., Saravanan, M., Kumar, S.A., Swedheetha, C.: Energy efficient high performance adder/subtractor circuits. In: 2022 3rd International Conference on Smart Electronics and Communication, pp. 333–338. IEEE, Trichy (2022)
24. Prakash, M.D., Krsihna, B.V., Satyanarayana, B.V.V., Vignesh, N.A., Panigrahy, A.K., Ahmadsaidulu, S.: A study of an ultrasensitive label free silicon nanowire FET biosensor for cardiac troponin I detection. SILICON **14**, 5683–5690 (2022)
25. Vesterbacka, M.: A 14-transistor CMOS full adder with full voltage-swing nodes. In: 1999 IEEE Workshop on Signal Processing Systems. SiPS 99. Design and Implementation, pp. 713–722. IEEE (1999)
26. Yano, K., Yamanaka, T., Nishida, T., Saito, M., Shimohigashi, K., Shimizu, A.: A 3.8-ns CMOS 16*16-b multiplier using complementary pass-transistor logic. IEEE J. Solid-State Circ. **25**(2), 388–395 (1990)
27. Aguirre-Hernandez, M., Linares-Aranda, M.: CMOS full-adders for energy-efficient arithmetic applications. IEEE Trans. Very Large Scale Integr. (VLSI) Syst. **19**(4), 718–721 (2011)
28. Chowdhury, S.R., Banerjee, A., Roy, A., Saha, H.: A high speed 8 transistor full adder design using novel 3 transistor XOR gates. Int. J. Electron. Circ. Syst. **2**(4), 217–223 (2008)

29. Wang, J.M., Fang, S.C., Feng, W.S.: New efficient designs for XOR and XNOR functions on the transistor level. IEEE J. Solid-State Circ. **29**(7), 780–786 (1994)
30. Naseri, H., Timarchi, S.: Low-power and fast full adder by exploring new XOR and XNOR gates. IEEE Trans. Very Large Scale Integr. (VLSI) Syst. **26**(8), 1481–1493 (2018)

Single-Phase Grid and Solar PV Integration with 15-Level Asymmetrical Multilevel Inverter

Asapu Siva[1][✉] ⓘ, Y. T. R. Palleswari[1] ⓘ, Kalyan Sagar Kadali[1] ⓘ,
Ramu Bhukya[1] ⓘ, Mamatha Deenakonda[2] ⓘ, and V. V. Vijetha Inti[2] ⓘ

[1] Department of Electrical and Electronics Engineering,
Shri Vishnu Engineering College for Women (Autonomous), Bhimavaram, Andhra Pradesh,
India
asivaeee@svecw.edu.in
[2] Department of Electrical and Electronics Engineering, Vishnu Institute of Technology,
Bhimavaram, Andhra Pradesh, India

Abstract. In the modern era of renewable energy generation and distribution, injecting solar power into the utility grid has gained universal recognition, also solar energy plays a crucial role for smart cities development. Grid-connected asymmetrical multilevel inverters have undergone significant development for integrating solar into the utility grid. This article aims to implement a 15-level asymmetrical inverter in a single-phase grid-integrated PV system and analyses the proposed system working with a 15-level inverter. This inverter topology has fewer switching devices, design with three DC sources and can generate fifteen output voltage levels; in this proposed system, these three DC sources are designed by three PV modules and three conventional boost converters. P&O (Perturb and Observe) Algorithm was implemented on each boost converter to extract the maximum power from three PV modules. The proposed system is simulated in MATLAB 2023a with toolboxes. The simulation results show whether the inverter can export the power to the grid under various solar irradiance conditions. The solar-based proposed system is well suited for groups of houses in smart cities.

Keywords: Photovoltaic (PV) · Multilevel inverter (MI) · MPPT · Smart cities

1 Introduction

Day by day, smart cities grow, and their power demand also increases day by day. As cities try to become "smart," clean energy sources like solar and wind can be a big part of helping them reach their goals and demand [1]. Solar photovoltaic (PV) systems are growing in popularity as distributed generators around the world due to their low environmental impact. PV panels have become much more affordable, but the DC power they typically produce is unstable. Therefore, DC-DC or DC-AC conversions are necessary before the PV power can be fed to the output load or connected to the grid power [2, 3].

© ICST Institute for Computer Sciences, Social Informatics and Telecommunications Engineering 2024
Published by Springer Nature Switzerland AG 2024. All Rights Reserved
P. Pareek et al. (Eds.): IC4S 2023, LNICST 537, pp. 37–47, 2024.
https://doi.org/10.1007/978-3-031-48891-7_3

Solar energy can be integrated with the load or grid in one of two ways: a single-stage [4] or two-stage process [5, 6]. In [7], a two-stage, single-phase PV system is presented that can function independently of batteries. In order to regulate a two-stage power system, a proportional-integral (PI) design was presented in the article [8]. Inverters play a crucial role in transforming the DC output of solar panels into the AC current used by the grid [9]. In-depth research has been done on the asymmetrical inverters for PV systems in the literature [10, 11].

This article aims to implement a 15-level asymmetrical inverter in a single-phase grid-integrated PV system and observe the performance of the proposed system when 15-level inverters are used in the single-phase grid-integrated PV system. This 15-level inverter was proposed [12] by Marif Daula Siddique. A grid-connected PV system (GCPVS) with a two-power stage, single phase, and no transformer is depicted in circuit diagram form in Fig. 1.

The proposed system, including the boost converter design and PV system design, is presented in Sect. 2. Section 3 of the paper explains a maximum power point tracking (MPPT) and inverter control. Simulation results for the proposed multilevel inverter are presented in Sect. 4. Section 5 provides a conclusion.

2   Proposed System

2.1   Main Circuit

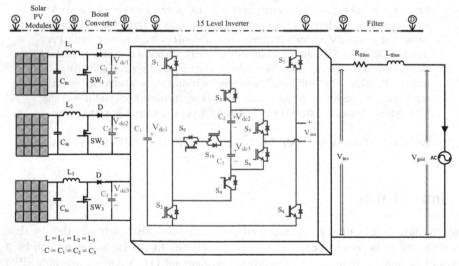

Fig. 1. Proposed System

Figure 1 depicts a schematic diagram of a two-stage conversion system. The circuit mainly consists of three DC-DC converters, 15 level inverter and a solar PV system. They are made to last long and withstand the harshest environmental conditions. In this system, three PV modules are connected to 15 level inverter as DC voltage sources via three boost converters and an inverter connected to the grid through filters.

2.2   15-level Inverter

The 15-level inverter switching states are shown in Table 1 (Table 2).

Table 1. Switches ON states

S.NO	ON state Switches				Vout
1	S_2	S_3	S_7	S_6	350V
2	S_1	S_3	S_6	S_7	300V
3	S_1	S_3	S_6	S_8	250V
4	S_2	S_6	S_7	S_9	200V
5	S_1	S_6	S_7	S_9	150V
6	S_1	S_6	S_8	S_9	100V
7	S_2	S_4	S_6	S_7	50V
8	S_1	S_3	S_5	S_7	0V
9	S_1	S_3	S_5	S_8	–50V
10	S_2	S_5	S_7	S_9	–100V
11	S_2	S_5	S_8	S_9	–150V
12	S_1	S_5	S_8	S_9	–200V
13	S_2	S_4	S_5	S_7	–250V
14	S_2	S_4	S_5	S_8	–300V
15	S_1	S_4	S_5	S_8	–350V

Table 2. PV Module parameters.

S.NO	PV Module's parameters	Values
1	current (short-circuit)	7.84A
2	voltage (open-circuit)	41 V
3	PV Current (at maximum point)	7.35 A
4	PV Voltage (at maximum point)	33 V
5	Power (at maximum point)	242.5 W
6	current (short-circuit)	7.84 A

2.3 The PV Array

The inverter in this system needs three DC power sources, and they all have different voltage ratings. Three Solar PV modules and three DC-DC boost converters were used to make these three DC power sources. As a DC source, each DC-DC boost converter is linked to a single 250 W PV module. Solar PV module design is carried out by simulating the environment's natural conditions, which are then processed to produce the voltage vs current at different irradiance and voltage vs power at different irradiance graphs. Table 1 shows the parameters of the solar module. Figure 2 shows the curves of the solar module panel.

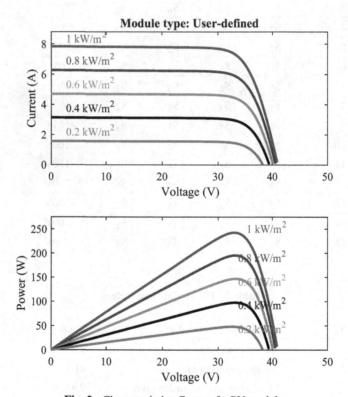

Fig. 2. Characteristics Curve of a PV module

2.4 Design of the DC-DC Converter

In the proposed system, three boost converters are used, and each boost converter's input voltage ranges from 25 V to 35 V. However, each boost converter's output voltage depends on its duty ratios and inverter input voltages (Vdc1 = 50 V, Vdc2 = 150 V, and Vdc3 = 150 V). Consider the boost converter's switching frequency, current ripple, voltage ripple and load capacity to be 50 kHz, 5%, 5% and 250 W, respectively. The inductance and capacitance are calculated as:

$$L = \frac{V_{in}(V_{out} - V_{in})}{f_{sw} \times \Delta I \times V_{out}} = 10.2\,mH$$

$$C = \frac{I_{out}(V_{out} - V_{in})}{f_{sw} \times \Delta V \times V_{out}} = 520\,\mu H$$

3 The Control System

The proposed system contains control strategies. One control strategy is for controlling boost converter output voltages, and the second is for controlling the inverter to export the power to the grid. The inverter control strategy should, however, include the following: 1) grid synchronization and 2) active power transfer by managing grid-injected current.

3.1 MPPT Algorithm

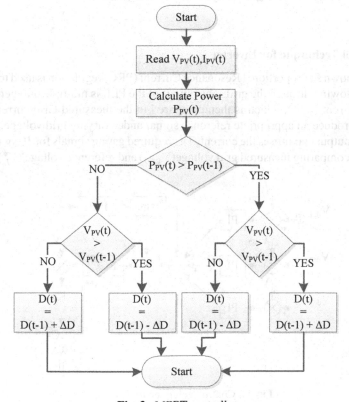

Fig. 3. MPPT controller

To track the maximum power from each solar module, the perturb and observe (P&O) algorithm is used for each module. Figure 3 is a flowchart illustrating the MPPT algorithm used to track the PV modules' maximum power point. It compares the sampled value of the voltage to the previous value in the same way and sets the maximum point's power duty according to the matching power's result (Fig. 4).

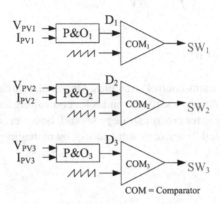

Fig. 4. Duty cycle control algorithm.

3.2 Control Technique for Inverter

A method known as Proportional Resonance Current (PRC) regulation is used to regulate the current flowing through the grid. The output of the PLL is multiplied to generate the reference current (i^*_{grid}), which is then compared to the measured Grid current (i_{grid}). In order to produce an appropriate reference signal under varying grid voltage, the PRC controller's output maximises the current. The required gating signals for 10 switches are obtained by comparing measured grid voltage(V_{grid}) and reference voltage (V^*_L) (Fig. 5).

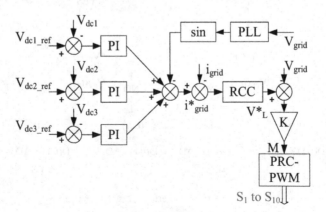

Fig. 5. Control technique for inverter

4   The Result Analysis

The process model and technique shown in Fig. 1 were verified through simulation studies. To examine the performance of the proposed system, two scenarios were simulated based on the model shown in Figs. 6, 7, 8 and 9

4.1   Proposed System Performance when Solar Irradiance Increases from 200 W/m² to 600 W/m²

At t = 2 s, The solar power generation increased from 140 w to 290 w as the solar irradiance increased from 200 w/m² to 400 w/m². The asymmetrical inverter can export 140 w power at 200 w/m² irradiance and 290 w power at 400 w/m² to the grid shown in Fig. 6(d). The boost converters maintain the output voltages as input to the inverter for successful operation. At t = 3s, with an increase in solar irradiance from 400 W/m² to 600 W/m², solar power generation increased from 290 w to 440 w and the same amount of power was export to the grid by the inverter. It has been observed that as the solar irradiance increases, the inverter can export power to the utility grid.

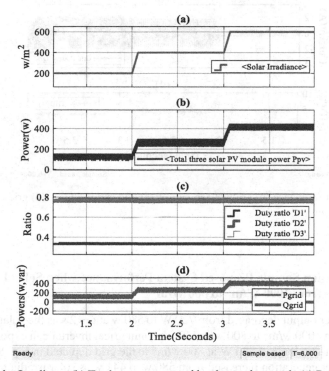

Fig. 6. (a) Solar Irradiance (b) Total power generated by three solar panels (c) Boost converter's Duty ratios (d) Active(P_{grid}) and reactive(Q_{grid}) power exporting to the grid

To export the power to the grid, the THD value of the current is necessary to maintain within IEEE standards. Figure 7(d) shows that at 200 w/m², 400 w/m², and 600 w/m² of solar irradiance, the THD value of the current is 4.5%, 4%, and 3.5%, respectively. The asymmetrical inverter can export the 0.45 A of peak current at 200 w/m^2 irradiance, 1.7 A of peak current at 400 w/m^2 and 1.9 A of peak current at 600 w/m^2 to the grid shown in Fig. 7(c).

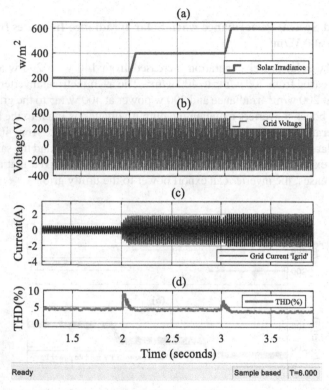

Fig. 7. (a) Solar Irradiance (b) grid voltage 'V$_{grid}$' (c) grid current 'I$_{grid}$' (d) grid current THD value in percentage

4.2 A Subsection Sample Proposed System Performance when Solar Irradiance Increases from 1000 W/m^2 to 600 W/m^2

The solar power output decreased from 720 W to 580 w at t = 2 s as the solar irradiance decreased from 1000 w/m^2 to 800 w/m^2. The asymmetrical inverter can export 720 W at 1000 w/m^2 of radiation and 580 W at 800 w/m^2 to the grid depicted in Fig. 8(d). At t = 3 s, solar power generation increased from 580 w to 441 w as solar irradiance decresed from 800 w/m^2 to 600 w/m^2, and the inverter exported the same amount of power to the grid. It has been noted that the inverter can export power to the utility grid as the solar irradiance rises.

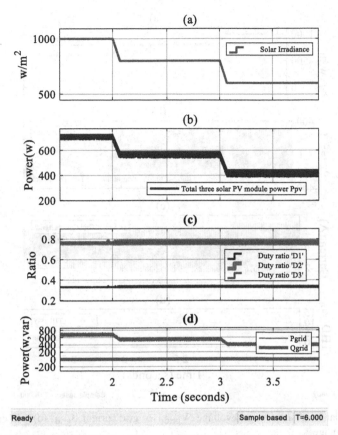

Fig. 8. (a) Solar Irradiance (b) Total power generated by three solar panels (c) Boost converter's Duty ratios (d) Active (P_{grid}) and reactive(Q_{grid}) power exporting to the grid.

To export the power to the grid, the THD value of the current is necessary to maintain within IEEE standards. The THD value of the current is 3%, 3.2%, and 3.5%, respectively, at 1000 w/m^2, 800 w/m^2, and 600 w/m^2 of solar irradiance, as shown in Fig. 9(d). The grid in Fig. 9(c) can receive peak currents of 2.75 A at 1000 w/m^2 irradiance, 2.35 A at 600 w/m^2, and 1.9 A at 600 w/m^2 from the asymmetrical inverter.

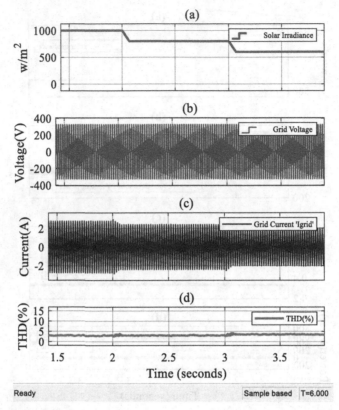

Fig. 9. (a) Solar Irradiance (b) grid voltage 'V$_{grid}$' (c) grid current 'I$_{grid}$' (d) grid current THD value in percentage

5   Conclusion

This work proposed the implementation of a fifteen-level inverter for PV applications. Conventional boost converters generate the required DC voltages, which are fed to the inverter for an AC stepped output waveform. The boost converters are designed to maintain the required output DC voltages for the inverter to function properly. The proposed system was tested under various solar irradiance conditions, and under all of them, the inverter performed better and could export power to the utility grid. When the solar irradiance increased and decreased, the inverter performed better. The inverter produces higher output voltage levels with fewer circuit components and low THD. The proposed system can export power to the grid while meeting IEEE standards in all solar irradiance conditions.

References

1. Pareek, P., Maurya, N.K., Singh, L., Gupta, N., Reis, M.J.C.S.: Study of smart city compatible monolithic quantum well photodetector. In: Gupta, N., Pareek, P., Reis, M. (eds.) Cognitive Computing and Cyber Physical Systems 2022, LNCS, vol. 472, pp. 215–224. Springer, Cham (2022)
2. Siva, A., Rajendran, V.: A novel auxiliary unit based high gain DC-DC converter for solar PV system with MPPT control. Int. J. Power Electron. Drive Syst. (IJPEDS) **13**, 2386–2395 (2022)
3. Bouaouaou, H., Lalili, D., Boudjerda, N.: Model predictive control and ANN-based MPPT for a multi-level grid-connected photovoltaic inverter. Electr. Eng. **104**, 1229–1246 (2022)
4. Siva, A., Rajendran, V.: A Novel asymmetric multilevel inverter with reduced number of switches for grid-tied solar PV system. Recent Adv. Electr. Electr. Eng. **15**, 379–389 (2022)
5. Sathiyanathan, M., Jaganathan, S. and Josephine, R.L.: Multi-input and multi-output bi-directional power converter for solar photovoltaic system. Electr. Eng. **103**, 3201–3216 (2021)
6. Marimuthu, M., Vijayalakshmi, S., Shenbaga, L.R.: A novel non-isolated single switch multilevel cascaded DC–DC boost converter for multilevel inverter application. J. Electr. Eng. Technol. **15**, 2157–2166 (2020)
7. Chappa, A., Gupta, S., Sahu, L.K., Gupta, K.K.: Resilient multilevel inverter topology with improved reliability. IET Power Electron. **13**, 3384–3395 (2020)
8. Hwang, J., Lim, S., Choi, M.: Reactive power control method for grid-tie inverters using current measurement of DG output. J. Electr. Eng. Technol. **14**, 603–612 (2019)
9. Zhao, L., Song, W. and Feng, J.: A compensation method of dead-time and forward voltage drop for inverter operating at low frequency. J. Electr. Eng. Technol. **14**, 781–794 (2019)
10. Liu, G., Caldognetto, T., Mattavelli, P.: Power-based droop control in DC microgrids enabling seamless disconnection from AC grids. In: IEEE Second International Conference on DC microgrids (ICDCM). IEEE, pp. 523–528 (2017)
11. Sharma, S., and Deshpande, A.: Design and development of maximum power point tracking algorithm using field programmable gate array. In: 2017 2nd IEEE International Conference on Recent Trends in Electronics, Information & Communication Technology (RTEICT), pp. 1560–1563 (2017)
12. Mantilla, M., Quinones, G., Castellanos, C., Petit, J., Orddonez, G.: Analysis of maximum power point tracking algorithms in DC–DC boost converters for grid-tied photovoltaic systems. In: Annual Conference of the IEEE Industrial Electronics Society, pp. 1971–1976 (2014)

A Review on Charging Control and Discharging Control of Plug-in Electric Vehicles in the Distribution Grid

Mekapati Suresh Reddy[1(✉)], Nalin Behari Dev Choudhury[1], Idamakanti Kasireddy[2], and Satyaki Biswas[1]

[1] National Institute of Technology Silchar, Assam, India
mekapatireddy@gmail.com, nalin@ee.nits.ac.in, biswassatyaki9@gmail.com
[2] Vishnu Institute of Technology, Bhimavaram, India
kasireddy.nit@gmail.com

Abstract. The potential of electric automobiles to discharge their batteries to the power grid (namely V2G technology) is discussed in this article as a way to lessen pollution and provide extra services, including income generation and grid reliance. The widespread use of electric cars might provide difficulties for the electrical infrastructure, such as power outages, overloading of transformers, and variations in bus voltage. The electrical grid and electric cars are connected through a middleman called an aggregator. This article investigates novel technologies and coordination mechanisms to control the discharging and charging of electric cars. In addition to that, Various optimization techniques are reviewed and also try to explain about challenges and future advances in the EV charging process and discharging process.

Keywords: EV Technology · Aggregator · V2G technology

1 Introduction

The use of electric vehicles (EVs) is crucial for reducing pollution and advancing the use of renewable energy sources. EVs may be a solution to environmental issues as the globe may experience oil scarcity in the future. The distribution grid can be stabilized with their assistance when used in conjunction with additional infrastructure, despite the fact that they might not be as successful when used alone as other technologies. With increasing numbers of electric vehicles (EVs) on the road, the distribution grid has to pay closer attention to how these vehicles are being charged and discharged. A component of the energy system, the distribution grid transports power from the transmission grid to users, including homes, businesses, and EV charging stations.

In order to coordinate EVs and manage their bi-directional charging process and discharging process processes, [1] proposed the idea of this EV aggregator. Electric vehicles (EVs) can assist in shifting energy use to periods when it is more affordable and less demanding by acting as an energy storage system using

P. Pareek et al. (Eds.): IC4S 2023, LNICST 537, pp. 48–62, 2024.
https://doi.org/10.1007/978-3-031-48891-7_4

V2G technology. This can reduce power costs and benefit the grid. EVs will boost the grid and lower demand on the electricity grid by moving the load from peak hours to off-peak hours to take part in demand size management [2].

V2G is also one other option for faster and more efficient than other storage options, making it a good choice for managing the power grid. The objective of V2G systems is to coordinate the charging process and discharging process and to achieve a reduction in dependency on the grid and acts as another source rather than generating power at remote locations and power balance to avoid problems with the grid [3–5].

For the charging process, efficient EV coordination is essential. Potential problems may be reduced by implementing group decision-making and centralised coordination within the scope of Vehicle to Grid (V2G) technology. It can be challenging to establish V2G coordination in big power systems, though. The present infrastructure, which was created only for loads without EV integration in the grid, could have problems, and EV owners might not want to immediately give electricity to the grid. This may result in issues with grid dependability and battery efficiency. Research is required to develop transdisciplinary computational models that may address these problems. Recent research focuses on improving the functionality of V2G systems as well as strategies to use V2G for commercial advantage and the provision of additional services. Some researchers are working together to find answers. This makes studying how to control the charging process and discharging process of V2G systems an exciting area of study [6–10].

This study attempts to illustrate how EV charging process and discharging process operations may be managed by associations of EV owners (aggregators). It compares different methods for making decisions about charging process and discharging process and looks at ways to reduce the pressure on the grid.

The use of technology namely (V2G) Vehicle-to-Grid, is the main emphasis since it may help to increase power quality and efficiency while lowering the cost of energy (CoE). It talks about ways to implement this approach, taking into account the capacity of the existing infrastructure. The paper provides a lot of information about how to set up networks for different sizes of EV fleets to help to reduce the load on the electrical grid.

2   Exploring the Literature on Charging-Discharging Techniques for Electric Vehicle Aggregators

The charging process and discharging process techniques of EVs are classified into two types controlled and uncontrolled, respectively. The uncontrolled method described does not involve the transmission of any details about the system from the user to the grid operator, which can potentially lead to issues such as instability of the grid, poor power quality, operational inefficiencies, and uncertainties regarding the battery's state of charge. To analyze the effect on electric power systems and establish control over the charging process and discharging process of EVs, it is necessary to conduct load modeling of EV charging [7,8].

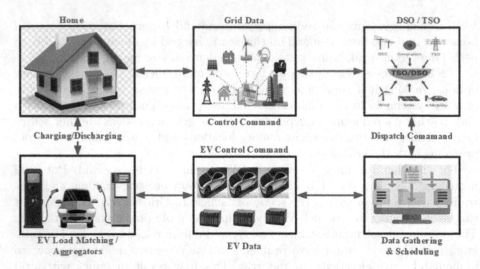

Fig. 1. EV Charge - Discharge using Control Center

The technique of controlled EV charging-discharging, also referred to as coordinated charging-discharging, has received significant attention in recent years. The coordinated and centrally regulated charge-discharge model is shown in Fig. 1, and it is carried out using a schematic algorithm. This method makes it easy for the person in charge to track and control how electric vehicles are charged and used. They can plan the schedule to avoid problems with electricity and make sure it works well for both the drivers and the company. The person in charge's goals includes making sure the system is working well and making improvements. Depending on how the system is controlled, there are different ways of charging process and discharging process using electric vehicles such as indirect control, intelligent control, bidirectional control, and multi-stage hierarchical control method and it has been referred to in Table 1.

Table 1. Control methods of charging process and discharging process

Method	Control	Suitable
Indirect control [11–15]	Energy cost, and Preventing grid overload	Small Scale EV
Intelligent control [17–27]	Reducing grid overload, Electricity costs and balancing the power grid	Large-scale EV
Bidirectional control [28–32]	EV as Energy storage to reduce peak load	Big-scale EV
Multi-stage hierarchical control [33–38]	EV owners, local aggregators, and a central coordinator	Large-scale EV

The Indirect Controlled Approach is a way of managing how electric vehicles are charged and used where there are no rules on how the charging should happen, such as how the charger is controlled, how long the charging takes, and how much the vehicle is charged.The emphasis is instead placed on minimizing grid overload and regulating extraneous variables like energy costs. EV owners make the charging schedule based on their own needs, while the operator's role is limited to providing information and incentives to influence their decisions. This approach is more straightforward and flexible but less effective in balancing the power grid. It may be used as a starting point for constructing more sophisticated techniques and is appropriate for small-scale EV charging stations [11–16].

The intelligent managed method is a technique for managing the charging process and discharging process of electric cars. In order to optimize the charging-discharging schedule, the operator actively manages the charging-discharging process. This approach is considered more complex than the Indirect Controlled Approach. Still, it can provide more accurate and efficient results, such as reducing grid overload, balancing the power grid, and reducing electricity costs. The operator uses techniques like real-time pricing, game theory, and optimization to determine the best charging-discharging schedule. This process is suitable for large-scale EVs charging systems and can significantly benefit the grid and EV owners. Based on the waiting times of EVs and their battery capacity, an individual EV-charging plan with a clever coordinating technique reduces overall charging costs. Based on the network's maximum daily power usage, the maximum number of electric cars that may be connected to the distribution grid at once is determined [17–27].

A method for managing the charging process and discharging process of EVs is named a bi-directional controlled approach (V2G). With this strategy, the EV is able to serve as energy storage at the dispersed end, enhancing the electric grid's capabilities and lowering peak demand. The grid and EV owners may benefit from additional benefits even though this strategy is thought to be more difficult than the indirect and intelligent controlled approaches. The V2G technology is appropriate for the large energy demands in the power distribution system, aids in managing the flow of power during variations in demand, and offers additional services to the power grid. This approach is appropriate for large-scale EV systems and advantageous to both the grid and EV owners [28–32].

An approach for organising the charging and usage of electric cars is the multistage hierarchical controlled technique.It incorporates a multi-level hierarchy of organisations (such as EV owners, local aggregators, and a central coordinator) making choices regarding charging process and discharging process, with each level of the hierarchy being in charge of various parts of coordination. The process is referred to as "multistage" because it entails a number of levels or phases of decision-making. For instance, at the beginning, EV owners will choose when to charge and discharge their cars. In the second step, local aggregators may plan the charging process and discharging processes of several EVs in a certain region. To guarantee the best possible functioning of the power grid as a whole, a central coordinator in the third stage coordinates the charging process and discharging process of EVs across several geographical locations. The approach is known as "hierarchical" because it uses a hierarchy of decision-making levels, with each level in charge of a distinct component of coordination. It is now feasible for EV collaboration to be more effective and efficient due to the increased alignment of the interests of numerous parties. The multistage hierarchical controlled charge-discharge approach may be applied in a variety of contexts. According to the demand and how the other groups of electric cars are being utilized, the power network may be adjusted using this technique [33–38].

3   Strategic Electric Vehicle Charging Management

The goal of this study is to compare various charging and operating techniques in order to determine the most effective strategy. It aims to improve the performance of electric cars by managing the way they use and store energy. This includes techniques like predicting the vehicle's future energy needs and correctly managing power usage or finding ways to optimize the charging process and discharging process of the battery. The ultimate goal of our study is to increase the performance and range of EVs so that they are more extensively used. The study will review earlier research on EMS optimization for EVs and discuss the various methods and ideas proposed. The authors of the study stress that adding EVs to the electrical grid may have negative consequences. Effective optimisation methods for EV charging process and discharging process are provided in Table 2, which can be used to reduce these problems. This essay aims to provide a detailed analysis of various approaches and the effects they have on the grid.

Fuzzy logic-based two-stage charging is a technique for managing the charging of electric vehicles (EVs) in a way that balances the needs of EV owners and the power utility grid. The initial step of decision-making in this strategy is concerned with the EV's urgent charging needs. This method uses fuzzy logic control to choose the appropriate charging rate while considering the driver's preferences, the battery's current state of charge, and factors like remaining driving distance into account. The second stage accounts for the grid's and the EV's long-term charging requirements, adjusting the charging schedule in accordance with the grid's overall load, the time of day, the availability of renewable sources, and other elements. This approach is considered to be more advanced than others since it allows for a flexible and responsive charging procedure while taking into consideration the needs of both EV owners and the power grid. While the second stage enables real-time modifications to the charging schedule to match the changing demands of the power grid, the first stage uses fuzzy logic control to produce a more effective and precise charging process. This approach is appropriate for large-scale EV charging stations because it has several advantages, including cutting electricity prices, balancing the power grid, minimising grid overload, and increasing grid reliability [24,39–43].

Model Predictive Control (MPC), a technology used to control EV charging in the Real-time Charging approach, aims to strike a balance between the

Table 2. Optimizatin Techniques for EV charging process and discharging process

Approach	Optimization	Impact
Two-stage [24, 39–43]	Fuzzy logic control	Reducing grid overload, balancing the power grid, lowering electricity costs, and improving grid reliability
Real-time [44–47]	Model Predictive Control	
Metaheuristic Control [48–51]	Advanced than others and used for Solving complex, Non-linear optimization Problems	
Heuristic Control [52–56]	Greedy algorithm, Dynamic Programming, Tabu search, Scatter search	

needs of EV owners and the power grid. MPC is a very advanced control technique that uses a mathematical model of the system in issue to predict its future behavior. In the context of EV charging, this technique uses a model of the grid and the EV battery to predict their future states. Based on this prediction, the algorithm optimizes the charging schedule to fulfill both the demands of the EV owner and the power grid, making it a more advanced charging strategy than others. It is a resource-use approach that is effective and better able to manage dynamic and unpredictable systems because of its flexibility and reactivity to the changing needs of the power grid in real-time. Large-scale EV charging systems can benefit from MPC because it allows the charging schedule to be changed to accommodate the changing needs of the power grid, reducing grid overload, balancing the power grid, lowering electricity costs, improving power grid reliability, and providing supplemental services to the grid [44–47].

A technique for managing how electric vehicles (EVs) charge and discharge that makes use of sophisticated optimisation algorithms based on metaheuristics is called metaheuristic charge-discharge control optimisation. Metaheuristics are a subset of optimisation algorithms that draw their inspiration from events in nature, such as animal behaviour or species development. The complicated, non-linear optimisation issues that arise during the charging process and discharging process of EVs are ideally suited for these algorithms' solution. In this method, various metaheuristics are used to find the best charging process and discharging process schedule. The algorithm considers factors like the current state of the EVs battery, the remaining driving distance, the current demand on the grid, and the availability of renewable energy sources.

In comparison to previous charging-discharging techniques, the Metaheuristic Charge-Discharge Control Optimisation approach is seen to be more advanced since it enables a more adaptable and responsive charging-discharging process while taking the changing demands of the power grid and the EV owner into account. With the help of metaheuristics, the charging-discharging schedule may be modified in real-time to take into account the fluctuating demands of the power grid and lessen the burden of the EV on it. Due to its many benefits, including its capacity to reduce grid overload, balance the power grid, lower energy prices, increase power grid dependability, and provide ancillary services to the grid, this approach may be appropriate for large-scale EV charging systems [48–51].

An strategy for controlling the charging process and discharging process processes of electric vehicles (EVs) that makes use of cutting-edge optimisation algorithms based on heuristics is called the Heuristic Control Optimisation based Charge and Discharge Method. In contrast to mathematical proofs or comprehensive system knowledge, heuristics are optimisation algorithms that depend on practical knowledge, rules of thumb, and basic mathematical models. This approach uses a variety of heuristics, such as the greedy algorithm, tabu search, dynamic programming, scatter search, etc., to determine the best charging process and discharging process schedule for the EV, taking into account elements like the battery's current condition, the amount of remaining driving time, the current demand on the power grid, and the availability of renewable energy sources. This approach is more sophisticated than previous charging process and discharging process techniques since it enables a more adaptable and quick procedure for charging process and discharging process while taking into consideration the altering requirements of both the power grid and the EV owner. This technique is appropriate for large-scale EV charging systems since the use of heuristics allows the charging process and discharging process schedule to be changed in real-time. This strategy also has a number of positive effects, including lowering electricity prices, balancing the power system, increasing grid resilience, and delivering useful services to the grid [52–56].

4   New Advances and Challenges in EV Charging Process and Discharging Process Techniques

4.1   New Advances

As the market for electric vehicles expands, a number of attractive future research areas might increase the contribution of EVs to the grid. In one of these regions, there is no need for physical connections because wireless charging is used. As a result, EV owners won't need to plug in their cars to charge, which makes the process quicker and more practical. Because wireless charging pads may be deployed in a variety of settings, including parking garages, public areas, and even on the street, this may further enhance the number of charging alternatives for EV drivers [57–63].

Due to its capacity to shorten charging times and ease EV drivers' range anxiety, rapid charging for electric cars (EVs) has become more popular. This is accomplished by using high power levels of 50 kW to 350 kW, which may charge EVs up to 80% in as short as 30 min [64–66].

Lightweight PV cells that are mounted to the steel of all vehicle body panels facing upward will be used to wirelessly charge EV batteries when the vehicle is in motion or while it is parked outside. The best wireless charging method for increasing the range of electric vehicles (EVs) uses free, environmentally friendly solar energy to charge the batteries [67–70].

Another possible area for development is the application of blockchain technology in systems for voice-to-text communication. Due to its ability to provide secure, decentralised, and transparent transactions, blockchain technology is the greatest option for managing energy transfers between EVs and the grid. The efficiency and cost-effectiveness of energy trading between EV owners and energy providers may subsequently increase, and the security and transparency of energy exchange between EVs and the grid may be enhanced [71,72].

Concerns about privacy and security must be addressed as EVs become more integrated into the power system. With the rise of connected vehicles, there is a growing concern about the collection and sharing of personal data by EV manufacturers and energy providers. To handle these issues, clear rules and laws about how personal information is gathered, kept, and shared must be implemented [73–77].

Moreover, the optimization of smart charging is a crucial area that needs to be further developed to balance the energy demand of EVs owner and the distribution grid. Smart charging optimization involves managing the charging process and discharging process of EVs in a way that minimizes the impact on the power grid, reduces electricity costs, and improves the grid's reliability. This can be achieved through advanced control methods, optimization algorithms, and energy management systems [78–81].

Many electricity providers offer TOU rates, which provide lower rates during off-peak hours when electricity demand is lower. By charging their electric vehicle during off-peak hours, owners can take advantage of these lower rates and minimize their overall charging costs [82–86].

4.2 Challenges

The following are the challenges in EV charging process and discharging process Techniques [24,87–92].

- Availability and accessibility of charging infrastructure, particularly in remote areas or regions with limited resources. This can limit the use of electric vehicles for long-distance travel or in areas with insufficient charging options.
- When implementing EV charging-discharging methods, it is important to consider factors such as the ability to charge EVs, the expense of batteries and other parts, and the effect on the power grid are important factors to be considered. To ensure the safe and secure collection, storage, and exchange of personal data, privacy and security issues must also be taken into account.
- As EVs are driven on the road increases, there is a risk of grid frequency issues caused by the sudden injection of large amounts of power into the grid during periods of high charging demand. This can give quality issues such as flicker, distortion, and voltage sag, which can negatively impact the performance of other devices connected to the grid.
- Advanced methods for EV charging and discharge optimisation must be used in order to efficiently manage the EVs (electric vehicles) being integrated into the distribution grid. In order to lessen the load on the grid and increase

efficiency, this involves using control techniques, optimisation algorithms, and energy management systems.

- The need to commercialise V2G technology is growing as interest in the technology rises. designing a market for V2G energy and designing a business model that will enable energy companies to offer V2G services are two examples of what is involved.
- With the use of the advantage of government incentives, such as rebates and tax credits which can helps to reduce the upfront cost of purchasing an electric vehicle or installing a home charging station. These incentives can also help to offset the long-term cost of owning and operating an electric vehicle.
- Pricing thing also needs to think about dynamic pricing and cost minimization. One challenge is the structure of electricity rates. Some utility companies offer variable rates that fluctuate throughout the day, which can make it difficult for electric vehicle owners to predict the cost of charging their vehicles. In addition to that, some utilities charge higher rates during peak usage times, which can coincide with the times when electric vehicle owners need to charge their vehicles
- Adoption of V2G technology in military applications poses several challenges, such as the lack of standardization in the EVs charging process and discharging process technologies, the high cost of batteries and other EV components, and the lack of widespread EV charging infrastructure. Additionally, the military's unique requirements for mobility, reliability, and security present additional challenges for V2G adoption.

5   Conclusion

This article offers a thorough analysis of the latest charging process and discharging process techniques, optimization methodologies, and objectives for electric vehicles. It examines various approaches to charging process and discharging process electric vehicles in upcoming Vehicle-to-Grid systems, including optimization techniques in V2G control, maintaining power grid stability, and managing high energy demand. The paper focuses on the role of an aggregator in V2G integration, current research on hierarchical EV optimization methods, and emerging multi-objective techniques for multistage hierarchy in charging process and discharging process for commercial use now and in the future. The article also introduces the basic concepts of charging-discharging planning and operation and suggests new advances in the areas for future research.

References

1. Kempton, W., Letendre, S.: Electric vehicles as a new power source for electric utilities. Transp. Res. Part D Transp. Environ. **2**, 157–175 (1997)
2. Daina, N., Sivakumar, A., Polak, J.: Electric vehicle charging choices: modelling and implications for smart charging services. Transp. Res. Part C Emerg. Technol. **81**, 36–56 (2017)
3. Kempton, W., Tomic, J., Letendre, S., Brooks, A., Lipman, T.: Vehicle-to-grid power: battery, hybrid, and fuel cell vehicles as resources for distributed electric power in California (2001)
4. Wang, Q., Liu, X., Du, J., Kong, F.: Smart charging for electric vehicles: a survey from the algorithmic perspective. IEEE Commun. Surv. Tutor. **18**, 1500–1517 (2016)
5. Nunes, P., Brito, M.: Displacing natural gas with electric vehicles for grid stabilization. Energy **141**, 87–96 (2017)
6. Saldanha, J., Dos Santos, E., De Mello, A., Bernardon, D.: Control strategies for smart charging process and discharging process of plug-in electric vehicles. Smart Cities Technol. **1**, 121–141 (2016)
7. Micari, S., Polimeni, A., Napoli, G., Andaloro, L., Antonucci, V.: Electric vehicle charging infrastructure planning in a road network. Renewable Sustain. Energy Rev. **80**, 98–108 (2017)
8. Kuppusamy, S., Magazine, M., Rao, U.: Electric vehicle adoption decisions in a fleet environment. Eur. J. Oper. Res. **262**, 123–135 (2017)
9. Kong, P., Karagiannidis, G.: Charging schemes for plug-in hybrid electric vehicles in smart grid: a survey. IEEE Access. **4**, 6846–6875 (2016)
10. Dharavat, N., et al.: Impact of plug-in electric vehicles on grid integration with distributed energy resources: a review. Front. Energy Res. **10**, 1099890 (2023)
11. Davis, L.: Electric vehicles in multi-vehicle households. Appl. Econ. Lett. **30**, 1–4 (2022)
12. Yu, J., Lin, J., Lam, A., Li, V.: Coordinated electric vehicle charging control with aggregator power trading and indirect load control. ArXiv Preprint ArXiv:1508.00663 (2015)
13. Kong, L., Han, J., Xiong, W., Wang, H., Shen, Y., Li, Y.: A review of control strategy of the large-scale of electric vehicles charging process and discharging process Behavior. IOP Conf. Ser. Mat. Sci. Eng. **199**, 012039 (2017)
14. Divshali, P., Choi, B.: Efficient indirect real-time EV charging method based on imperfect competition market. In: 2016 IEEE International Conference On Smart Grid Communications (SmartGridComm), pp. 453–459 (2016)
15. Hu, J., Si, C., Lind, M., Yu, R.: Preventing distribution grid congestion by integrating indirect control in a hierarchical electric vehicles' management system. IEEE Trans. Transp. Electrification **2**, 290–299 (2016)
16. Imthias Ahamed, T.P., Devaraj, D.: Optimized charge scheduling of plug-in electric vehicles using modified placement algorithm. In: 2019 International Conference on Computer Communication and Informatics (ICCCI), pp. 1–5. IEEE
17. Moghaddami, M., Sarwat, A.: A three-phase ac-ac matrix converter with simplified bidirectional power control for inductive power transfer systems. In: 2018 IEEE Transportation Electrification Conference and Expo (ITEC), pp. 380–384 (2018)
18. Tan, K., Ramachandaramurthy, V., Yong, J.: Bidirectional battery charger for electric vehicle. In: 2014 IEEE Innovative Smart Grid Technologies-Asia (ISGT ASIA), pp. 406–411 (2014)

19. Hofmann, M., Schafer, M., Ackva, A.: Bi-directional charging system for electric vehicles: a V2G concept for charging process and discharging process electric vehicles. In: 2014 4th International Electric Drives Production Conference (EDPC), pp. 1–5 (2014)
20. Lambert, G., et al.: Bidirectional charging system for electric vehicle. (Google Patents, 2017), US Patent App. 15/307,255
21. Habib, S., Khan, M., Abbas, F., Tang, H.: Assessment of electric vehicles concerning impacts, charging infrastructure with unidirectional and bidirectional chargers, and power flow comparisons. Int. J. Energy Res. **42**, 3416–3441 (2018)
22. Dehaghani, E., Cipcigan, L., Williamson, S.: The role of electric vehicles in smart grids. Planning and Operation of Active Distribution Networks, pp. 123–151 (2022)
23. Tachikawa, K., Kesler, M., Danilovic, M., Esteban, B., Atasoy, O., Yeung, K.: Bi-Directional wireless power transfer for vehicle-to-grid: demonstration and performance analysis. SAE Technical Paper (2019)
24. Solanke, T., Ramachandaramurthy, V., Yong, J., Pasupuleti, J., Kasinathan, P., Rajagopalan, A.: A review of strategic charging-discharging control of grid-connected electric vehicles. J. Energy Storage **28**, 101193 (2020)
25. Reddy, K., Meikandasivam, S.: Load flattening and voltage regulation using plug-in electric vehicle's storage capacity with vehicle prioritization using anfis. IEEE Trans. Sustain. Energy **11**, 260–270 (2018)
26. Akil, M., Dokur, E., Bayindir, R.: Optimal scheduling of aggregated electric vehicle charging with a smart coordination approach. In: 2022 11th International Conference on Renewable Energy Research and Application (ICRERA), pp. 546–551 (2022)
27. Akil, M., Dokur, E., Bayindir, R.: Optimal scheduling of on-street EV charging stations. In: 2022 IEEE 20th International Power Electronics and Motion Control Conference (PEMC), pp. 679–684 (2022)
28. Li, F., Ji, F., Guo, H., Li, H., Wang, Z.: Research on integrated bidirectional control of EV charging station for V2G. In: 2017 2nd International Conference on Power and Renewable Energy (ICPRE), pp. 833–838 (2017)
29. Tang, Y., Chen, Y., Madawala, U., Thrimawithana, D., Ma, H.: A new controller for bidirectional wireless power transfer systems. IEEE Trans. Power Electron. **33**, 9076–9087 (2017)
30. Zhang, M.: Battery charging process and discharging process research based on the interactive technology of smart grid and electric vehicle. AIP Conf. Proc. **1971**, 050004 (2018)
31. Samanta, S., Rathore, A.: A single-stage universal wireless inductive power transfer system with V2G capability. In: 2018 International Conference on Power, Instrumentation, Control and Computing (PICC), pp. 1–5 (2018)
32. Verma, A., Singh, B.: Three phase off-board bi-directional charger for EV with V2G functionality. In: 2017 7th International Conference on Power Systems (ICPS), pp. 145–150 (2017)
33. Singh, A., Pathak, M.: A multi-functional single-stage power electronic interface for plug-in electric vehicles application. Electric Power Compon. Syst. **46**, 135–148 (2018)
34. Zhu, X., Han, H., Gao, S., Shi, Q., Cui, H., Zu, G.: A multi-stage optimization approach for active distribution network scheduling considering coordinated electrical vehicle charging strategy. IEEE Access. **6**, 50117–50130 (2018)
35. Wang, Y., Bai, H., Li, W., Bu, F., Hua, Y., Han, D.: Function system and application scenario design of energy big data application center. In: 2019 IEEE 3rd

Conference on Energy Internet and Energy System Integration (EI2), pp. 1788–1792 (2019)

36. Shukla, A., Verma, K., Kumar, R.: Multi-stage voltage dependent load modelling of fast charging electric vehicle. In: 2017 6th International Conference on Computer Applications in Electrical Engineering-Recent Advances (CERA), pp. 86–91 (2017)

37. Luo, L., et al.: Optimal planning of electric vehicle charging stations comprising multi-types of charging facilities. Appl. Energy **226**, 1087–1099 (2018)

38. Wang, Y., Thompson, J.: Two-stage admission and scheduling mechanism for electric vehicle charging. IEEE Trans. Smart Grid. **10**, 2650–2660 (2019)

39. ERDOGAN, N., ERDEN, F., KISACIKOGLU, M.: A fast and efficient coordinated vehicle-to-grid discharging control scheme for peak shaving in power distribution system. J. Modern Power Syst. Clean Energy **6**(3), 555–566 (2018). https://doi.org/10.1007/s40565-017-0375-z

40. Bandpey, M., Firouzjah, K.: Two-stage charging strategy of plug-in electric vehicles based on fuzzy control. Comput. Oper. Res. **96**, 236–243 (2018)

41. Wang, B., et al.: Predictive scheduling for Electric Vehicles considering uncertainty of load and user behaviors. In: 2016 IEEE/PES Transmission And Distribution Conference And Exposition (T&D), pp. 1–5 (2016)

42. Yan, D., Yin, H., Li, T., Ma, C.: A two-stage scheme for both power allocation and EV charging coordination in a Grid-Tied PV-battery charging station. IEEE Trans. Industr. Inf. **17**, 6994–7004 (2021)

43. Yan, D., Ma, C., Chen, Y.: Distributed coordination of charging stations considering aggregate EV power flexibility. IEEE Trans. Sustain. Energy **14**, 356–370 (2023)

44. Fetene, G., Kaplan, S., Sebald, A., Prato, C.: Myopic loss aversion in the response of electric vehicle owners to the scheduling and pricing of vehicle charging. Transp. Res. Part D Transp. Environ. **50**, 345–356 (2017)

45. Ji, Z., Huang, X., Xu, C., Sun, H.: Accelerated model predictive control for electric vehicle integrated microgrid energy management: a hybrid robust and stochastic approach. Energies **9**, 973 (2016)

46. Li, T., Liu, H., Wang, H., Yao, Y.: Multiobjective optimal predictive energy management for fuel cell/battery hybrid construction vehicles. IEEE Access. **8**, 25927–25937 (2020)

47. Cai, S., Matsuhashi, R.: Model predictive control for EV aggregators participating in system frequency regulation market. IEEE Access. **9**, 80763–80771 (2021)

48. Eldeeb, H., Faddel, S., Mohammed, O.: Multi-objective optimization technique for the operation of grid tied PV powered EV charging station. Electric Power Syst. Res. **164**, 201–211 (2018)

49. Aziz, M., Budiman, B.: Extended utilization of electric vehicles in electrical grid services. In: 2017 4th International Conference on Electric Vehicular Technology (ICEVT), pp. 1–6 (2017)

50. Yong, W., Haihong, B., Chunning, W.: Research on charging process and discharging process dispatching strategy for Electric Vehicles. Open Fuels Energy Sci. J. **8** (2015)

51. Gang, J., Lin, X.: A novel demand side management strategy on electric vehicle charging behavior. In: 2018 IEEE 15th International Conference on Networking, Sensing and Control (ICNSC), pp. 1–5 (2018)

52. Lee, J., Park, G.: Revenue analysis of a lightweight V2G electricity trader based on real-life energy demand patterns. Int. J. Multimedia Ubiquitous Eng. **10**, 9–18 (2015)

53. Panwar, L., Reddy, K., Kumar, R., Panigrahi, B., Vyas, S.: Strategic Energy Management (SEM) in a micro grid with modern grid interactive electric vehicle. Energy Conversion Manag. **106**, 41–52 (2015)
54. Aljanad, A., Mohamed, A., Shareef, H., Khatib, T.: A novel method for optimal placement of vehicle-to-grid charging stations in distribution power system using a quantum binary lightning search algorithm. Sustain. Cities Soc. **38**, 174–183 (2018)
55. Dogan, A., Bahceci, S., Daldaban, F., Alci, M.: Optimization of charge/discharge coordination to satisfy network requirements using heuristic algorithms in vehicle-to-grid concept. Adv. Electr. Comput. Eng. **18**, 121–130 (2018)
56. Turker, H., Bacha, S.: Optimal minimization of plug-in electric vehicle charging cost with vehicle-to-home and vehicle-to-grid concepts. IEEE Trans. Veh. Technol. **67**, 10281–10292 (2018)
57. Zhai, H., Pan, H., Lu, M.: A practical wireless charging system based on ultra-wideband retro-reflective beamforming. In: 2010 IEEE Antennas and Propagation Society International Symposium, pp. 1–4 (2010)
58. Beh, T., Imura, T., Kato, M., Hori, Y.: Basic study of improving efficiency of wireless power transfer via magnetic resonance coupling based on impedance matching. In: 2010 IEEE International Symposium on Industrial Electronics, pp. 2011–2016 (2010)
59. Inamdar, S., Fernandes, J.: Review of wireless charging technology for electric vehicle. In: 2022 IEEE 10th Power India International Conference (PIICON), pp. 1–5 (2022)
60. Mahesh, A., Chokkalingam, B., Mihet-Popa, L.: Inductive wireless power transfer charging for electric vehicles-a review. IEEE Access. **9**, 137667–137713 (2021)
61. Fan, Z., Jie, Z., Yujie, Q.: A survey on wireless power transfer based charging scheduling schemes in wireless rechargeable sensor networks. In: 2018 IEEE 4th International Conference on Control Science and Systems Engineering (ICCSSE), pp. 194–198 (2018)
62. Vinoth Kumar, K., Maruthi, B., Rahul, R., Santhosh Melvin, D., Sathish, S.: A review of dynamic wireless power transfer system technology used in solar wireless electric vehicle charging stations. In: 2022 International Conference on Automation, Computing And Renewable Systems (ICACRS), pp. 198–201 (2022)
63. Lu, X., Wang, P., Niyato, D., Kim, D., Han, Z.: Wireless charging technologies: fundamentals, standards, and network applications. IEEE Commun. Surv. Tutor. **18**, 1413–1452 (2016)
64. Konara, K., Kolhe, M.: Charging coordination of opportunistic EV users at fast charging station with adaptive charging. In: 2021 IEEE Transportation Electrification Conference (ITEC-India), pp. 1–6 (2021)
65. Tu, H., Feng, H., Srdic, S., Lukic, S.: Extreme fast charging of electric vehicles: a technology overview. IEEE Trans. Transp. Electrification **5**, 861–878 (2019)
66. Tan, J., Wang, L.: Real-time charging navigation of electric vehicles to fast charging stations: a hierarchical game approach. IEEE Trans. Smart Grid. **8**, 846–856 (2017)
67. Tanveer, M.S., Gupta, S., Rai, R., Jha, R.N.K., Bansal, M.: Solar based electric vehicle charging station. In: 2019 2nd International Conference on Power Energy, Environment and Intelligent Control (PEEIC), Greater Noida, India, 2019, pp. 407–410 (2019). https://doi.org/10.1109/PEEIC47157.2019.8976673
68. Fathabadi, H.: Plug-in hybrid electric vehicles: replacing internal combustion engine with clean and renewable energy based auxiliary power sources. IEEE Trans. Power Electron. **33**(11), 9611–9618 (2018)

69. Abdelhamid, M., Pilla, S., Singh, R., Haque, I., Filipi, Z.: A comprehensive optimized model for on-board solar photovoltaic system for plug-in electric vehicles: energy and economic impacts. Int. J. Energy Res. **40**(11), 1489–1508 (2016)
70. Mobarak, M., Kleiman, R., Bauman, J.: Solar-charged electric vehicles: a comprehensive analysis of grid, driver, and environmental benefits. IEEE Trans. Transp. Electrification **7**, 579–603 (2021)
71. Shen, J., Zhou, T., Wei, F., Sun, X., Xiang, Y.: Privacy-preserving and lightweight key agreement protocol for V2G in the social Internet of Things. IEEE Internet Things J. **5**, 2526–2536 (2017)
72. Zhou, Z., Tan, L., Xu, G.: Blockchain and edge computing based vehicle-to-grid energy trading in energy internet. In: 2018 2nd IEEE Conference on Energy Internet and Energy System Integration (EI2), pp. 1–5 (2018)
73. Li, D., Yang, Q., Yu, W., An, D., Zhang, Y., Zhao, W.: Towards differential privacy-based online double auction for smart grid. IEEE Trans. Inf. Forens. Secur. **15**, 971–986 (2020)
74. Islam, S., Badsha, S., Sengupta, S., Khalil, I., Atiquzzaman, M.: An intelligent privacy preservation scheme for EV charging infrastructure. IEEE Trans. Industr. Inf. **19**, 1238–1247 (2023)
75. Almarshoodi, A., Keenan, J., Campbell, I., Hassan, T., Ibrahem, M., Fouda, M.: Security and privacy preservation for future vehicular transportation systems: a survey. In: 2023 IEEE 12th International Conference On Communication Systems And Network Technologies (CSNT), pp. 728–734 (2023)
76. Chavali, S., Cheema, H., Delgado, R., Nolan, E., Ibrahem, M., Fouda, M.: A review of privacy-preserving authentication schemes for future internet of vehicles. In: 2023 IEEE 12th International Conference on Communication Systems and Network Technologies (CSNT), pp. 689–694 (2023)
77. Eiza, M., Shi, Q., Marnerides, A., Owens, T., Ni, Q.: Efficient, secure, and privacy-preserving pmipv6 protocol for V2G networks. IEEE Trans. Veh. Technol. **68**, 19–33 (2019)
78. Roszczypala, D., Batard, C., Poitiers, F., Ginot, N.: Implementation of dynamic programming algorithms for electric vehicle smartcharging in a real parking lot with supervision. In: 2020 IEEE 29th International Symposium On Industrial Electronics (ISIE), pp. 886–891 (2020)
79. Tao, M., Ota, K., Dong, M., Qian, Z.: AccessAuth: capacity-aware security access authentication in federated-IoT-enabled V2G networks. J. Parallel Distrib. Comput. **118**, 107–117 (2018)
80. Gao, F., Zhu, L., Shen, M., Sharif, K., Wan, Z., Ren, K.: A blockchain-based privacy-preserving payment mechanism for vehicle-to-grid networks. IEEE Network **32**, 184–192 (2018)
81. Li, Y., Zhang, P., Wang, Y.: The location privacy protection of electric vehicles with differential privacy in V2G networks. Energies **11**, 2625 (2018)
82. Kenneth, N., Logenthiran, T.: A novel concept for calculating electricity price for electrical vehicles. In: 2017 IEEE PES Asia-Pacific Power And Energy Engineering Conference (APPEEC), pp. 1–6 (2017)
83. Yu, N., et al.: Research on dynamic pricing strategy of electric vehicle charging based on game theory under user demand service scheme. In: 2022 International Conference On Manufacturing, Industrial Automation And Electronics (ICMIAE), pp. 94–99 (2022)

84. Hongli, L., Xuxia, L., Kaikai, W., Jingyu, Z., Qiang, L.: Day-ahead optimal dispatch of regional power grid based on electric vehicle participation in peak shaving pricing strategy. In: 2022 IEEE 5th International Electrical And Energy Conference (CIEEC), pp. 1265–1270 (2022)
85. Yan, Q., Manickam, I., Kezunovic, M., Xie, L.: A multi-tiered real-time pricing algorithm for electric vehicle charging stations. In: 2014 IEEE Transportation Electrification Conference and Expo (ITEC), pp. 1–6 (2014)
86. Ma, H., Lai, L., Sun, J.: New energy double-layer consumption method based on orderly charging of electric vehicles and electricity price interaction. In: 2022 7th Asia Conference On Power And Electrical Engineering (ACPEE), pp. 476–481 (2022)
87. Cheng, A., Tarroja, B., Shaffer, B., Samuelsen, S.: Comparing the emissions benefits of centralized vs. decentralized electric vehicle smart charging approaches: a case study of the year 2030 California electric grid. J. Power Sources **401**, 175–185 (2018)
88. Kester, J., Noel, L., Rubens, G., Sovacool, B.: Promoting Vehicle to Grid (V2G) in the Nordic region: expert advice on policy mechanisms for accelerated diffusion. Energy Policy **116**, 422–432 (2018)
89. Shareef, H., Islam, M., Mohamed, A.: A review of the stage-of-the-art charging technologies, placement methodologies, and impacts of electric vehicles. Renewable Sustain. Energy Rev. **64**, 403–420 (2016)
90. Khonji, M., Chau, S., Elbassioni, K.: Challenges in scheduling electric vehicle charging with discrete charging rates in AC power networks. In: Proceedings of the Ninth International Conference on Future Energy Systems, pp. 183–186 (2018)
91. Singh, M., Kumar, P., Kar, I., Kumar, N.: A real-time smart charging station for EVs designed for V2G scenario and its coordination with renewable energy sources. In: 2016 IEEE Power and Energy Society General Meeting (PESGM), pp. 1–5 (2016)
92. Fang, C., Lu, H., Hong, Y., Liu, S., Chang, J.: Dynamic pricing for electric vehicle extreme fast charging. IEEE Trans. Intell. Transp. Syst. **22**, 531–541 (2021)

Area Efficient and Ultra Low Power Full Adder Design Based on GDI Technique for Computing Systems

T. Saran Kumar[1] , I. Rama Satya Nageswara Rao[1] , Y. Satya Vinod[1] ,
P. Harika[1] , B. V. V. Satyanarayana[2](✉) , and A. Pravin[1]

[1] ECE Department, BVC Engineering College (A), Odalarevu 533210, India
[2] ECE Department, Vishnu Institute of Technology, Bhimavaram, India
vvsatya.b@gmail.com

Abstract. The relevance of ultra-low-voltage (ULV) operation for attaining minimal energy usage has increased recently. The fundamental building element of computational arithmetic in many computer and signal/image processing applications is the full adder. An innovative 1-bit hybrid adder circuit that uses both multi-threshold voltage (MVT) transistor logic and GDI (gate-diffusion input) logic is disclosed. The suggested motivation of the Multi Threshold Voltage-gate-diffusion input hybrid adder design is to furnish low energy efficient utilization with a small footprint. Standard 45 nano-meter CMOS process technology is used to simulate the suggested hybrid architecture with a ULV of 0.2 V. The suggested design made considerable improvements in contrast to the previous published designs, yielding >57% and 92% reductions Using only 14 transistors in the Energy and Delay Product respectively, according to the post-layout simulation findings. The suggested design technique produces full functionality, which shows resistance against the processes of global, local variations. The suggested design offers >57% energy efficient compared to the current efforts, according to energy measures that are adjusted for 32 and 22 nm technologies.

Keywords: MVT · GDI · ULV · TGA · TFA · Energy Delay Product

1 Introduction

There are numerous applications needing fast speed, compact space, and low power consumption due to technological improvements and the rise in the usage of portable communication systems including computers, smart phones, Internet of Things (IoT) devices, i-Pads and others. Therefore, low-energy circuits are required for the creation of components of systems and processors with specialized use [1]. The implementation of digital systems is a demanding area of interest for circuit design experts due to this need. Operating digital circuits at ULV, or at close to or below the transistor's One of the most efficient strategies to decrease energy consumption is to use threshold voltage [2].

P. Pareek et al. (Eds.): IC4S 2023, LNICST 537, pp. 63–75, 2024.
https://doi.org/10.1007/978-3-031-48891-7_5

The most prevalent and often utilized arithmetic operations in various DSP systems and very large-scale integration (VLSI) are further, subtraction, multiplication, and accumulation. High-performance DSPs and application-specific processors were made possible by when algorithms with standards such as correlation-convolution-digital filtering are executed; these arithmetic operations are effectively used [3]. The one bit complete full adder cell serves as the fundamental implementing element for carrying out these arithmetic operations. Therefore, improving the entire adder cell's performance is crucial for improving the performance of the system/architecture as a whole. Numerous complete adder ideas using various logic and technological philosophies have been documented in the literature. While other systems employ numerous logic styles, some are built on a single logic style (hybrid designs). Each complete adder design has a similar functionality, but they all have advantages and drawbacks in terms of performance variables including size, speed, and power usage.

The most common method is the full adder architecture for static C-CMOS complementary metal-oxide semiconductors [4]. The 28 transistor architecture resembles a typical N-MOS and P-MOS PU and PD transistors in a CMOS framework. This structure's key benefit is that it is resistant to transistor size and supply voltage scaling. Additionally, it has complete swing logic, which is necessary for constructing intricate structures. Due to the huge PMOS transistors used in this configuration, it has a high input capacitance and takes up more space. Regarding power use and transistor count, mirror adders are among the clever designs that resemble static CMOS full adders, although they have shorter carry propagation latency than CCMOS [5].

Another common design using 32 transistors is the entire adder using CPL logic [6–9]. CMOS does not use supply lines, rather, pass transistor source in PTL is coupled to certain input signals. Although this adder logic's numerous intermediary switching nodes provide efficient voltage swing restoration and higher transistor count make it an unsuitable option for low power applications. Transmission gate logic complete adders are suggested as a solution to the pass transistor logic problem with voltage deterioration [10]. Alioto and associates suggested Shams et al. [11] provided the Transmission Gate Adder (TGA), which employs twenty transistors, in contrast to the Transmission Function Adder - (TFA), which uses sixteen transistors. The transmission function theory and transmission gates serve as the foundation for these adders. The parallel coupling of NMOS and PMOS pass transistors creates the transmission gate structure. The major benefit of its low power efficient is strength of its transmission gate logic arrangement, however due to its weak driving capabilities, Due to cascading of either TFA or TGA; this logic is not advised to creating complicated systems.

To decrease the area, latency, and power, since then, there have been a number of hybrid full adder works described [12–14] a hybrid adder with 14 transistors (14T) was presented by Vesterbacka [15]; however it had pass logic transistors with no complete swing. In Hung et al. [16], another comparable hybrid adder with only 10 transistors was presented. Poor driving capabilities affect both the 14T and 10T adders, which have 10 transistors each [17], suggested a HPSC. It concurrently generates the XOR and XNOR functions using a six transistor pass logic network. Full swing logic is produced by the HPSC full adder, albeit at the expense of additional latency and transistor use. The majority-based adder is another adder that employs hybrid logic [18]. It requires capacitors and static inverters to produce the superior of capabilities because of less number of transistors. Complete adder with 24 transistors (24T) based on a 3-input XOR architecture provided by Tung et al. [19] also employs two distinct types of logic: pass transistor and CMOS. Goel et al. [20] suggested The FA-Hybrid, a different CPL-based hybrid full adder, which employs an unique XOR/XNOR architecture utilizing cross-coupled PMOS transistors with NMOS transistors increase speed. It offers superior driving capabilities but consumes more power when the output, static CMOS inverters are employed.

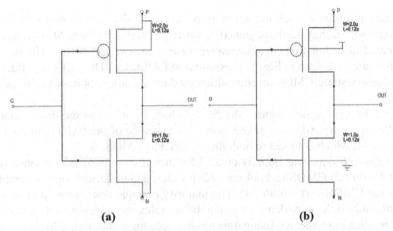

(a)   **(b)**

Fig. 1. a). Originally Proposed, b). CMOS Compatible

DP Logic Full Adder (DPTL-FA) and Swing Restored PT Logic Full Adder (SRPL-FA) are two hybrid designs that Aguirre et al. [21] suggested (DPL-FA). For more energy-efficient computing, these complete adders were created employing ground-less/powerless pass transistors. In contrast to DPL-FA, which uses complementary transistors, SRPL-FA uses PMOS restoration transistors to achieve full swing logic. Bhattacharyya et al. [22] suggested another hybrid 16T full adder (16T Hybrid). Weak inverters in the sum generating module and robust carry module which consists of TG gates were used in this complete adder design to lower the PDP. Full swing outputs are provided by Gate-diffusion-input based OR, AND and XOR gates in a different full adder architecture. Because swing restoration transistors are used for each individual gate in this system, more transistors are needed. Recently, a brand-new, 10-transistor complete Adder circuit (10T) for applications requiring minimal power was suggested [23]. If buffers are not employed, this architecture falls short in terms of energy efficiency and complete logic swing. The majority of hybrid full adders designs do, however, demonstrate gain in only power, speed, or area at the price of the other performance criteria.

There is a need to investigate novel design techniques since none of the entire adder designs that have hitherto been published in-survey show reliable working in furnishing full logic- swing logic with a space- and energy-saving ULV result. The innovative energy-efficient complete adder cell described in this study was created by combining multiple threshold voltage transistors with the GDI approach. The section of this essay is categorized into the following. In this Sect. 2, the GDI approach is briefly summarized. The design strategy for the suggested complete adder circuit is shown in Sect. 3. In Sect. 4, simulation findings and yield comparisons of the suggested design with several ones already in use are covered. Sect. 5 presents yield of 32-bit CP adder, and Sect. 6 summarizes the results.

2 GDI Approach Overview

This section provides a quick introduction to Gate-Diffusion Input, one of the modern digital logic approaches that have gained popularity [24]. Using the GDI approach, very sophisticated logic functions may be realized using just two transistors. The usage i.e. straightforward cell, seen in Fig. 1, is essential to GDI logic. The cell's construction is comparable to a static-CMOS inverter, although there certain important variations to be aware of.

The GDI cell has three inputs: the N-i/p, which is linked to the drain, source of the NMOS, the P-i/p, which is linked to the drain, source of the PMOS, and the G-i/p common input, which is linked to both the PMOS and NMOS.

The GDI technique was initially created for manufacturing in Silicon-on-Insulator (SOI) and twin-tub CMOS technologies [25]. Later, a Gate DI cell that is compatible with normal CMOS was introduced. The majority of logic operations, such as AND, OR, XOR, and MUX, were demonstrated to be complex and necessitate the use of 6–12 transistors when implemented Using transmission gate logic and static CMOS. However, the same operations could be implemented by simply changing the inputs in GDI cells, requiring only two transistors.. Comparison of the number of transistors used in Gate DI and conventional CMOS designing of various logic functions is shown in Table 2 along with Table 1 contains the Boolean table for leveraging GDI to implement various Boolean functions. Unlike the ubiquitous NAND and NOR logic gates, GDI's F1 and F2 universal logic functions may be utilized to implement additional complicated functions more effectively.

3 Proposed Model of Hybrid Full Adder Design

On the whole, a simple 1-bit complete adder's logic operations be described as

$$\text{Sum Output} = (A \text{ xor } B) \text{ xor } C \tag{1}$$

$$\text{Carry\_out} = AB + Cin.(A \text{ xor } B) \tag{2}$$

Only 14 transistors are used in the suggested complete adder architecture, as seen in Fig. 2. Five logic blocks that were created utilizing the MVT-GDI approach make up the majority of it.

$$\text{Sum Output} = Cin (A \text{ xor } B) + Cin (A \text{ x} - \text{nor } B) \tag{3}$$

One Swing Restored Transmission Gate (SRTG), two multiplexers, one XOR/XNOR, two XOR/XNOR, and technology are the other components. They were then altered near Table 1: application of various logic operations using GDI cells (Table 3).

Table 1. Utilizing GDI, some Boolean functions

N	P	G	OUT	Function
0	B	A	A'B	F1
B	1	A	A' + B	F2
0	1	A	A'	NOT
B	0	A	AB	AND
1	B	A	A + B	OR
C	B	A	A'B + AC	MUX

Table 2. Comparison of Transistors Count

Function	Required Transistors Number	
	CMOS	GDI
f1	six	two
f2	six	two
or	six	two
not	two	two
and	six	Two
xor	twelve	Four
mux	twelve	Two

Table 3. Transistor Sizes for Proposed Design

Transistor Name	Length (nm)	Width(nm)
T1, 5	45	240
T2, 6	45	120
T3, 7	45	360
T4, 8	45	120
T11	45	480
T9, 10, 12, 13, 14	45	120

Block of SRPTs or swing restored pass transistors. The GDI approach is used in the creation of the XOR/XNOR block. The XOR/XNOR blocks' inverters are integrated with typical VT devices since the route of the inverters has no voltage loss. In order to achieve the sum function, The Gate DI Multiplexer-1 multiplexes the control input (Cin) with the result of the X-OR (A B) and the X-NOR (A X-NOR B). As a result, the (1) can also be represented as in (3).

The GDI MUX-2 produces the carry output (C_{out}), multiplexing the inputs C-in and B by connecting a decision line to the X-NOR logic's output (A X-NOR B). Consequently, the (2) can also be shown as in (4)

$$Carry\,Out = (A\,x - nor\,B)\,.\,C_{in} + (A\,x - nor\,B)B \tag{4}$$

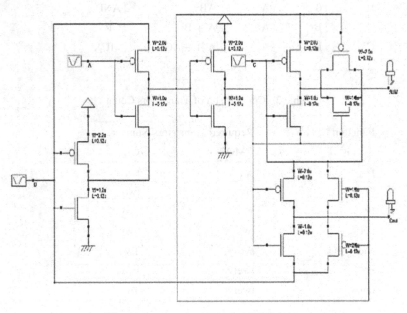

Fig. 2. Proposed 14 T Hybrid Full Adder

The suggested structure resembles other earlier XOR/XNOR logic-based systems as well as the authors' earlier GDI-based approach, although none of them achieves complete logic swing with just 14 transistors. A SRTG at the sum o/p and SRPTs at the carry o/p's are used in the suggested design to assure the entire swing (Cout). Table 4 illustrates how this complete adder operates with relation to the transistor states. When the sum o/p generation GDI MUX1 and the carry o/p generation Gate D input MUX-2 experience a VT drop The swing restoration transistors (T11, T12, and T13) are visible., and T14) are "ON" to give complete swing logic. Since, as indicated in Table 4, there is frequently no VT drop at the output; the transistors (T11, T12, T13, and T14) are also integrated with regular VT transistors.

Table 4. Functional Table

A	B	C	SUM	Cout
0	0	0	0	0
0	0	1	1	0
0	1	0	1	0
0	1	1	0	1
1	0	0	1	0
1	0	1	0	1
1	1	0	0	1
1	1	1	1	1

4 Comparisons of Results and Performance

The outcomes of the simulations and performance evaluations of the suggested complete adders are presented in this section. Cadence 45 nm CMOS technology is used for the simulations, with a Ultra LV of 0.2 V. The simulations' results for, Energy Delay Product (EDP), power, speed, energy and layout area performance metrics are contrasted with those of other designs described in the written word. Table 5 shows the simulation results for all adders, along with their number of transistor and area of the layout. The entire full –adder design shown in Table 5 are run at a speed of 20 kHz and a temperature of 300 K to preserve consistency in comparisons. For any of the complete adder architectures, no additional buffers are required to produce fair results. The observations of different adders performance for each criteria is shown in further detail below Table 5.

Table 5. Simulation Results Comparison

Design	Average Power	Transistor Count	Reference
C-CMOS	2.568	28	[4, 6]
16T Hybrid	2.506	16	[18]
CPL	6.236	32	[8, 9]
TFA	2.98	16	[10]
Proposed	3.053	14	Present

From Table 5, it is clear that the count of transistors and greater switching frequency, the CPL design consumes the most energy. The load's charge and discharge capacitances resulted in this power consumption, which may be stated as follows: (5)

$$P = \alpha C_L V_{DD}^2 \cdot f \tag{5}$$

where f is the circuit's operating frequency, CF is the switching frequency, and CL load capacitance.. Due to its lower transistor density, the GDI architecture is determined to utilize the least amount of power. However, although having the same number of transistors, the GDI design uses less power than the 10T version because it has less leakage current [26–28]. The introduction of extra Low VT devices and swing repaired gates, which increase the switching activity and leakage components, respectively causes the suggested design to use more power than necessary. Because there is less delay, which is a crucial factor while running at ULV, this cannot be a problem in terms of energy usage.

Due to the exponential growth of the delay with supply voltage scaling in accordance with [29], it is crucial to include the delay while determining the circuit's energy metric when it is running at ULV. For strong inversion operation, the (5) can be represented as in (6).

$$C_L V_{DD} / I_{0e}^{((VDD-VT)/nV_{th})} \tag{6}$$

The latency in the sub-threshold zone of operation is exponentially larger due to the exponentially falling ON-current, contrary to what is predicted, where there is no major dependence on the VDD during strong inversion operation (Ion). It is evident from (5) that when the circuit operates at ULV, speed will exponentially decline, narrowing application range to low- to medium-frequency ranges [30, 31].

By subtracting delay equals 50% (input) of the input minus 50% (output)of the voltage swing calculated for each input transition. The speed is significantly increased since In the design proposed, the carry input (C_{in}) is propagated via a one GDInput-MUX, minimizing the carry delay propagated. When compared to previous designs in the literature, it can be shown that the suggested design has a substantially lower latency and achieves >64% savings. This is accomplished by utilizing more swing restored gates and low VT components, which enhance output driving capabilities. The 10T design's longer delay is mostly caused by its lower driving power at lower supply voltage.For digital computational systems, the two key performance indicators that assess a circuit's effectiveness are energy and the Energy Delay Product (EDP). Table 5 makes it evident that, when compared to the other designs, the suggested design has the best energy and EDP parameters.

The suggested complete adder circuit's area is determined by the Fig. 3 shows architecture created for 45 nm technology. Due to the higher transistor density of the CPL architecture, it requires more space, density, and it also has a layout that is more complicated with the existence of additional metal-rails used to implement the logic. Despite having more transistors, C^2MOS & mirror full-adder designs use about the same space in the TGA design because of their straightforward and consistent architectures. With the exception of the 10T and GDI designs, the suggested design takes up considerably less space than the other concepts.

Additionally, the suggested design offers consistent performance TT, FF, FS, SF and SS are just a few of the process corners that they can be used for. Figure 2 displays the differences in the suggested design's power consumption and delay. As anticipated, the largest power consumption and delay are shown at the FF and SS corners, while the least values are seen at the SS and FF process corners.

The energy metric of the proposed design in 45 nm technology is normalized to 32 and 22 nm and compared with the recently proposed CMOS hybrid full adder designs in

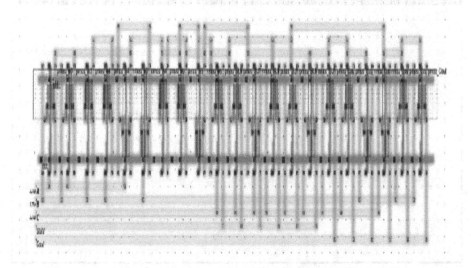

Fig. 3. Proposed 14 T Hybrid Full Adder

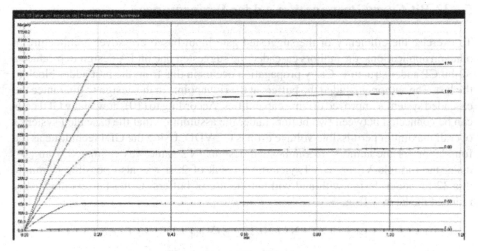

Fig. 4. I_{ds} and V_{ds} Responses

order to determine the efficiency of a suggested development with the latest technology improving trends adhering to the (ITRS) and (IRDS). According to Fig. 4, 5 the normalized energy (EN) meter is derived from the technological scaling trends of the energy measure and VDD.

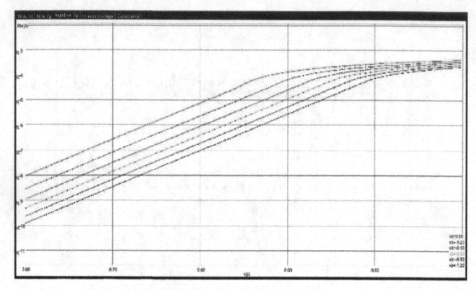

Fig. 5. Log I_d and V_g Responses

5 32-Bit Carry Propagation Adder Performance

To exercise the efficiency of the suggested gate input-based crossover architecture in hand on applications, the authors cascaded the suggested one bit FA design to produce a 32-bit CP adder structure. Carry propagation occurs in this from the first adder block to the last. The authors included the buffers at the right points in the design to guarantee the cascaded system's improved driving capabilities. According to (10), how many GDI cells can be connected between two buffers under the assumption that the cascaded system's overall maximum permissible voltage drop is 0.2 VDD. Using the GDI technique as the foundation for the authors' recommended designs, N's value was estimated from (10) using the formula Vdrop = VT, yielding a result of 2. It was also possible to mimic this 32-bit adder design with and without the use of buffers. Using buffers drastically reduced the latency (Fig. 6).

$$N = 0.2\,V_{DD}/V_{drop} \tag{7}$$

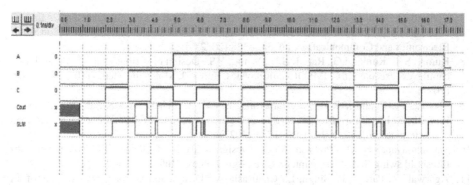

Fig. 6. Simulation Results

6 Conclusion

Here, the authors present the 14TMVT-GDI, a brand-new complete adder circuit. The simulations were run in cadence mode using a 0.2 V and 45 nm technology ULV. The outcomes are contrasted with other published hybrid, 10T, and GDI full adder designs as well as typical such as CMOS, CPL, TGA, and others, are full adder architectures. The suggested Designs are resistant to regional and global change changes, according to Monte-Carlo simulations. According to developments in ITRS technology scaling, normalized energy consumption also reveals that the suggested design delivers greater energy savings of 57% during the time before the current works. The suggested architecture was expanded upon to incorporate 32-bit full adders at the relevant stages (after two stages), both with and without the usage of buffers. Therefore, the majority of applications for energy- and space-efficient computing might adopt the suggested complete adder circuit architecture.

References

1. Alioto, M.: Enabling the Internet of Things: From Integrated Circuits to Integrated Systems, 1st edn. Springer, New York (2017)
2. Wang, A., Calhoun, B.H., Chandrakasan, A.P.: Sub-Threshold Design for Ultra Low-Power Systems, 1st edn. Springer, New York (2006)
3. Prakash, M.D., Krsihna, B.V., Satyanarayana, B.V.V., Vignesh, N.A., Panigrahy, A.K., Ahmadsaidulu, S.: A study of an ultrasensitive label free silicon nanowire FET biosensor for cardiac troponin I detection. SILICON **14**, 5683–5690 (2022)
4. Parvathi, M.: High-accurate, area-efficient approximate multiplier for error-tolerant applications. In: Pant, M., Kumar Sharma, T., Arya, R., Sahana, B., Zolfagharinia, H. (eds.) Soft Computing: Theories and Applications. Advances in Intelligent Systems and Computing, vol. 1154, pp. 91–102. Springer, Singapore (2020)
5. Balan. N.S., Murugan, B.S.: An high speed area efficient implementation of prime field based twisted edwards curve point multiplication using FPGA architecture. In: International Conference on Distributed Computing and Electrical Circuits and Electronics, pp. 1–5 (2022)
6. Rao, I.R.S.N., Krishna, B.M., Shameem, S., Khan, H., Madhumati, G.L.: Wireless secured data transmission using cryptographic techniques through FPGA. Int. J. Eng. Technol. (IJET), 0975–4024 (2016)

7. Ramesh, G., Manikandan, P., Naveen, P., Saravanan, M., Kumar, S.A., Swedheetha, C.: Energy efficient high performance adder/subtractor circuits. In: 3rd International Conference on Smart Electronics and Communication, pp. 333–338 (2022)
8. Kumar, S.T., Kumar, I.G., Rao, R.S.N., Vinod, Y.S.: Design of various shaped MEMS based cantilevers executed in COMSOL multiphysics. Europ. J. Mol. Clin. Med. 7(10), 1972–1979 (2020)
9. Meriga, C., Ponnuri, R.T., Satyanarayana, B.V.V., Gudivada, A.A.K., Panigrahy, A.K., Prakash, M.D.: A novel teeth junction less gate all around FET for improving electrical characteristics. SILICON 14, 1979–1984 (2022)
10. Satyanarayana, B.V.V., Prakash, M D.: Design analysis of GOS-HEFET on lower sub-threshold swing SOI. Analog Integr. Circ. Sign. Process. 109, 683–694 (2021)
11. Agarwal, V., Pareek, P., Singh, L., Chaurasia, V.: Design and performance analysis of all optical half adder based on carrier reservoir SOA-Mach Zehnder interferometer (MZI) configuration. In: International Conference on Numerical Simulation of Opto-electronic Devices, pp. 43–44 (2022)
12. Navi, K., Moaiyeri, M.H., Mirzaee, R.F., Hashemipour, O., Nezhad, B.M.: Two new low-power full adders based on majority-not gates. Microelectron. J. 40(1), 126–130 (2009)
13. Bui, H.T., Wang, A., Jiang, Y.: Design and analysis of low-power 10-transistor complete adders employing innovative XOR-XNOR gates. IEEE Trans. Circ. Syst. II Analog Digit. Sign. Process. 49, 25–30 (2002)
14. Vesterbacka, M.: A 14-transistor CMOS full adder with full voltage-swing nodes. In: 1999 IEEE Workshop on Signal Processing Systems. Design and Implementation SiPS 99, pp. 713–722. Taipei, Taiwan (1999)
15. Vinod, Y.S., Kumar, T.S., Nageswara Rao, I.R.S.: Ultra power optimization of full-adder using adiabatic logic. Test Eng. Manag. 83, 19076–19079 (2020)
16. Tung, C.K., Hung, Y.C., Shieh, S.H., Huang, G.S.: A low-power high-speed hybrid CMOS full adder for embedded system. In: 2007 IEEE Design and Diagnostics of Electronic Circuits and Systems, pp. 1–4. Poland's Krakow, Poland (2007)
17. Satyanarayana, B.V.V., Prakash, M.D.: device and circuit level design, characterization and implementation of low power 7T SRAM cell using heterojunction tunneling transistors with oxide overlap. Microprocess. Microsyst. 77 (2020)
18. Bhattacharyya, P., Kundu, B., Ghosh, S., Kumar, V., Dandapat, A.: Performance analysis of a low-power high-speed hybrid 1-bit full adder circuit. IEEE Trans. Very Large Scale Integr. (VLSI) Syst. 23(10), 2001–2008 (2014)
19. Shams, A.M., Darwish, T.K., Bayoumi, M.A.: Performance analysis of low-power 1-bit CMOS full adder cells. IEEE Trans. Very Large Scale Integr. (VLSI) Syst. 10(1), 20–29 (2002)
20. Shoba, M., Nakkeran, R.: GDI based complete adders for energy efficient arithmetic applications. Eng. Sci. Technol. Int. J. 19(1), 485–496 (2015)
21. Dokania, V., Verma, R., Guduri, M., Islam, A.: Design of 10T full adder cell for ultralow-power applications. Ain Shams Eng. J. 9(4), 2363–2372 (2018)
22. Satyanarayana, B.V.V., Prakash, M.D.: Lower subthreshold swing and improved miller capacitance heterojunction tunneling transistor with overlapping gate. In: Materials Today Proceedings, vol. 45, pp.1997–2001. Springer (2021)
23. Morgenshtein, A., Shwartz, I., Fish, A.: November. gate diffusion input (GDI) logic in standard CMOS nanoscale process. In 2010 IEEE 26-th Convention of Electrical and Electronics Engineers, pp. 776–780, IEEE, Israel (2010)
24. Narendra, S.G., Chandrakasan, A.P.: Leakage in Nanometer CMOS Technologies. Springer Science & Business Media, 1st edition. Springer, USA, (2006)

25. Sanapala, K., Sakthivel, R.: Analysis of GDI logic for minimum energy optimal supply voltage. In: 2017 International Conference on Microelectronic Devices, Circuits and Systems, pp. 1–3. IEEE, Vellore (2017)
26. Lee, P.M., Hsu, C.H., Hung, Y.H.: Novel 10-T full adders realized by GDI structure. In: 2007 International Symposium on Integrated Circuits, pp. 115–118. IEEE (2007)
27. Satyanarayana, N.V., Kumar, R.B.: Review on compressors based approximate multipliers design. In 2021 5th International Conference on Computing Methodologies and Communication, pp. 1016–1021. IEEE (2021)
28. Sakthimohan, M., Deny, J.: An efficient design of 8 * 8 wallace tree multiplier using 2 and 3-bit adders. In: Shakya, S., Balas, V.E., Haoxiang, W., Baig, Z. (eds.) Proceed-ings of International Conference on Sustainable Expert Systems. Lecture Notes in Networks and Systems, vol. 176, pp. 23–39. Springer, Singapore (2021)
29. Misra, A., Birla, S., Singh, N., Dargar, S.K.: High-performance 10-transistor adder cell for low-power applications. IETE J. Res. 1–19 (2022)
30. Bhuvaneswary, N., Prabu, S., Karthikeyan, S., Kathirvel, R., Saraswathi, T.: Low power reversible parallel and serial binary adder/subtractor. In: Balas, V.E., Solanki, V.K., Kumar, R. (eds.) Further Advances in Internet of Things in Biomedical and Cyber Physical Systems. Intelligent Systems Reference Library, vol. 193, pp. 151–159. Springer, Cham. (2021)
31. Zimmermann, R., Fichtner, W.: Low-power logic styles: CMOS versus pass-transistor logic. IEEE J. Solid-State Circ. **32**(7), 1079–1090 (1997)

IoT-Based Hi-Tech Battery Charger for Modern EVs

V. V. Vijetha Inti[1]([✉]) [iD], Mamatha Deenakonda[1] [iD], N. Bhanupriya[1] [iD], RajyaLakshmi.Ch[2] [iD], TRPalleswari Yalla[3] [iD], and Asapu Siva[3] [iD]

[1] Department of EEE, Vishnu Institute of Technology, Bhimavaram, Andhra Pradesh, India
vijetha.i@vishnu.edu.in
[2] Department of Basic Science, Vishnu Institute of Technology, Bhimavaram, Andhra Pradesh, India
[3] Sri Vishnu Engineering College for Women, Bhimavaram, Andhra Pradesh, India

Abstract. These days, we find many batteries used in Electric Vehicles are burning or blasting due to overcharging or long time charging. During such condition, the battery body temperature may rise, to avoid these types of mishaps, here this special type of battery charger is designed using the latest technology such that the battery can be charged using wireless technology and at the same time the battery voltage and its body temperature data will be monitored continuously through an embedded system. An important feature added to the system is that if the battery body temperature raises more than the threshold value, immediately supply to the battery will be disconnected automatically and an alarm will be energized. Once the alarm is energized it remains in energized condition until the reset button is activated. The battery condition will be monitored digitally and it will be displayed through an LCD interfaced with Arduino board. Another important feature added to the system is that entire information will be transmitted to the concerned mobile phone through a WiFi module using IOT technology. Wireless charging is a new technology for charging batteries that allows charging over short distances without cables. The advantage of wireless charging is that charging is quicker and easier, we need not have to plug and unplug each time, simply by placing the vehicle in its parking place where the power transmitting coil is installed under the ground, the battery starts charging automatically.

Keywords: Electric Vehicle · Arduino · IoT · Wireless Charger · Wireless Power Transfer Systems

1 Introduction

A wireless charger for electric vehicles is a new subject for discussion. Many scientists across the world are conducting many experiments on this technology to improve the efficiency and distance between the power-transmitting coil and the power receiving coil. Presently the main drawback of this system is, poor efficiency and poor distance, once this is improved this technology will become popular and we can find these chargers everywhere. In practice, in this method of charging [1], when considered to charge an

P. Pareek et al. (Eds.): IC4S 2023, LNICST 537, pp. 76–85, 2024.
https://doi.org/10.1007/978-3-031-48891-7_6

electric car battery, the power receiving coil must be arranged below the chassis of the vehicle. Here the vehicle is not constructed, but a demo piece can be simulated. With the help of a digital monitoring system constructed with an Arduino board, battery body temperature, and its voltage can be monitored during charging or in idle condition. 12 V–2 Ah rechargeable battery is used and its terminal voltage is monitored continuously through an LCD interfaced with the Arduino board. When the EV batteries are charging [2–4] in the hot environment, especially in summer, battery cathodes are the main cause of the heat release. Effective thermal management is crucial for battery protection, especially in our country where the ambient temperature is high. Therefore, it is thought that the purpose of this effort is to prevent battery burning caused by charging-related overheating. A power transmitting coil and a power receiving coil frequently play a prominent role in wireless chargers [5–7] when it comes to charging batteries. Without using any conducting wires, the energy will be transferred from one coil to another coil using this manner. The electro-dynamic induction technique, also known as resonant inductive coupling, is used. Near-field and far-field wireless power techniques fall into these two groups. In this instance, we used near-field or non-radioactive approaches. Power is delivered over short distances by magnetic fields using inducting coupling between coils of wire or by electric fields using capacitive coupling between metal electrodes. The most used wireless technology is inductive coupling. When a secondary coil is brought near to this magnetic field, maximum energy will be grabbed which is converted as a pure DC source and which can be used to charge the battery. The output is not regulated there by voltage up and downs can be monitored by varying the distance between two coils. The Arduino processor used in the project work is having built-in-with ADC, so additional ADC is not required for converting the analog data generated by the temperature sensor. In the same manner, the battery terminal voltage is also monitored and displayed through an LCD interfaced with Arduino.

2 Wireless Power Transfer Systems

The proposed work will leverage the power of the Internet of Things [7, 8] to enable remote monitoring, control, and management of the charging process. The key strategies for increasing the effectiveness of wireless charging for electric vehicles as well as the issues with electromagnetic interference and radiation are examined. The two methods of charging an electric vehicle are conductive (or wired) charging and wireless charging.

The vehicle's charge inlet and the electric supply are connected during wired charging. Even while cable charging [9–12] is common, it has several drawbacks, including untidy cords and safety concerns in moist environments. Since a few years ago, there has been a significant increase in interest in providing electric loads over a field without using any physical connections to the grid [13–16]. Wireless power transfer systems (WPTSs) are the devices that operate the through-the-field supply. Their deployment has begun for the purpose of recharging the batteries installed in grid-detachable machinery. The equipment is supplied with power while it is moving, to remove the batteries or at the very least diminishing their capacity. Recharging occurs when the equipment is standing in a purposeful configuration. In comparison to its wired version, wireless charging of EV batteries offers several benefits [17, 18], including the elimination of the need for any plugs, cables, or outlets, friendlier charging procedures, fearless energy transfer in any setting, and more. For these reasons, WPTSs are anticipated to have a significant impact on how EVs will be charged in the future. Finally, the primary issues with and fixes for electric vehicle wireless charging technologies are examined. Reduced petrol prices and reduced greenhouse gas emissions are two benefits of electrified transportation. A variety of charging networks must be built in a user-friendly setting to promote the adoption of electrified transportation. Systems for wirelessly charging electric vehicles may prove to be a practical substitute for current plug-in charging methods.

2.1 Wireless Power Transfer (WPT)

Wireless power transfer, also known as inductive charging or wireless charging, eliminates the need for physical cables by using magnetic fields to transfer power between a charging pad or ground-based transmitter and a receiver installed in the EV. WPT technology is still evolving, and its efficiency varies based on the specific system and implementation.

2.2 Conductive Charging (Conventional Chargers)

Conductive charging involves physically connecting the EV to a charging station using a cable. This method has been widely used and is the most common way of charging EVs today. The efficiency of conductive charging depends on various factors, including the charging infrastructure, cable quality, and the onboard charger of the EV. Generally, conductive charging has an efficiency range of 85% to 95%, with modern chargers typically achieving efficiency levels toward the higher end of that range.

3 Block Diagram of HI-TECH Battery Charger

As per the block and circuit diagrams shown in this chapter, the process or functional description begins with the Remote control unit. Little energy will be transported over a distance of about 35 cms because this project work incorporates electromagnetic field coils, which demonstrate the fundamental theory of wireless energy source (Figs. 1 and 2).

During our trial runs, we found that around 300milli amps of current is obtained at a source voltage of 12 approximately. This power is enough to charge the battery arranged in the electric vehicle. The battery used here is rated for 12 V−2 Ah (Ampere hour), when this battery is charged with a 300 mA power source, then the charging time can be defined as battery rating/charging current rating, I.e. 2/0.3 = 6.6 h. So if a high power source is used, then the battery can be charged in less time.

Charging Time = Battery Rating/Charging Current Rating

Charging Time = 2A h/0.3 A

Charging Time = 6.67 h

Therefore, with a 300mA power source, it would take approximately 6.67 h to fully charge the 12 V, 2 Ah battery. Keep in mind that this calculation assumes ideal charging conditions and does not account for factors like charging efficiency or any safety mechanisms that may be implemented in the charging process.

The following chapter includes a thorough functional explanation, and the key components of this system are listed below.

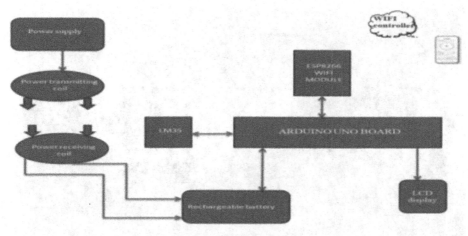

Fig. 1. Block Diagram of Hi-Tech Battery Charger

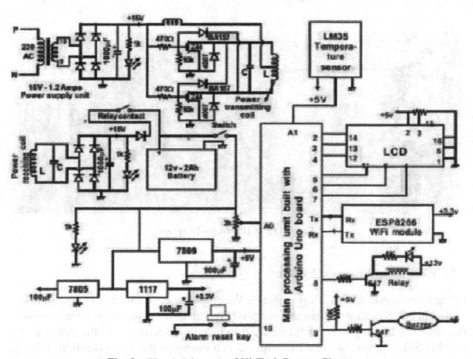

Fig. 2. Circuit Diagram of Hi-Tech Battery Charger

3.1 PCB Hardware Circuit

The PCB is designed at the first level of construction, with component, and circuit mounted on it, and sensors are connected to it (Fig. 3 and Fig. 4).

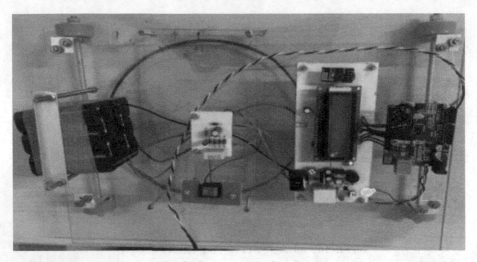

Fig. 3. Power transferring circuit

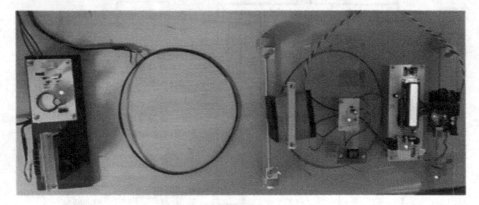

Fig. 4. PCB Hardware Diagram

3.2 Working of Proposed System

Design Implementation of Cost Effective Wireless Power Transmission system is presented in this project work, since it is a prototype module low power transmitter circuit is constructed with power MOSFETs. To induce more current into the primary coil, two MOSFETs are used and are configured in a Push-pull mode of operation. With the help of a diode connected in feedback mode to both power MOSFETs and switches in a sequence (alternatively one after another) such that both MOSFETs will not conduct at a time. As the primary coil is made as center tapped and is divided into two sections, each section will be energized individually through corresponding MOSFET. If the top MOSFET energized bottom remains in de-energized condition, similarly if the bottom MOSFET energized top remains in off condition. In this manner of switching, the switching frequency depends on the fast recovery diodes connected in the feedback loop.

4 Results

(Fig. 5, Fig. 6, Fig. 7 and Fig. 8).

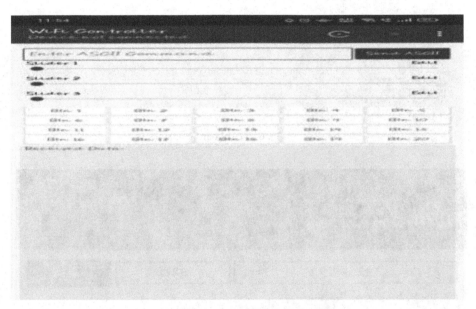

Fig. 5. Homepage display

Through Website we can see the output. When the system get started a current is given to the kit and converts to DC and then arduino and WIFI gets on. The Website was designed for the Internet of Things. It can control hardware remotely, it can display current data, it can store data, visualize it and do many other cool things. Instead, it's supporting the hardware of your choice. Whether your Arduino or Node MCU is linked to the Internet over Wi-Fi, App will get you online and ready for the Internet of your Things. The website serves as a sophisticated and user-friendly platform, offering a multitude of functionalities that make it an indispensable tool for anyone exploring the realm of the Internet of Things (IoT). When the system is activated, a current is supplied to the IoT kit, seamlessly converting it to direct current (DC), thereby powering up the Arduino and WIFI components. As a dedicated IoT platform, the website has been meticulously designed to cater to the diverse needs of IoT enthusiasts and professionals alike. Its primary purpose is to provide a seamless and intuitive interface through which users can remotely control their connected hardware. Whether you are at home, in the office, or halfway around the globe, the website empowers you to interact with your IoT devices with ease and efficiency.

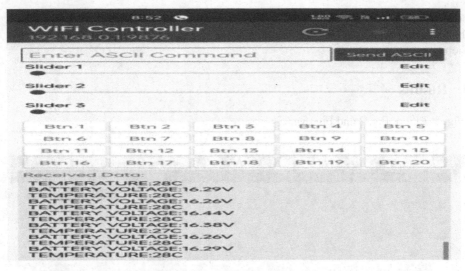

Fig. 6. Dashboard

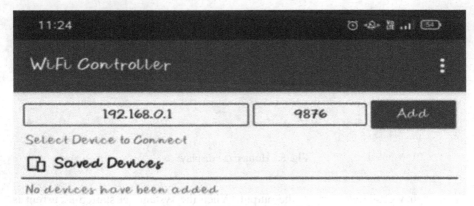

Fig. 7. Login page

Beyond remote control, the website functions as a comprehensive data hub. It allows users to access and display real-time data generated by their IoT devices. From sensor readings to performance metrics, the website presents this information in an easily understandable and visually appealing format. This real-time data visualization gives users valuable insights into their IoT devices' behavior, performance, and environment, enabling them to make informed decisions and optimizations. Additionally, the website offers a powerful data storage feature. Users can securely store historical data, enabling them to track trends and conduct in-depth analyses over extended periods. The ability to access historical data ensures that users have a comprehensive understanding of their IoT devices' long-term performance, facilitating data-driven decision-making and strategic planning.

Fig. 8. Signup page

The website's versatility is a standout feature, accommodating a wide range of IoT hardware choices. It readily supports both Arduino and NodeMCU, making it a truly agnostic platform. No matter the user's preference for IoT hardware, the website adapts effortlessly, seamlessly integrating with the chosen device and establishing a seamless connection to the Internet over Wi-Fi. Moreover, the website does not stop at mere data representation and control; it also enables users to perform various other cool and innovative tasks. The possibilities are boundless, limited only by the imagination and creativity of the user. From triggering automated actions based on sensor data to setting up alerts and notifications, the website empowers users to customize their IoT experience according to their unique needs and preferences.

5   Conclusion

The development of a hi-tech battery charger for modern electric vehicles using the Internet of Things technology is a significant step towards making EVs more convenient and efficient for users. The field of wireless transmission of electrical energy between two magnetically coupled coils requires lot of experiments to obtain better results. In our trail runs we have winded many types of magnetic coils, we have tried with different Gauge wires, with different turn's ratio, different sized coils, etc. Finally we have focused and concluded on one set of coils winded with 21 SWG wire with a ring size of $8''$, both primary and secondary coils are having six turns each. With these coils we found that the range increased slightly when compared with other coils. Finally the distance between the two coils is defined as 50mm, at this distance the battery is charged with less current, when the distance is decreased by less than 30 mm, we observed that the battery is charging at around 300 ma. Since it is a prototype module, low power transmitter is constructed because of restricted power source at primary side. Not only power source, economy is also criteria, but after conducting so many experiments we are very confident that we can build a high power transmitter and range also can be increased accordingly.

The development of a hi-tech battery charger for modern electric vehicles (EVs) using Internet of Things (IoT) technology marks a significant milestone in advancing the convenience and efficiency of EV usage. By incorporating IoT, the charger gains the ability to communicate and exchange data with other devices, ensuring seamless integration with the vehicle and enabling smart charging capabilities.

References

1. Lanza, F., Allotta, M., Rizzoni, G.: Smart charging infrastructure for electric vehicles with IoT integration. In: Proceedings of IEEE Intelligent Transportation Systems Conference (ITSC) (2015)
2. Rahim, N.A., Sulaiman, S., Saad, N.H.M.: Smart charging strategies for electric vehicles integrated with IoT-based grid management. In: Proceedings of IEEE International Conference on Innovative Research and Development (ICIRD) (2019)
3. Gomez-Moreno, P.I., Sanchez-Lopez, F.M., León-de-Mora, Cand J. J. Durillo-Ruiz.: Design and implementation of a smart EV charger using IoT technologies. In: Proceedings of IEEE International Conference on Consumer Electronics (ICCE) (2019)
4. Lanza, F.M., Allotta, M., Rizzoni, G.: Smart charging infrastructure for electric vehicles based on the Internet of Things. In: Proceedings of IEEE Transactions on Intelligent Transportation Systems (2015)
5. Teo, T.S., Gan, L.C., Tong, T.W.: Smart charging for electric vehicles using IoT a review. In: Proceedings of IEEE International Conference on Sustainable Energy, Engineering, and Technologies (SEETech) (2019)
6. Le, DN, Parvathy, V.S., Gupta, D., et al.: IoT enabled depth wise separable convolution neural network with deep support vector machine for COVID-19 diagnosis and classification. Int. J. Mach. Learn. Cybern. **12**, 3235–3248 (2021). https://doi.org/10.1007/s13042-020-01248-7
7. Wang, H., Gao, J., Liu, X., Wang, L.: A smart charging method for electric vehicles based on IoT technology. In: Proceedings of IEEE Power & Energy Society General Meeting (PESGM) (2018)
8. Goorha, S.G., Gupta, S.K., Vishwakarma, B.R.: IoT based battery charger for electric vehicles. In: Proceedings of International Conference on Machine Learning, Big Data, Cloud and Parallel Computing (COMITCon) (2019)
9. Mehta, V.N., Jadeja, P.K., Shah, N.C.: Design and development of IoT based EV charging infrastructure. In: Proceedings of IEEE International Conference on Intelligent Sustainable Systems (ICISS) (2021)
10. Medapati, P.K., Tejo Murthy, P.H., Sridhar, K.P.: Lamstar: For IoT-based face recognition system to manage the safety factor in smart cities. Trans. Emerg. Telecommun. Technol. **31**(12), e3843 (2019). https://doi.org/10.1002/ett.384
11. Sunitha, K.V.N., et al.: Identification and remediation of vulnerabilities in IoT based health monitor. Int. J. Innov. Technol. Explor. Eng. (2019). https://doi.org/10.35940/ijitee.B7805.129219
12. Rambabu, K., Chinnaiah, M.C., Dubey, S., Yellanki, S.: Remote accessible security system with IoT using labview. In: Proceedings of Cognitive Science and Technology (2023). https://doi.org/10.1007/978-981-19-2358-6_36
13. Vijetha Inti, V.V., Vakula, V.S.: Design and implementation of dual switch inverter fed from renewable energy source, 2020/3. Test Eng. Manag. **83**,13993–13999 (2020)
14. Vijetha Inti, V.V., Vakula, V.S.: Design and Matlab/Simulink implementation of four switch inverter for microgrid utilities. Energy Procedia **117**, 615–625 (2017)
15. Chakravarthi, B.N.C.V., Hari Prasad, L., Chavakula, R.L., Vijetha Inti, V.V.: Solar energy conversion techniques and practical approaches to design solar pv power station. In: Pal, D.B., Jha, J.M. (eds.) Sustainable and Clean Energy Production Technologies, pp. 179–201, Singapore: Springer Nature Singapore (2022). https://doi.org/10.1007/978-981-16-9135-5_8
16. Inti, V.V.V., Lakshmi, C.R., Manoj, V.S., Swamy, G.R.: Analysis of DC-DC power converter for fuel cell applications. In: Proceedings of 6th International Conference on Communication and Electronics Systems (ICCES), Coimbatre, India, pp. 63–66 (2021). https://doi.org/10.1109/ICCES51350.2021.9489135(2021)

17. Chen, Y., Zhang, H., Yang, Z., Hu, Z.: Internet of Things-based intelligent electric vehicle charging infrastructure for smart cities. In: Proceedings of IEEE International Conference on Communications (ICC) (2017)
18. Prasanthi, L.S., Bharathi, B.L., Subrahmanyam, M.R.V.: Smart grid based electric vehicle charging infrastructure using IoT. In: Proceedings of International Conference on Inventive Systems and Control (ICISC) (2017)

Comparative Analysis of PWM Methods for Three Level Neutral Point Clamped Inverter

Mamatha Deenakonda[1]([✉]) [iD], V. V. Vijetha Inti[1] [iD], G. Rama Swamy[2] [iD], M. Lavanya[1] [iD], P. Bhavani[3] [iD], and B. Tirumala Rao[2] [iD]

[1] Department of EEE, Vishnu Institute of Technology, Bhimavaram, Andhra Pradesh, India
mamatha.d@vishnu.edu.in
[2] Department of Basic science, Bhimavaram, Andhra Pradesh, India
[3] Department of Basic science, SRKR Engineering College, Bhimavaram, Andhra Pradesh, India

Abstract. In present scenario renewable energy sources are crucial and they taking lead to supply power to the consumers. These renewable energy sources mostly generate dc power, which are to be converted into alternating supply for consumer's use. So there is an impelling need of inverter to convert direct current supply into alternating supply. These inverters are classified into numerous types and depending upon application, the type of inverter is used. In general, a normal inverter will produce square wave output where as a multi-level Inverter will produce staircase waveform which is nearer to sinusoidal. By the level upgrading, the output waveform is very near to the pure sine nature. In this paper, a comparative analysis between the basic Sine PWM and advanced Third Harmonic Injection method for a three level Neutral Point Clamped Inverter (NPC).

Keywords: Neutral Point Clamped (NPC) Inverter · Pulse Width Modulation (PWM) · Third Harmonic Injection PWM (THIPWM · Sine PWM (SPWM)

1 Introduction

The multilevel inverters have been highly penetrated into power industry due to their remarkable features such as modularity, enhanced power quality, reactive power compensation and need of fewer filters etc. The power and voltage rating of a MLI can be increased by proper selection of configuration. The number of levels being increased decreases the voltage strains on the devices, boosts the inverter's ability to handle more power, and considerably enhances the output power quality. Three general categories can be used to classify multilevel inverters [1, 2]. They are the Neutral Point Clamped Multilevel Inverter (NPCMLI) or Diode Clamped Multilevel Inverter, the Flying Capacitor Multilevel Inverter (FCMLI) and the Cascaded H-Bridge Multilevel Inverter. The three-level Diode Clamped multilevel inverter (NPCMLI), which was suggested by Nabae, Takahashi, and Akagi in 1981. By using the appropriate Pulse Modulation Techniques (PWM), these converters' output can be significantly altered [3].Conventionally, sinusoidal pulse width modulation (SPWM) [4] has been applied to regulate the inverter

P. Pareek et al. (Eds.): IC4S 2023, LNICST 537, pp. 86–95, 2024.
https://doi.org/10.1007/978-3-031-48891-7_7

output voltage. Even though, the SPWM technique is simple to implement, its DC bus utilization ratio is 79%. Further, space vector pulse width modulation (SVPWM) [5–8] techniques are used to control the inverter which improved the DC bus utilization by 15.5%. However, the implementation of SVPWM sophisticated digital controller to perform various calculations.

In this paper, third harmonic injection pulse width modulation (THIPWM) in which benchmark wave is a combination of sinusoidal signals with fundamental and third harmonic frequency is used to control the three level NPCMLI. Implementation of THIPWM is simple and provides additional 15.5% DC bus utilization which is similar performance that of SVPWM.

Sinusoidal Pulse Width Modulation (SPWM):
The sine wave is used as the reference wave and the high frequency triangle wave is used as the carrier wave in the sinusoidal PWM approach.

Third Harmonic Injection Pulse Width Modulation (THIPWM):
The fundamental and third harmonic components of the sinusoidal signal are added to create the reference wave in the Third Harmonic Injection PWM technique, and the high frequency triangle wave is used as the carrier wave.

2 Sinusoidal PWM For Three-Level Inverter

The sinusoidal PWM for three-level NPC inverter shown in Fig. 1 is implemented on the basis of unipolar switching scheme. Comparatively the implementation of sine PWM for three-level inverter is little bit difficult to that of two-level inverter. The Table 1 describes the operation of the sine PWM for three-level inverter. The carrier triangle wave has a frequency of 1050 Hz and the benchmark sine wave has a frequency of 50 Hz. Compared to the conventional MLIs, there are different configurations are been developed recently. But the traditional converters are more feasibility than these non-conventional converters [9, 10] (Fig. 1).

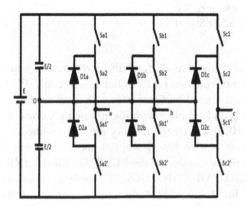

Fig. 1. Three Level NPC Inverter [11]

Table 1. Operation Cycle of the Three-Level Neutral Point Clamped Inverter.

Output Voltage V_{ao} (V)	Switch States			
	S_{a1}	S_{a2}	S_{a1}'	S_{a2}'
$V_{ao} = + V_d/2$	1	1	0	0
$V_{ao} = 0$	0	1	1	0
$V_{ao} = -V_d/2$	0	0	1	1

There are two circumstances. The switches are provided with switching pulses based on these circumstances. First, the sinusoidal wave magnitude must be more than zero, and second, it must be greater than the triangular wave magnitude. The two switches Sa1 and Sa2 are turned ON if the two requirements are met. The switches Sa2 and Sa1' turn ON if any one of the conditions is broken. The two switches Sa1' and Sa2' are switched ON if none of the conditions are met.

Assume $Vr > 0$; $Vr > Vc$; as conditions A and B then for switch Sa1(A AND B), Sa2 (A OR B), Sa3 (A NAND B), and Sa4 (A NOR B) are the appropriate logic. The logic for switches Sa1 and Sa2 are implemented and their inverse is applied to switches Sa3 and Sa4 respectively. The individual Switching scheme for each switch is shown in Table 2.

Table 2. Three Level NPC Inverter with Sine PWM

$V_r > 0$ (A)	$Vr > Vc$ (B)	Sa1 (A AND B)	Sa2 (A OR B)	S_{a1}' (A NAND B)	S_{a2}' (A NOR B)	Output Voltage (V)
0	0	0	0	1	1	$-V_{dc}$
0	1	0	1	1	0	0
1	0	0	1	1	0	0
1	1	1	1	0	0	$+ V_{dc}$

Vr - Voltage magnitude Reference Signal
Vc - Voltage magnitude Carrier Signal

In this work, the analysis is made for NPC inverter with and without filters at the output AC side. So it is noticed that, Fig. 2 indicates a 3-level NPC Inverter without filter and the Fig. 3 indicates a 3-level NPC Inverter with filter circuit.

The selected filter is an L-C filter with the following values: L = 21 mH; C = 470 uF. Figures 8, 9 and 10, respectively, display the line-to-line output voltages for the three-level NPC inverter of sine PWM without and with filters. The aforementioned circuits were successfully simulated in the MATLAB-SIMULINK software. Using the POWERGUI tool in MATLAB-SIMULINK, FFT analysis is performed to determine the harmonics. The individual harmonics along total harmonic distortion for 3-Level NPC inverter without and with filters are shown in Fig. 4 and Fig. 5 respectively.

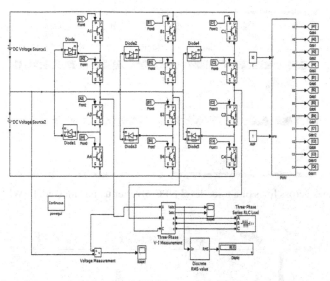

Fig. 2. Diagram of a three-level NPC inverter in SIMULINK without a filter

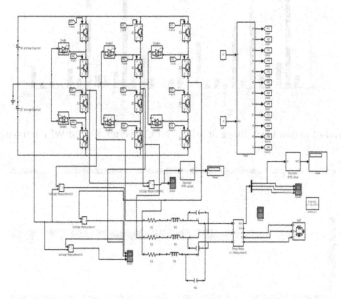

Fig. 3. Schematic of Three level NPC inverter with filter in SIMULINK

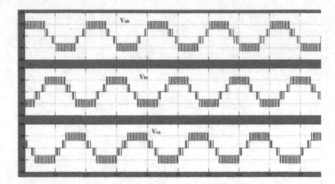

Fig. 4. Voltages across lines of a three-level NPC inverter using sinusoidal PWM and no filter

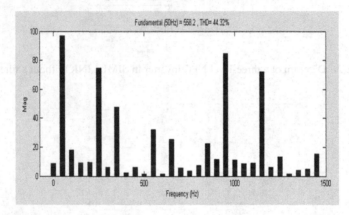

Fig. 5. THD of output line voltage of 3-level NPC inverter with SPWM and without filter

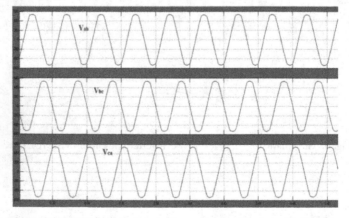

Fig. 6. Three-Level NPC inverter line to line voltages with sinusoidal PWM and filter

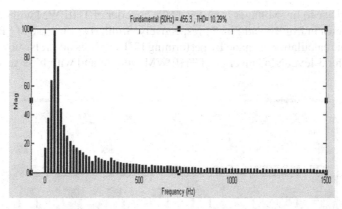

Fig. 7. THD of the 3-level NPC inverter's output line voltage while using SPWM and a filter

3 Third Harmonic Injection PWMFor Three-Level Inverter

The THIPWM [12, 13] methodology, which is akin to the selective harmonic injection method, uses a modified reference signal that includes a fundamental and third harmonic component in addition to the sinusoidal reference signal, resulting in a 15.5% greater voltage amplitude. As a result, THIPWM utilises the DC bus more effectively than SPWM.

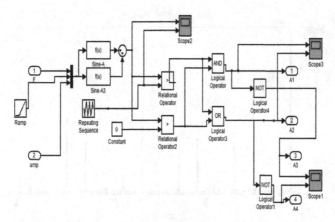

Fig. 8. Diagram of phase A third harmonic injection PWM in SIMULINK

While developing a PWM generation circuit for producing THIPWM pulses, reference signal is created by combining fundamental and third harmonic content of the sinusoidal signal as shown Fig. 8. It is shown that f (u): sin-A and f (u): sin-A3 both is combined to generate the reference signal and the remaining pulse generation process is same as that of sinusoidal PWM generation circuit.The generated THIPWM pulses are given to the 3-Level NPC inverter without/with filter circuits as shown in Fig. 6 and 7.

The obtained line to line voltages for 3-level NPC inverter of THIPWM without and with filters are shown in Fig. 10 and Fig. 11 respectively. Similarly as in case of SPWM, here also harmonic calculation is made by performing FFT analysis and it is shown in Fig. 9 and Fig. 10 for 3-level NPC inverter of THIPWM without and with filter respectively.

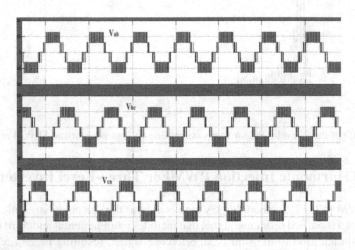

Fig. 9. Voltages across lines of a 3-level NPC inverter with THIPWM and without a filter

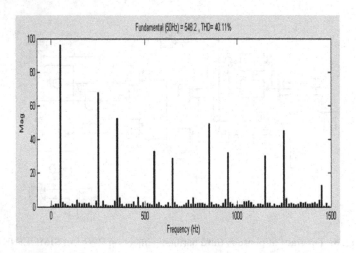

Fig. 10. THD of line voltage of 3-level NPC inverter with THIPWM and without filter

3.1 Comparison Between Sine PWM and THIPWM

From Table 3 it is observed that, the third harmonic injection PWM [14] not only reduces the amount of voltage THD, but also reduces the usage of input DC voltage. It can produce

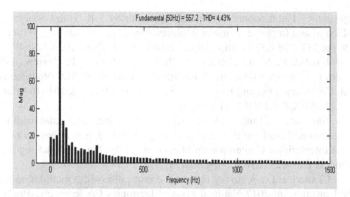

Fig. 11. FFT analysis of line voltage of 3-level NPC inverter with THIPWM and with filter

Table 3. Comparisons between Modulation Techniques

Parameter	Sinusoidal PWM			THIPWM		
	DC I/P	Output line voltage	Voltage % THD	DC I/P	O/P	Voltage % THD
Line voltage without filter	620	420	44.32	510	428	40.11
Line voltage with filter	620	408	10.29	510	415	4.43

nearly same output Line Voltage of 415 V even if 15.5% input voltage is not available as compared with sinusoidal PWM. It is noticed that, the simulation process is conducted at modulation index of 0.75. The power quality [15–19] is mainly improved with the reduction of total harmonic distortion.

4 Conclusion

The three-level Neutral Point Clamped inverter is simulated in MATLAB-SIMULINK utilising sinusoidal PWM and third harmonic injection techniques. Third harmonic injection PWM has been shown to lower an SPWM's output THD from 44.32% to 40.11%. Additionally, the THIPWM approach is lowering the output voltage THD for the same inverter with filter from 10.29% to 4.43%. Finally, it is clear from the data that the THIPWM approach reduces overall harmonic distortion while also making appropriate use of the DC input voltage.

References

1. Prasad, K.N.V., Kumar, G.R., Kiran, T.V., Narayana, G.S.: Comparison of different topologies of cascaded H-Bridge multilevel inverter. In: 2013 International Conference on Computer Communication and Informatics, pp. 1–6 (2013). https://doi.org/10.1109/ICCCI.2013.646 6135

2. Dhanamjayulu, C., Padmanaban, S., Ramachandaramurthy, V.K., Holm-Nielsen, J.B., Blaab-jerg, F.: Design and Implementation of Multilevel Inverters for Electric Vehicles. In: IEEE Access, **9**, pp. 317–338 (2021). https://doi.org/10.1109/ACCESS.2020.3046493

3. Chakravarthi, B.N.C.V., Narasimharao, P.V., Paul, O.R., Paturi, R.P.: Review of Pulse Modulation Controllers for reduced swiches Multil-level converters. In: 2021 6th International Conference on Communication and Electronics Systems (ICCES), pp. 161–164 (2021). https://doi.org/10.1109/ICCES51350.2021.9489165

4. Hussin, H., Saparon, A., Muhamad, M., Risin, M.D.: Sinusoidal pulse width modulation (SPWM) design and implementation by focusing on reducing harmonic content. In: 2010 Fourth Asia International Conference on Mathematical/Analytical Modelling and Computer Simulation, pp. 620–623 (2010). https://doi.org/10.1109/AMS.2010.125

5. Kumar, A., Chatterjee, D.: A survey on space vector pulse width modulation technique for a two-level inverter. In: 2017 National Power Electronics Conference (NPEC), pp. 78–83 (2017). https://doi.org/10.1109/NPEC.2017.8310438

6. Patro, G.M., Rao, P.M.: An approach for design of space vector PWM inverter. 2019 4th International Conference on Electrical, Electronics, Communication, Computer Technologies and Optimization Techniques (ICEECCOT), pp. 95–100 (2019). https://doi.org/10.1109/ICE ECCOT46775.2019.9114588(2019)

7. Rao, B.N., Narukullapati, D.S.M., Kasireddy, B.K., Kuma, D.G.: Selective harmonic elimination based THD minimization of a symmetric 9-level inverter using ant colony optimization. Mathematical Modelling of Engineering Problems **8**(5), 769–774 (2021). https://doi.org/10.18280/mmep.080512

8. Kasireddy, I., Ramakrishnareddy, K.: An efficient selective harmonic based full bridge DC-DC converter for LED lighting applications. In: 2019 National Power Electronics Conference (NPEC), Tiruchirappalli, India, pp. 1–6 (2019). https://doi.org/10.1109/NPEC47332.2019.9034817(2019)

9. Swamy, B., Veeraiah, N., Nagaraju, E., Chakravarthi, B.N.C.V.: Review of Non Conventional power electronic converters for Solar PV with Non-linear loads. In: 2021 6th International Conference on Communication and Electronics Systems (ICCES), pp. 358–361 (2021). https://doi.org/10.1109/ICCES51350.2021.9489077

10. Chakravarthi, B.N.V., Naveen, P., Pragaspathy, S., Raju, V.S.N.N.: Performance of Induction Motor with hybrid Multi level inverter for Electric vehicles. In: 2021 International Conference on Artificial Intelligence and Smart Systems (ICAIS), pp. 1474–1478 (2021). https://doi.org/10.1109/ICAIS50930.2021.9395885

11. Krug, H., Kume, T., Swamy, M.: Neutral-point clamped three-level general purpose inverter - features, benefits and applications. In: 2004 IEEE 35th Annual Power Electronics Specialists Conference (IEEE Cat. No.04CH37551), **1**, pp. 323–328 (2004). https://doi.org/10.1109/PESC.2004.1355764

12. Iqbal, W., Qureshi, I.M., Majeed, H.B.A., Khan, A.: Performance evaluation of third harmonic injection PWM technique for three phase multilevel inverter. In: 2021 International Conference on Frontiers of Information Technology (FIT), pp. 78–83 (2021). https://doi.org/10.1109/FIT53504.2021.00024

13. Colak, I., Bayindir, R., Kabalci, E.: A modified harmonic mitigation analysis using Third Harmonic Injection PWM in a multilevel inverter control. Proceedings of 14th International Power Electronics and Motion Control Conference EPE-PEMC, pp. T2–215-T2–220 (2010). https://doi.org/10.1109/EPEPEMC.2010.5606607(2010)

14. Rahimi, A.S., Subki, A.: Comparative study of Sinusoidal PWM and third harmonic injected PWM on three phase cascaded H-bridge multilevel inverter at various amplitude modulation indices. In: 2018 IEEE PES Asia-Pacific Power and Energy Engineering Conference, pp. 520–525 (2018). https://doi.org/10.1109/APPEEC.2018.8566478

15. Chakravarthi, B.N.V., Siva Krishna Rao, G.V.: Impact of power quality issues in grid connected photovoltaic system. In: 2020 4th International Conference on Electronics, Communication and Aerospace Technology (ICECA), pp. 155–158 (2020). https://doi.org/10.1109/ICECA49313.2020.9297618
16. Vijetha Inti, V.V., Vakula, V.S.: Design and Matlab/Simulink implementation of four switch inverter for microgrid utilities. Energy Procedia 117, 615–625 (2017)
17. Vijetha, V.V., Lakshmi, C.R., Surya Manoj, V., Rama Swamy, G.: Analysis of DC-DC power converter for fuel cell applications. 2021 6th International Conference on Communication and Electronics Systems (ICCES). IEEE (2021)
18. Chakravarthi, B.N.V., Siva Krishna Rao, G.V.: Optimal real power penetration to solar PV fed double boost integrated multilevel converter with improved power quality. Journal of Circuits, Systems and Computers (2020)
19. Pareek, P., Maurya, N.K., Singh, L., Gupta, N., Reis, M.J.: Study of smart city compatible monolithic quantum well photodetector. In International Conference on Cognitive Computing and Cyber Physical Systems, pp. 215–224 (2022). Springer Nature Switzerland, Cham

Smart City Eco-System
and Communications

Single Use Plastic Bottle Recognition and Classification Using Yolo V5 and V8 Architectures

Venkata Durgarao Matta(✉) ⓘ, K. A. Venkata Ramana Raju Mudunuri ⓘ,
B. Ch. S. N. L. S. Sai Baba ⓘ, Kompella Bhargava Kiran ⓘ,
C. H. Lakshmi Veenadhari ⓘ, and B. V. Prasanthi ⓘ

Computer Science and Engineering Department, Vishnu Institute of Technology, Bhimavaram, Andhra Pradesh, India
durgaraomatta@gmail.com

Abstract. Improper disposal of single use plastic bottles leads to many problems including danger to marine life and land pollution. Burning of plastic in turn releases dioxins and polychloride biphenyls. They are very harmful if inhaled and are threat to vegetation too. Manual sorting of plastic bottles and safe disposal is not an easy task. A lot of recycling initiatives use manual sorting for plastic recycling, which depends on plant staff visually identifying and selecting plastic bottles as they move along the conveyor belt. Automatic sorting of plastic bottles has advantage of non-intrusive sorting, speed, consistency, cost effectiveness in long run and even prevents health hazards to workers working in recycling environment. As a result, it is imperative to replace human sorting systems with intelligent automated systems. In this study, convolutional neural network architectures such as YOLOv5 and YOLOv8 were utilized to detect plastic bottles in images. Despite YOLOv8 having more parameters and requiring more computation time, it was found that YOLOv8 outperformed YOLOv5 in accurately identifying plastic bottles in the images.

Keywords: CNN · Computer Vision · Plastic bottles · Yolo V5 and Yolo V8

1 Introduction

It is estimated that 2.5 million tons of CO_2 is being released annually by improper water bottle disposals. Approximately, 1.1 million ocean creatures die because of plastic pollution caused by plastic water bottles. Beside this, many toxic gases like polychlorinated biphenyls are released by burning plastic bottles [1]. Determining the amount of plastic bottle waste is a big issue. Continuous usage of plastic bottles and no proper disposal might lead to many disasters and disturbs the harmony of nature. Determining the plastic bottles waste in public areas is an important area of research. In future, the cities may be ranked based on plastic management. One of the solutions to find the solution is to use computer vision models which are built using convolutional neural network (CNN) to

© ICST Institute for Computer Sciences, Social Informatics and Telecommunications Engineering 2024
Published by Springer Nature Switzerland AG 2024. All Rights Reserved
P. Pareek et al. (Eds.): IC4S 2023, LNICST 537, pp. 99–106, 2024.
https://doi.org/10.1007/978-3-031-48891-7_8

find out the plastic bottle usage in an area. These models can be built on techniques like Histogram of Oriented Gradients (HOG), Region-based Convolutional Neural Networks (R-CNN), Faster R-CNN. The models used above gave less accuracy and performance is not up to the mark in many cases.

The basic concept of this paper is to use some good object detection techniques for uncovering plastic bottles in an given area. In this paper a we have used YOLO V5, V8 as they are Single shot detection techniques to find the targe image with a better accuracy. The output by using these techniques have bounding boxes along with probability of classification. The reason for selecting these models is for its speed. They are one short detection techniques. Means, the image is passed only once [3]. These features make these models methodical and are better algorithms. On the other hand, region-based CNN find the approximate region and recognizes the bounding box in a separate stage but are more accurate. The YOLO has gone through five levels of major iterations. Minimal requirement of data, simple architecture and easy implementation make these architectures unique.

The paper is organized as follows. The related work is first reviewed. The architectures are then described. The experimental outcomes are then emphasized. Finally, the paper is concluded.

2   Related Work

Walden et al. [4] proposed a paper in which the plastic bottles are converted into hue, saturation and (HSV). Filters are then applied along with binary thresholding. These images are then converted into gray pixels. Blob analysis is then utilised to determine the number of plastic bottles. Here, no CNN based approach is used and there is no way to increase the accuracy of models. Dhokley et al. [5] proposed a similar method to detect plastic waste and uses YOLO V3. But the proposed approach uses much advanced models. Christopher et al. [6] proposed a paper "PET-Bottle-Recognizer" to detect Polyethylene-Terephthalate Based- Bottles. This accuracy of the model is 85.70%. They have sued mean average pooling to calculate the accuracy. Jungiu et al. [7] have come with an approach based on improved yolo V3. The system extracts the features by using shuferNet network. The screening accuracy is around 91.3% provided the detection rate is set at 26 frames per second. An approach proposed by keqiong et al. uses stochastic configuration network and yolo v5 to detect the plastic bottles[8]. They have dataset is generated using Hikvision MV CE050-30GM camera. Gilroy et al. [9] have performed the detection of plastic bottles in river by using yolo V5 algorithm which is focused on custom dataset. The model has an accuracy of 84%, a precision rate of 79.14%, and a recall rate of 57.37% when deployed on raspberry pie.

Most of the researchers have used models like VGG 16, YOLO V3, YOLO V5 and YOLOV5s and the accuracy of the models were around 85%. This paper uses the latest version YOLO V5 and YOLO V8. Our model is tested on Plastic Bottles in the wild Image Dataset from Kaggle.

3 Architectures Used

This dataset has 8000 images out of which 70% is used for training, and 20% & 10% for testing and validation respectively. The rationale behind this choice is to balance the trade-off between having enough data for training, assessing model performance on unseen data, and tuning hyperparameters. [10]. The model built is a binary classifier, where we check for plastic bottle in an image. In this paper, we are using YOLO for object detection. Earlier, Object detection was performed using sliding window method. Later more faster versions like, Region based Convolutional neural network(R-CNN) [11], fast Region based convolutional neural network (Fast R-CNN) [12], faster region based convolutional neural network (Faster R-CNN) [13]. In 2016, YOLO (you only look once) were invented which outperformed all the previous pervious object detection algorithms. In case of image classification, we just look if the object is present in an image. But, in the case of object detection, we exactly look for object inside an image using bounding boxes. This is referred to as object localization. In terms of a bounding box, we have a vector of minimal elements [PC, BX, BY, BW, BH C1, C2]. Pc is the probability of a given class, Bx, By is the center coordinate, and Bw and Bh are the width and height of the bounding box, C1 and C2 are the class labels. If there is no object in an image, the value of Pc is 0 and the rest of the values in the vectors do not have any meaning. Here if we need to detect multiple objects, the vector size will be increased accordingly. For example, if 10 objects need to be detected, then vector size will be 70. In case of yolo algorithm, the image is divided into grids. There is no rule for dividing the data into specific number of grids. If an image is divided into a 4×4 grid, then each grid is individually searched for the object based on the coordinates of center. And, if each grid is represented by a vector of size 7, the image will have $4 \times 4 \times 7$ volume of information. So, the training attribute is images with grid and bounding boxes and training labels would be a three-dimensional vector. While predicting for objects in an image, the output would be 16 vectors in case of 4×4 grid.

Basic YOLO algorithm has some limitations, at first it can detect multiple bounding rectangles for a given object. One approach to solve this issue is to select the bounding box with highest probability. This approach works in case of single object detection. In case of multiple object detection, another approach called Intersection over union might work well. This method finds the rectangles with overlapping area

$$IOU = \frac{Intersect\ area}{Union\ area} \tag{1}$$

The Intersection over union (IOU) method is also known as Non max suppression. The larger the value of IOU, the better the accuracy. In some cases, an object can be inside another object, this scenario can be handled by concatenation of 2 vectors resulting in a vector of size 14.

The first version of YOLO was released in year 2016 [14]. The concept of YOLO was related to regression. It was able to predict images at 45 frames per second. The similar version of yolo known as lighter YOLO could predict at a speed of 144 frames per second but with lesser layers. The next version of YOLO as known as YOLO 2 was able to input of different sizes and was able to balance between speed and accuracy [15]. In 2018, YOLO V3 was released which was based on Darknet-53 architecture

[16]. The YOLO versions up to three were proposed by Joe Redmon. Later version V4 was having features like Weighted-Residual-Connections (WRC) Cross Stage Partial connections (CSP), Cross Mini-Batch Normalization (CmBN), Self-adversarial training (SAT), Mish activation, Mosaic data augmentation, DropBlock regularization, CIoU loss. Later YOLO V5 was released and was the first version from yolo which was developed using Pytorch and removed the drawbacks of Darknet framework [17]. In year 2023, YOLO v7 was released. The Extended Efficient Layer Aggregation Network (E-ELAN) stands for the computing block in the YOLOv7 backbone [18]. By employing "expand, shuffle, merge cardinality" to accomplish the capacity to constantly increase the learning ability of the network without breaking the original gradient route, the YOLOv7 E-ELAN architecture helps the model learn better. Both YOLO v5 and V8 architectures use CSPDarknet53 architecture and the detection accuracy is improved by using anchor boxes.

The main advantage of YOLO v5 and YOLO v8 is their simplicity, single forward pass, lightweight and opensource. In order to reduce detection of same object multiple times, we use non-Maximum suppression technique. They use Adam and Mish as their optimizer and activation function respectively. [19]. Yolo architectures suffer with accurately localizing small objects. These are overcome using good resolution input images, more data argumentation and adjusting the anchor boxes is done. Using these approaches help to reduce false positive and false negative.

4   Experimental Results

We have ploted F1- Confidence curve, precision Recall curve, precision confidence curve and recal confindence curve to check the performance of the model. The relationship between a binary classifier's precision and recall as the decision threshold changes is depicted graphically by the F1 confidence curve. The F1 score, which is the harmonic mean of precision and recall, is plotted versus the confidence threshold. The trade-off between recall and precision is depicted by the curve, which displays how the f1 score changes as the confidence threshold is altered. The precision typically rises while the recall falls as the threshold rises, and vice versa. For a certain classification task, the f1 confidence curve can assist in determining the best threshold by balancing precision and recall. A high F1 score means that the classifier is successfully striking a balance between recall and precision. Using the f1 confidence curve,

An illustration of a binary classifier's performance at various classification thresholds is the precision-recall curve. The accuracy and recall values for various threshold values are plotted, Precision is the proportion of true positives among all projected positives, while recall is the proportion of true positives among all real positives. The precision-recall curve can be used to visualize the trade-off between precision and recall at different decision thresholds. A perfect classifier would provide a point in the top-right corner of the curve with precision and recall both equal to 1.0. On the curve, a random classifier would generate a straight line from (0,0) to (1,1). A decent classifier's curve ought to be as near the top-right as possible. A binary classifier's recall as a function of the degree of certainty or confidence in its predictions is shown graphically as the recall-confidence curve. It displays a visualization of the recall values at various classifier confidence levels.

The recall-confidence curve can also be used to assess the effects of several feature sets or hyperparameters on the performance of a single classifier or to compare the effectiveness of various classifiers. In general, better performance of the classifier is indicated by a higher recall value for a particular confidence level. It should be emphasized that the recall-confidence curve cannot give a thorough assessment of a classifier's effectiveness because it only examines recall and ignores accuracy or other metrics. Additional evaluation metrics, such as the precision-recall curve, F1 score, or area under the ROC curve, must be considered in addition to the recall-confidence curve for a more full assessment of the classifier's performance. The curve makes determining the confidence level at which a classifier performs well and the confidence level over which that performance begins to decline easy. It enables the selection of the confidence level that optimizes remembrance, which might be useful when the goal is to achieve high recall at the expense of lower precision.

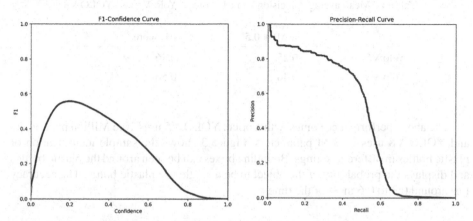

Fig. 1. F1 confidence curve for YOLO V8 (left). Precision Recall curve for YOLO V8 (Right)

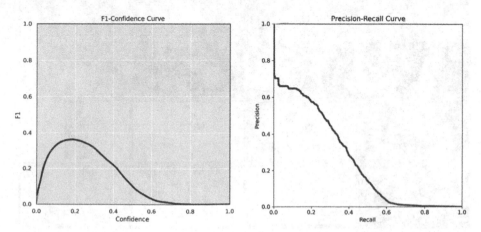

Fig. 2. F1 confidence curve for YOLO V5 (left). Precision Recall curve for Yolo V5 (Right)

Single-use plastic bottles come in various shapes, sizes, and materials. Some bottles may have unique designs or labels that differ significantly from the original training data. This problem can be reduced to an extent by using multiple data argumentation techniques.

As we can see above Fig. 1 and 2, the mAp value of models build using YOLO V5 and V8 stands at 0.252 and 0.46 respectively. Similarly, F1 score of both models stands 0.56 and 0.36. The same is displayed in the below table. Yolo V5 was run for 40 epochs and Yolo V8 was run for 25 epochs. Overfitting was observed after increasing the number of epochs. The images were resized to 416×416 and the batch size was 16 in both environments.

The experiment was performed on machine with NVIDIA QuADro GV100 having 5120 CUDA cores and 640 Tensor cores (Table 1).

Table 1. Mean average precision and F1 score of Yolo V5 and YOLO V8

	mAP@ 0.5	F1 Score
Yolo V5	0.252	0.36
Yolo V8	0.46	0.56

The above performance comes with a price. YOLO V5 uses 10.6 Million parameters and YOLO V8 uses 61.3 M parameters. Figure 3 shows the sample identification of plastic bottles in different settings. Bounding boxes can be seen around the plastic bottles and displays the probability of the object to be a single use plastic bottle. The accuracy lies around 0.3 to 0.6 most of the time.

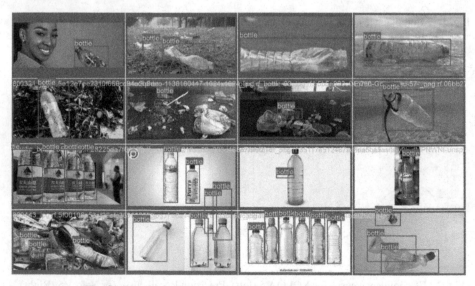

Fig. 3. Sample detection of plastic bottles by YOLO V5 and V8.

5   Conclusion

Given the abundance of plastic bottle waste in the environment, it is easy to come across scenes and photos of this trash. The main goal of this paper is to detect single-use plastic bottles in each image by employing cutting-edge convolutional neural networks. The dataset consists of 8000 images of pre-annotated single use plastic bottles collected from publicly available sources. This paper contrasts the performance of models build using YOLO v5 and v8 architectures to detect plastic bottles. It can be observed that efficiency of model created using YOLO V8 has outperformed the model build using YOLO V5. The models are validated using mean average precision and F1 score. Mean average precision of YOLO V8 is 0.56. Whereas the F1 Score is approximately 0.36.

References

1. Yuan, Z., Nag, R., Cummins, E.: Human health concerns regarding microplastics in the aquatic environment - from marine to food systems. Sci. Total. Environ. **823**, 153730 (2022). https://doi.org/10.1016/j.scitotenv.2022.153730
2. Cao, Y., Wang, H.: Object detection: algorithms and prospects. In: 2022 International Conference on Data Analytics, Computing and Artificial Intelligence (ICDACAI), Zakopane, Poland, pp. 1–4 (2022). https://doi.org/10.1109/ICDACAI57211.2022.00031
3. Diwan, T., Anirudh, G., Tembhurne, J.V.: Object detection using YOLO: challenges, architectural successors, datasets and applications. Multimed Tools Appl **82**, 9243–9275 (2023). https://doi.org/10.1007/s11042-022-13644-y
4. Walden, K., Mehrubeoglu, M.: Quantifying plastic bottle debris in waterways using image processing (2020). In: Proceedings of the 2020 IEEE 7th International Conference on Engineering Technologies and Applied Sciences (ICETAS), Kuala Lumpur, Malaysia, pp. 1658–1663 (2020). https://doi.org/10.1109/CSCI51800.2020.00305
5. Dhokley, W., Khambati, H., Kapadia, K., Dubey, A.: Identification of products in plastic waste using object detection. Int J Creative Res Thoughts **8**, 6 (2020). ISSN 2320–2882
6. Cunanan, C.F.: PET-Bottle-recognizer: a machine vision recognition of polyethylene-terephthalate based-bottle for plastic waste classification and recycling. In: Proceedings of the 2020 IEEE 7th International Conference on Engineering Technologies and Applied Sciences (ICETAS), Kuala Lumpur, Malaysia, pp. 1–5 (2020). https://doi.org/10.1109/ICETAS51660.2020.9484217
7. Xiao, J., Tang, Y., Zhao, Y., Yan, Y.: Design of plastic bottle image recognition system based on improved YOLOv3. In: Proceedings of the 2020 5th International Conference on Mechanical, Control and Computer Engineering (ICMCCE), Harbin, China, pp. 2047–2050 (2020). https://doi.org/10.1109/ICMCCE51767.2020.00445
8. Chen, K., An, J., Fang, Y., Bu, T.: Research on solid waste plastic bottle cognitive based on YOLOv5s and deep stochastic configuration network. In: Proceedings of the 2022 7th International Conference on Automation, Control and Robotics Engineering (CACRE), Xi'an, China, pp. 275–280 (2022). https://doi.org/10.1109/CACRE54574.2022.9834213
9. Sio, G.A., Guantero, D., Villaverde, J.: Plastic waste detection on rivers using YOLOv5 algorithm. In: Proceedings of the 2022 13th International Conference on Computing Communication and Networking Technologies (ICCCNT), Kharagpur, India, pp. 1–6 (2022). https://doi.org/10.1109/ICCCNT54827.2022.9984439
10. Shi, J., Zhou, Y., Zhang, W.X.Q.: Target detection based on improved mask rcnn in service robot. In: Proceedings of the 2019 Chinese Control Conference (CCC), Guangzhou, China, pp. 8519–8524 (2019). https://doi.org/10.23919/ChiCC.2019.8866278

11. Sah, S.: Plastic Bottles in the wild Image Dataset, Version 1. Retrieved 28 March 2023 (2023). https://www.kaggle.com/general/46091

12. Ullah, A., Xie, H., Farooq, M.O., Sun, Z.: Pedestrian detection in infrared images using fast RCNN. In: Proceedings of the 2018 Eighth International Conference on Image Processing Theory, Tools and Applications (IPTA), Xi'an, China, pp. 1–6 (2018). https://doi.org/10.1109/IPTA.2018.8608121

13. Wang, D., Wang, L., Peng, D., Qi, E.: Research on appearance defect detection of power equipment based on improved faster-RCNN. In: Proceedings of the 2021 6th International Conference on Power and Renewable Energy (ICPRE), Shanghai, China, pp. 290–295 (2021). https://doi.org/10.1109/ICPRE52634.2021.9635270

14. Redmon, J., Divvala, S., Girshick, R., Farhadi, A.: You only look once: unified, real-time object detection. In: Proceedings of the 2016 IEEE Conference on Computer Vision and Pattern Recognition (CVPR), Las Vegas, NV, USA, pp. 779–788 (2016). https://doi.org/10.1109/CVPR.2016.91

15. Bhuvaneshwary, N., Jayameenakshi, M., Lakshmi, S.A., Shrivalli, K.: People detection and identification using yolo. In: Proceedings of the 2021 5th International Conference on Electrical, Electronics, Communication, Computer Technologies and Optimization Techniques (ICEECCOT), Mysuru, India, pp. 85–87 (2021). https://doi.org/10.1109/ICEECCOT52851.2021.9708016

16. Bi, F., Yang, J.: Target detection system design and FPGA implementation based on YOLO v2 algorithm. In: Proceedings of the 2019 3rd International Conference on Imaging, Signal Processing and Communication (ICISPC), Singapore, pp. 10–14 (2019). https://doi.org/10.1109/ICISPC.2019.8935783

17. Ting, L., Baijun, Z., Yongsheng, Z., Shun, Y.: Ship detection algorithm based on improved YOLO V5. In: Proceedings of the 2021 6th International Conference on Automation, Control and Robotics Engineering (CACRE), Dalian, China, pp. 483–487 (2021). https://doi.org/10.1109/CACRE52464.2021.9501331

18. Balaji, R., Prabaharan, G., Singh, A.R., Athisayamani, S., Sarveshwaran, V., Dani-ya, S.: Multi-scale features fusion with YOLOv3 for detecting small and fine tumors in MRI images. In: Proceedings of the 6th International Conference on Electronics, Communication, and Aerospace Technology (ICECA 2022), Coimbatore, India, pp. 1545–1549 (2022). https://doi.org/10.1109/ICECA55336.2022.10009122

19. Kim J.H., Kim N., Won C.S.: High-speed drone detection based on Yolo-V8. In: Proceedings of the ICASSP 2023 - 2023 IEEE International Conference on Acoustics, Speech and Signal Processing (ICASSP), Rhodes Island, Greece, pp. 1–2 (2023). https://doi.org/10.1109/ICASSP49357.2023.10095516

Model Predictive Control (MPC) and Proportional Integral Derivative Control (PID) for Autonomous Lane Keeping Maneuvers: A Comparative Study of Their Efficacy and Stability

Ahsan Kabir Nuhel[1](✉), Muhammad Al Amin[1], Dipta Paul[1], Diva Bhatia[2], Rubel Paul[3], and Mir Mohibullah Sazid[1]

[1] American International University, Kuratoli, Bangladesh
nuhe17050@gmail.com
[2] Vellore Institute of Technology, Vellore, India
[3] Sonargaon University, Dhaka, Bangladesh

Abstract. The escalating frequency of fatal crashes has led to an enhanced focus on road safety, resulting in the creation of diverse driver assistance systems. Several instances of these systems encompass active braking, lane departure warning, cruise control, lane maintaining, and numerous additional examples. However, the primary objective of this research is to examine the effectiveness and reliability of a model predictive control (MPC) and a proportional integral derivative (PID) control in executing lane keeping maneuvers within an autonomous vehicle. In this paper, a custom controller for autonomous lane-changing maneuvers is developed by utilizing the Model Predictive Control (MPC) and Proportional-Integral-Derivative (PID) controllers. Different trajectory models are employed to assess the overall effectiveness of the designed model, showcasing its superiority over existing models.

Keywords: Autonomous car · Trajectory models · MPC · PID · Model Prediction

1 Introduction

Autonomous vehicle is one of key technological advancement to keep the road safe and decrease the number of fatal accidents [1]. Apart from than, most of the modern cars come with Advance Driving Assistance System (ADAS) to help the driver in the road. The main functions for these kinds of systems are autonomous breaking, lane keeping assist, object detection, lane departure warning and so on. In this project, a custom controller have designed to keep the car in the lane [2]. At first, Kinematic Bicycle model is discussed which is used to design the MPC controller.

P. Pareek et al. (Eds.): IC4S 2023, LNICST 537, pp. 107–121, 2024.
https://doi.org/10.1007/978-3-031-48891-7_9

While designing the MPC controller, as inputs, this method considers the current condition of the vehicle, including its position, velocity, and heading, as well as the positions of surrounding vehicles [3]. System dynamic is also taken into consideration to fulfill the control objective. Different trajectory models are generated to observe the models and checking the outputs with the pre-existing models [4]. The results are analyzed with reference trajectory, straight and curve trajectory for different attributes.

The paper is divided into five sections. In Sect. 2, the paper describes earlier fundamental research concepts. Section 3 describes the proposed methodology and experimental setup. Section 4 analyzes the results, and Sect. 5 discusses potential future applications.

2 Literature Review

In the research of Trajectory Tracking Control using MPC Controller for of Quadcopter [5], G. Ganga and co-author Meher Madhu Dharmana constructed a quadcopter by utilising the dynamic equation in 2017. They proceeded to devise a linear Model Predictive controller and a Proportional-Integral-Derivative controller to address the issue of trajectory tracking for the quadcopter. The findings of this study demonstrate the superiority of the Linear Model Predictive controller over the Proportional-Integral-Derivative controller in the context of design objectives.

In a paper Hengyang Wang, Biao Liu, Xianyao Ping, and Quan worked on Autonomous Vehicles Path Tracking Control Based on an Improved Model Predictive Controller. They proposed an enhanced MPC controller that incorporates fuzzy adaptive weight control. The objective of their study is to address the challenge of lane tracking in autonomous vehicles. The fuzzy adaptive control algorithm is employed to implement this controller, which primarily involves the cost function by dynamically increasing the weight in the classical model predictive control (MPC) approach.

In 2020 Shuping Chen, Huiyan Chen, and Dan Negrut worked on the study of Path Tracking for Autonomous Vehicles with the Implementation of Model Predictive Control Incorporating Three Vehicle Dynamics Models with Varying Fidelities. They proposed the practical application of path tracking for autonomous vehicles. In the study conducted by [7], an MPC controller was developed utilising three distinct models: an 8-DOF model, the bicycle model, and a 14-DOF model. The reference paths employed in the experiment consisted of a straight line and a circular trajectory. The researchers also conducted a comparative analysis of the performances exhibited by various models.

Ak Nuhel, MM Sazid and MNM Bhuiyan designed a level 5 autonomous car using machine learning, deep learning and CNN [8]. Apart from that, MPC controller is used to design the path of the car using different trajectory analysis. An overview of vehicle safety was also discussed.

In the paper, Eugenio Alcalá, Vicenç Puig, and Joseba Quevedo worked on "LPV-MPC Control for Autonomous Vehicles" [9] in 2019. They addressed a trajectory tracking issue pertaining to autonomous vehicles. The approach employs a cascade control strategy, wherein an external loop is utilised for position control through the implementation of a Learning Parameter Varying (LPV) Model Predictive Control (MPC) controller. The distinction between the LPV-MPC controller and the Nonlinear (NL) MPC controller is demonstrated, and the superior performance of the LPV-MPC controller is discussed.

3 Methodology

3.1 Vehicle Model

According to Limebeer and Massaro [10], the vehicle model serves as a fundamental basis for investigating the features of vehicle control. Consequently, an accurate vehicle model is crucial for developing a reliable and precise model of vehicle control. Both the kinematic bicycle model and the dynamic bicycle model are analyzed in detail in this research, for the purpose of implementing autonomous driving capabilities in our vehicle. The utilization of models is imperative for conducting an in-depth analysis of vehicles, particularly for the purpose of modeling the controller.

Kinematic Bicycle Model.
The motion of a bicycle can be conceptualized within the theoretical framework of the Kinematic Bicycle Model.

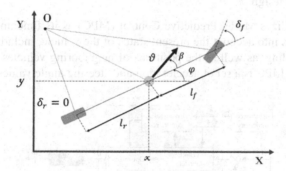

Fig. 1. Kinematic Bicycle model

The studies by Rajamani and Zhang [2] show that the kinematic bicycle model is commonly used in the study of vehicle features [11]. Figure 1 demonstrates the kinematic bicycle model. There are four wheels total, two in the front and two in the back, however in this model they function as one. The forward lumped wheel sits precisely in the geometric center of the front axle, whereas the rear lumped wheel is centered on the back axle. The kinematic model can be quantitatively expressed by Eqs. (1) through (5) if it is assumed that the front wheel is the only part responsible for steering.

$$\dot{x}\, vcos(\psi + \beta), \tag{1}$$

$$\dot{y}\, vsin(\psi + \beta), \tag{2}$$

$$\dot{\psi}\, \frac{v}{l_r}(\beta), \tag{3}$$

$$\dot{v} \, a \tag{4}$$

$$\beta \arctan\left(\frac{l_r}{l_f + l_r} + \tan(\delta_f)\right) \tag{5}$$

The inertial frame's center of mass is represented by the coordinates (X,Y), where X and Y are variables. The direction of the self-driving car's movement is indicated by ψ and its velocity is marked by the variable v. Distances from the center of mass to the front and rear wheels are represented by the l_f and l_r variables, respectively. The present model incorporates acceleration (a) and the stecring angle (δ) as control inputs. In order to simplify the analysis and facilitate practical application, it is common practice to assume that the rear wheels of a vehicle have zero steering angle $\delta_r = 0$, with the focus being solely on the steering of the front wheel. The present kinematic model exhibits a relatively elementary structure in comparison to alternative models that incorporate additional physical factors such as aerodynamic drag, gravitational force, and frictional resistance. The kinematic model can be identified with only two parameters, namely (l_f and l_r), rendering it applicable for both longitudinal and lateral control purposes.

3.2 Controller Design

The controller utilizes Model Predictive Control (MPC) as its fundamental approach. This method takes into account the present states of the vehicle, including its position, velocity, and heading, as well as the positions of neighboring vehicles, as inputs [12]. The output of the MPC is a set of acceleration and steering angle values (Fig. 2).

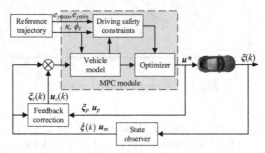

Fig. 2. Stabilization via Model-Predictive Controlling High-Speed Autonomous Ground Vehicle

System Dynamics.
The study utilizes a nonlinear kinematic bicycle model to depict the characteristics of vehicle dynamics [3]. The kinematics are rewritten here for completeness from Eq. 6 to 10:

$$\dot{x} = vcos(\psi + \beta) \tag{6}$$

$$\dot{y} = vsin(\psi + \beta) \tag{7}$$

$$\dot{\psi} = \frac{v}{l_r}sin(\beta) \tag{8}$$

$$\dot{v} = a \tag{9}$$

$$\beta = tan^{-1}\left(\frac{l_r}{l_f + l_r}tan(\delta)\right) \tag{10}$$

Cartesian coordinates (x, y) represent the location of the vehicle's center of mass, whereas the inertial heading ψ is the direction in which the vehicle is moving. The variable v stands for the speed of the car, while the letter a stands for the acceleration felt by the vehicle's center of mass in the direction of the speed. Distances from the vehicle's center to the front and rear axles are denoted by the l_f and l_r variables. Both the front wheel's steering angle (δ) and the vehicle's acceleration (a) are used as inputs for control. The following mathematical representation captures the essence of the discrete-time dynamical model derived by the Euler discretization method:

$$z(t + 1) = f(z(t), u(t)) \tag{11}$$

where $z = \begin{bmatrix} x & y & \psi & v \end{bmatrix}^T$ and $u = \begin{bmatrix} a & \delta \end{bmatrix}^T$ (12) for time t.

Control Objective.
The principal goal of the control system is to effectively integrate into the designated lane, while concurrently avoiding any potential collisions with other vehicles. We have a predilection for changing lanes at a prior intersection. Enhanced driving comfort is typically associated with a preference for smooth accelerations and steering. The presentation of the objective function formulation is as follows:

$$J = \sum_{\ell=t}^{t+T} \lambda_{div}(x(\ell \mid t); x_{end})D(\ell \mid t) \tag{12}$$

$$+ \sum_{\ell=t}^{t+T} \lambda_v\|v(\ell \mid t) - v^{ref}\|^2 \tag{13}$$

$$+ \sum_{\ell=t}^{t+T-1} \lambda_\delta \| \delta(\ell \mid t) \|^2 \tag{14}$$

$$+ \sum_{\ell=t}^{t+T-1} \lambda_a \| a(\ell \mid t) \|^2 \tag{15}$$

$$+ \sum_{\ell=t+1}^{t+T-1} \lambda_{\Delta\delta} \| \delta(\ell \mid t) - \delta(\ell - 1 \mid t) \|^2 \tag{16}$$

$$+ \sum_{\ell-t+1}^{t+T-1} \lambda_{\Delta a} \parallel a(\ell \mid t) - a(\ell - 1 \mid t) \parallel^2 \tag{17}$$

The temporal value of l is determined by the data collected at time t, and is denoted by the symbol $(l|t)$. The latitude coordinate of the road's terminus is represented by the symbol x_{end}. The ego vehicle's distance norm to the target lane at time l is denoted by $D(l|t)$. The symbol v^{ref} refers to the reference velocity. The regularization of each penalty is accomplished by employing λ_{div}, λ_v, λ_δ, λ_a, $\lambda_{\Delta\delta}$, and $\lambda_{\Delta a}$, correspondingly. The timely lane change is incentivized through the utilization of a dynamic weight, denoted as λ_{div}, which is expressed as a convex function. Specifically, λ_{div} is represented as $\parallel \frac{1}{x_{end} - x} \parallel$. The aforementioned expression denoted by (12) serves to penalize the deviation of the car's center from the vertical midpoint of the designated lane. The expressions denoted by (14) and (15) serve to impose a penalty on the exertion of control in relation to the steering angle and acceleration, respectively. To improve ride quality, the steering rate and disturbance is impacted by the equations labelled as (12) and (12), respectively.

Transitional Model Predictive based Controlling for Lane Changing.
To avoid rear-end crashes during overtaking maneuvers, autonomous trajectory path tracking using Model Predictive Control is implemented [14]. In Fig. 4 we see a model of the control architecture used by the autonomous vehicle. For making a lane change, the major focus for designing trajectories is to reduce the amount of yaw acceleration the vehicle experiences. The constraints pertaining to the dynamics of vehicles and the boundaries of the roadside are articulated as limitations within a set of convex optimization problems. The acquisition of reference positions and velocities is achieved through the utilization of a convex optimization algorithm. Model Predictive Control (MPC) controllers, like the one seen in Fig. 4, make use of mathematical models of autonomous cars to anticipate how the system would evolve in the future. This methodology is employed to enhance future vehicle performance and minimize the discrepancy between the planned trajectory route and the actual route (Fig. 3).

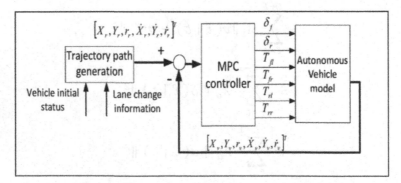

Fig. 3. Architecture for the management of systems in autonomous vehicles.

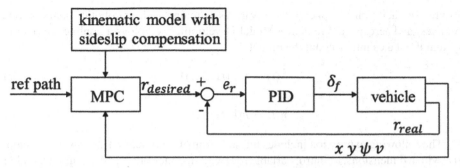

Fig. 4. The Lateral Control configuration of the proposed model

PID Controller

The study presents a PID controller that is specifically designed to ensure that the vehicle adheres to a desired yaw rate [13], $\dot{\psi}$, while simultaneously minimizing the sideslip, β. Equation (18) pertains to the pid control law. This consists of K_p, K_i, and K_d, the proportional, integral, and derivative gains, respectively. These three parts work together to form the G_c block of the controller transfer function.

$$G_c = K_p + \frac{K_i}{s} + K_d s \qquad (18)$$

K_p, K_i and K_d values were obtained using the built-in tuning tool in MATLAB SIMULINK and was imported in coding script., as presented in Table 1. In order to evaluate the efficacy and purpose of control techniques during vehicular maneuvers, distinct control parameters are established for each test velocity to enhance the system's response.

Table 1. PID Tuning Parameters

Control Parameters	Speed	
	Slow 30 km/h	High 80 km/h
K_d	3	0.03
K_p	7	0.81
K_i	5	8.10

3.3 Development of Trajectory Generator and Tracking Controller

The development of a trajectory tracking controller is a crucial area of research in the field of control systems. This controller aims to enable precise tracking of desired trajectories by a dynamic system. By utilizing advanced control algorithms and sensor feedback, the trajectory tracking controller enhances the system's ability to follow predefined paths accurately, facilitating applications in various domains such as robotics, autonomous vehicles, and aerospace. Prediction Model Development employs a nonlinear dynamic system that takes into account the output:

$$\dot{\xi}(t) = f(\xi(t), \mu(t)) \tag{19}$$

$$\eta(t) = h(\xi(t), \mu(t)) \tag{20}$$

The following expression includes the state transition function f(.,.), a state variable $\xi(t)$ with n dimensions, a control variable $\mu(t)$ with m dimensions, and an output variable $\eta(t)$ with p dimensions.

Convert the continuous systems of Eqs. (17) and (18) into a linear time-varying system.

$$\xi(k+1) = A_{k,t}\xi(k) + B_{k,t}\mu(t) + d_{k,t} \tag{21}$$

$$\eta(t) = C_{k,t}\xi(k) + D_{k,t}\mu(t) + e_{k,t} \tag{22}$$

Taking into account the following presumptions:

$$\acute{A}_{k,t} = \begin{bmatrix} A_{k,t} & B_{k,t} \\ 0_{m\times n} & I_m \end{bmatrix} \tag{23}$$

$$\acute{B}_{k,t} = \begin{bmatrix} B_{k,t} \\ I_m \end{bmatrix} \tag{24}$$

$$\acute{C}_{k,t} = [C_{k,t} \quad D_{k,t}] \tag{25}$$

$$\acute{D}_{k,t} = D_{k,t} \tag{26}$$

$$\acute{\xi}(k|t) = \begin{bmatrix} \xi(k|t) \\ \mu(k-1|t) \end{bmatrix} \tag{27}$$

$$\acute{d}(k|t) = \begin{bmatrix} d(k|t) \\ 0_m \end{bmatrix} \tag{28}$$

The symbol $0_{m\times n}$ represents a zero matrix with dimensions $m \times n$, while I_m denotes an identity matrix with dimensions m.

When designing the trajectory tracking controller, it is imperative to take into account the constraints imposed by the vehicle dynamics.

3.4   The Constraint of Sideslip Angle at the Mass Center

Significant changes in driving stability can occur when the mass center's sideslip angle β is outside the linear range of the lateral force, necessitating the imposition of constraints.

The mass center slip angle's constraint range is commonly represented by the arctangent function.

$$-arctan(0.02\mu g) \leq \beta \leq arctan(0.02\mu g) \tag{29}$$

The Constraint of Lateral Acceleration

The amount of grip provided by a car's tires on the road has a direct impact on the vehicle's performance. The presence of different coefficients of adhesion on the road results in the generation of distinct longitudinal and lateral forces that are applied to the Tire by the ground. This research establishes a connection between forward velocity, sideways velocity, and road adhesiveness, highlighting the presence of inequality in this relationship. The inequality of the given expression is as follows.

$$\sqrt{a_x^2 + a_y^2} \leq \mu g \tag{30}$$

can be presented, where the acceleration along the longitudinal axis is denoted by a_x and the acceleration perpendicular to it is denoted by a_y. The longitudinal velocity of the vehicle showed almost no variation during a relatively short time period. It is reasonable to hypothesize that the vehicle maintains a constant longitudinal speed. As a result, Eq. (30) can be simplified in the subsequent manner.

$$|a_y| \leq \mu g \tag{31}$$

The inability to accurately calculate can be ascribed to the constraint conditions being either overly inclusive or overly restrictive, depending on the specific road adhesion circumstances. A relaxation coefficient is included in to provide the adaptive modification of constraint conditions in response to the solution scenario of each control iteration, making the restriction on lateral acceleration a flexible constraint. This inequality holds for every value of the lateral acceleration $|a_y| \leq \mu g$:

$$a_{y,min} - \varepsilon \leq a_y \leq a_{y,max} + \varepsilon. \tag{32}$$

The maximum and minimum lateral accelerations, $a_{y,max}$ and $a_{y,min}$, are shown below, where ε stands for the relaxation factor.

3.5 Experimental Setup

The experiment was set up in simulation using the command prompt of Windows version 11 to execute the main file and a support file for backing up library functions. Python version 3.7, along with NumPy and Matplotlib libraries, was utilized for the process, while separate animation scripts were incorporated for visualizations. The experiment involved constants related to an autonomous vehicle and model parameters, which are detailed in Table 2, allowing researchers and engineers to analyze their impact on the vehicle's behavior and the model's performance.

Table 2. Dimensions of the car model and other environmental attributes

Parameters	Values
Mass (m)	1500 g
Mass moment of inertia (I_z)	3000 g
Front wheel cornering stiffness (C_{af})	19000 N/m
Back wheel cornering stiffness (C_{ar})	33000 N/m
The gap between the front wheel and the mass (L_f)	2 c.m
The gap between the back wheel and the mass (L_r)	3 c.m
Sampling time (T_s)	0.02 s
Lane width	7 m
Number of lanes	5
Reference Trajectory frequency	0.01 Hz

4   Results Analysis

The author of the study devised three distinct trajectory models to assess the efficacy and performance of the overall system under investigation (Fig. 5). These systems were meticulously engineered and implemented to evaluate different aspects and functionalities of the overarching system. The accompanying visual aids, presented below, depict specific instances where a car adeptly tracks and follows its designated reference trajectory. In these illustrations, the reference trajectory is represented by the blue color, while the actual position of the vehicle is denoted by the red color. These visual representations vividly showcase the successful execution of the implemented control algorithms and highlight the capability of the system to accurately adhere to the desired trajectory while maintaining the desired position).

The front wheel demonstrated smooth movement and effectively adjusted its position to accommodate both positive and negative slopes as required by the system, effectively avoiding overshooting. When analyzing the hybrid parabola scenario, a minor delay was observed during the initial response, and the front wheel exhibited aggressive behavior. This behavior can be attributed to the fact that, initially, the vehicle was oriented in the x-direction with a yaw angle of 0 radians, while the reference yaw angle was set to 0.5 radians at time 0 s. Consequently, there was a noticeable deviation in tracking the predetermined setpoint. As the vehicle gradually aligned itself with the desired setpoint, its velocity decreased, leading to a more consistent speed. Additionally, it was noted that the steering wheel angle reached its maximum limit of *pi/6* radians during this specific instance (showcases in Fig. 6).

From the analysis presented in Fig. 7, it becomes evident that the controller's performance deteriorates when operating at higher frequencies. This degradation can be attributed to the fact that the car's longitudinal velocity remains unchanged, leaving insufficient time for the system to stabilize, despite the amplitude of the input signal not being significantly marginal. To overcome this limitation, the longitudinal velocity was

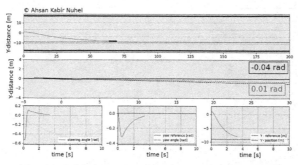

(a)   Car following reference trajectory in a straight line

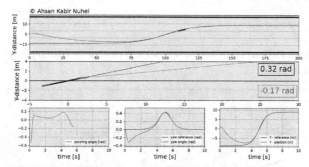

(b)   Car following reference trajectory in a curvy line

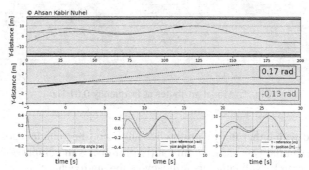

(c)   Car following reference trajectory in a hybrid (Mixed trajectory of sinusoidal and parabolic) line

Fig. 5. Car's attributes in following difference reference trajectories

subsequently adjusted to a value of 20 m/s. Particularly, the experiments show that the vehicle can successfully follow a reference input's optimally trajectory. Moving on to Fig. 8, we observe the output pertaining to variations in higher weight matrices within the cost function for both state and final horizon period outputs. The outcomes reveal a proportional relationship between the weight values and the minimization of specific

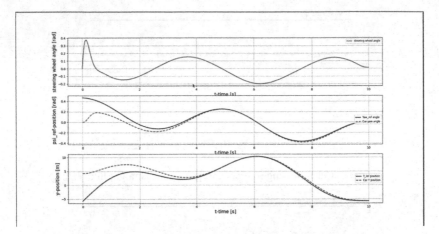

Fig. 6. Different attributes of car at following curvy line trajectory

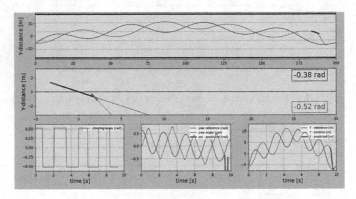

Fig. 7. Model attributes at frequency 0.01 Hz

errors associated with either the yaw reference angles or the vehicle's position. Consequently, the error in the yaw angle significantly diminishes in comparison to the larger positional error, thereby validating the efficacy of the approach.

According to the findings presented in Fig. 9, the incorporation of the Proportional-Integral-Derivative (PID) controller resulted in observable oscillation in both the steering wheel and yaw reference angle. This phenomenon arises due to the fact that the PID controller solely considers errors within a single sampling time and lacks the ability to take into account the entire time horizon, unlike the Model Predictive Control (MPC) controller. Therefore, the task of fully mitigating overshooting becomes a challenging attempt for the PID controller. On the other hand, due to the MPC controller's ability to make more informed decisions by utilising future predictions derived from the system model, the oscillations and error reduction achieved are significantly smoother in comparison to those obtained with the PID controller.

Furthermore, to assess the overall effectiveness of the proposed model, three distinct trajectories were formulated. Among these trajectories, only the exponential and

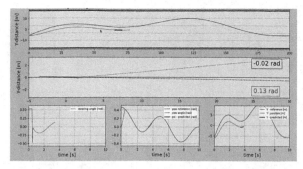

Fig. 8. Outputs concerning changing in the weight matrices by breaking the identity law (by prioritizing Yaw angle error)

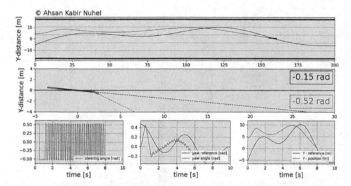

Fig. 9. Output concerning PID controller.

cubic polynomials were considered, and their corresponding outcomes are depicted in Figs. 10 and 11. The results exhibit a satisfactory tracking performance characterized by consistent and accurate adherence to the prescribed reference setpoints, while maintaining a desirable level of smoothness. It is worth noting that the vehicle's initial course angle is set to zero, whereas the initial course angle of the reference trajectory exceeds zero. Consequently, the vehicle demonstrates the ability to adjust its direction during this phase. This modification enables accurate monitoring of the reference trajectory, thereby improving both the resilience and precision. As the vehicle's speed increases and the adhesion coefficient of the road surface reduces, the aforementioned simulation findings show that noticeable deviations from the desired trajectory are noticed when using double-shifting. Moreover, such deviations may give rise to hazardous situations where the vehicle loses control over its direction.

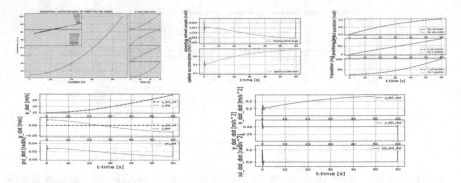

Fig. 10. Results pertaining exponential route

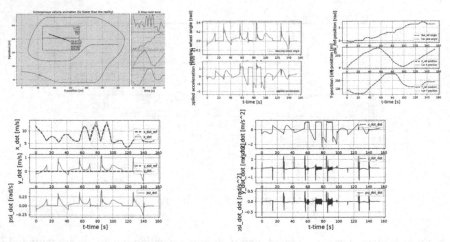

Fig. 11. Results pertaining cubic polynomial route

5 Conclusion

The paper uses a MPC controller to track the path of any cars. It can be used in ADAS or Autonomous cars for keeping the vehicle in lane. For the modern safety requirements, it important have ADAS in every modern vehicle [14]. This MPC controller can meet one of the key features in ADAS system. The paper analysis the different trajectory of the car using the custom made MPC controller and shows how it is better than pre-existing controller such as PID controller. The vehicle movement and speed also taken into consideration while doing the maneuvers so that the cars can keep their lane. Using the MPC controller, Computer vision and Deep learning, more advance ADAS or Autonomous Vehicle can be designed for which will make our roads safer.

References

1. Ahmed, H.U., Huang, Y., Lu, P., Bridgelall, R.: Technology Developments and Impacts of Connected and Autonomous Vehicles: An Overview. **1**, 382–404, Smart Cities 5, USA (2022)

2. Yun, H., Park, D.: Virtualization of Self-Driving Algorithms by Interoperating Embedded Controllers on a Game Engine for a Digital Twining Autonomous Vehicle. **17**, 2102, Electronics 10, USA (2021)
3. Cui, H., et al.: Deep kinematic models for kinematically feasible vehicle trajectory predictions. In: 2020 IEEE International Conference on Robotics and Automation (ICRA), pp. 10563–10569. IEEE, Paris (2020)
4. Sazid, M.M., Haider, I., Rahman, M.E., Nuhel, A.K., Islam, S., Islam, M.R.: Developing a solar powered agricultural robot for autonomous thresher and crop cutting. In: 2022 12th International Conference on Electrical and Computer Engineering (ICECE), pp. 144–147, IEEE Dhaka, Bangladesh (2022)
5. Ganga, G., Dharmana, M.M.: MPC controller for trajectory tracking control of quadcopter. In: 2017 International Conference on Circuit Power and Computing Technologies (ICCPCT), pp. 1–6. IEEE, Kollam, India (2017)
6. Zohu, Z., Rother, C., Chen, J.: MPC Controller for trajectory tracking control of quadcopter. In: ICCPCT, pp.1–6, IEEE (2017)
7. Chen, S., Chen, H., Negrut, D.: Implementation of MPC-based path tracking for autonomous vehicles considering three vehicle dynamics models with different fidelities. Automotive Innovation **3**, 386–399 (2020)
8. Nuhel, A.K., Sazid, M.M., Bhuiyan, M.N.M., Arif, A.I., Roy, P.H., Islam, M.R.: Developing a Self-Driving Autonomous Car using Artificial Intelligence Algorithm. In: 2022 6th International Conference on Electronics, Communication and Aerospace Technology, pp. 1240–1249. IEEE, India (2022)
9. Alcalá, E., Puig, V., Quevedo, J.: LPV-MPC control for autonomous vehicles. IFAC-PapersOnLine **52**(28), 106–113 (2019)
10. Massaro, M., Limebeer, D.J.N.: Minimum-lap-time optimisation and simulation. Veh. Syst. Dyn. **59**(7), 1069–1113 (2021)
11. Nuhel, A.K., Sazid, M.M., Bhuiyan, M.N.M., Arif, A.I.: Designing and performance- analysis of a 3 DOF robotic manipulator arm and its higher order integration for 7 DOF robotic arm. In: 2022 4th International Conference on Sustainable Technologies for Industry 4.0 (STI), pp. 1–6, IEEE, Dhaka, Bangladesh (2022)
12. Mozaffari, S., Al-Jarrah, O.Y., Dianati, M., Jennings, P., Mouzakitis, A.: Deep learning-based vehicle behavior prediction for autonomous driving applications: a review. IEEE Trans. Intell. Transp. Syst. **23**(1), 33–47 (2020)
13. Peicheng, S., Li, L., Ni, X., Yang, A.: Intelligent vehicle path tracking control based on improved MPC and hybrid PID. IEEE Access **10**, 94133–94144 (2022)
14. Williams, T., et al.: Transportation planning implications of automated/connected vehicles on Texas highways. No. FHWA/TX-16/0–6848-1. Texas A&M Transportation Institute (2017)

Performance Evaluation of DSR, AODV and MP-OLSR Routing Protocols Using NS-2 Simulator in MANETs

Hameed Khan[1]([✉]), Kamal Kumar Kushwah[2], Jitendra Singh Thakur[1], Gireesh Gaurav Soni[3], Abhishek Tripathi[1], and Sandeep Rao[1]

[1] Department of Computer Science and Engineering, Jabalpur Engineering College, Jabalpur, M.P., India
hameed.khan20@gmail.com
[2] Department of Applied Physics, Jabalpur Engineering College, Jabalpur, M.P., India
[3] Department of Applied Physics and Optoelectronics, SGSITS, Indore, M.P., India

Abstract. MANET (Ad-Hoc Mobile Network) is a systematic aggregation of identical types and varieties of nodes. These nodes are dynamically created as desirable and capable of communicating, barring a primary infrastructure-based system. As these nodes connect devices such as mobiles, tablets, etc., they can develop a range of provider delivery with an appreciation for network performance. Network traffic is an essential assignment in the ad-hoc mobile area network. Route agreements efficiently enhance carrier exceptionality in better access, partial delivery of packets, and minimal storage delays. The predominant motive of this analysis is to evaluate the legal method concerning the parameters of the various quality of service enhancement services. The simulation outcomes affirm that the proposed scenario affords a better dimension of the exceptional testing of the compliance offerings at MANET.

Keywords: MANET · AODV · DSR · MP-OLSR · NS-2 · Routing Protocols

1 Introduction

The well-known problems associated with wireless and mobile communications, such as bandwidth maximizing efficiency, strength control, and enhancing transmission quality, are carried over by mobile ad hoc networks (MANET). The multi-hop nature and lack of installed infrastructure have given rise to new research concerns such as ad hoc addressing, self-routing, configurations advertising, discovery, and preservation. Mobile ad hoc network architecture is very uncertain and dynamic. The nodes' distribution and ability to self-organize also have a significant impact. Except for a consistent architecture, MANET is a dynamic environment where numerous nodes may be freely distributed and connected to other nodes. Figure 1's representation of the basic architecture of a mobile ad-hoc network illustrates how various networking elements, such as a server, access point, GPS satellite, etc., interact with one another.

© ICST Institute for Computer Sciences, Social Informatics and Telecommunications Engineering 2024
Published by Springer Nature Switzerland AG 2024. All Rights Reserved
P. Pareek et al. (Eds.): IC4S 2023, LNICST 537, pp. 122–133, 2024.
https://doi.org/10.1007/978-3-031-48891-7_10

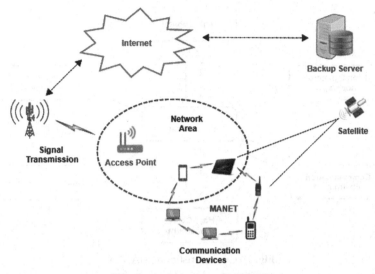

Fig. 1. Architecture of MANET

In MANET, the motive of this routing protocol is to decide how the nodes themselves determine how they can locate, connect, and transmit packets to different nodes [1]. These route strategies are divided primarily into functional and hybrid contracts. Active protocols preserve all the viable pathways between all current nodes equally. In inactive protocols, only the contact nodes related to the assisting nodes have required to switch packets. Hybrid Agreements comprise negotiating strategies to discover an excellent way to change packets to an ad-hoc cellular network. In MANET, nodes can regularly alternate locations with complete instructions to produce specific route issues [2]. The most challenging problem is finding an efficient way between two nodes with multiple hops in the network based on the quality of service parameters for proactive and reactive protocols such as throughput and partial delivery packages by altering the MANET's network load and dimension [3]. Figure 2 shows the essential characteristics of a mobile ad-hoc network.

2 Literature Survey

When more extensive networks are taken into consideration, tests on the Network Simulator (NS-2) have demonstrated that the Dynamic Source Routing protocol (DSR) is only slightly less efficient than the Ad-hoc On-demand Distance Vector (AODV). However, since AODV affects several networking websites, it is more prone to assault than DSR [4]. Additionally, it has been shown that the (DSR) increases the overall performance of vehicular ad hoc networks (VANET) in comparison to AODV protocols in terms of high power consumption, low packet loss, increased delivery rate, and decreased latency even in a wide variety of vehicular networks. However, primarily based on the simulation results of NS-2, it examines four quality parameters, end delays, termination, packet loss, and energy consumption [5]. The DSR agreements with the AODV under egocentric and

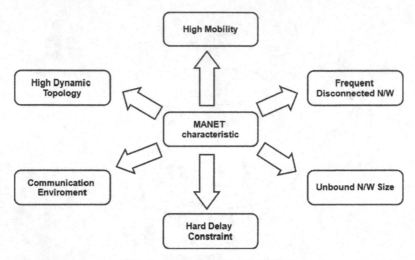

Fig. 2. Characteristics of MANET

dark attacks additionally produced comprehensive tests and outcomes that showed DSR used to be more affected by egocentric node attacks than AODV.

On the other hand, DSR performed better than AODV under black hole attacks. Depending on the individual parameters, such as widespread packing and installation, AODV has higher performance and much less packet delay than the more significant packet delay in DSR [6]. Similarly to other parameters like energy consumption per byte under the attack of egocentric nodes, DSR consumes less energy than AODV. However, the DSR consumes more energy under the black hole attack than the AODV. In short, each DSR and AODV technique is affected by this attack, and performance varies under different parameters [7]. Also, selecting the proper routes according to the network finally impacts the implementation of this network in secure and efficient ways.

In contrast, the protocol (MP-OLSR) hybrid segment was analyzed higher than AODV and DSR. However, its performance will affect trade with an exclusive community and variability in mobility [8]. Due to its potential to keep a connection through periodic information exchange, AODV performance is most desirable to DSR. AODV demonstrates its most influential and overall performance under greater mobility than DSR for real-time functions. To check product accuracy using the Analysis of Variance check (1-way ANOVA), AODV indicates better route overall performance (security and power optimization) than the preferred routing method, DSR [9]. AODV ensures an excessive packet delivery ratio (PDR) and several installations. During the simulation experiments in the MATLAB 2018a simulator, many amendments to network topology extend the computational complexity of current MP-OLSR routing processes as calculating new routes becomes more complicated [10]. With its extensive range of nodes and network statistics, DSR exceeds AODV in terms of performance, PDR, and packet loss ratio using the Netsim 10.2 simulator.

The Riverbed Simulator Modeler examined AODV, MP-OLSR, and DSR. Regarding E2E delays, records lowered and surpassed, and the MP-OLSR protocol fared better than the other agreements, AODV and DSR [11]. Analyze performance for all performance metrics utilizing Riverbed Simulator Modeler, AODV, DSR, OLSR, and GRP. In terms of stop delays, termination, and packet disposal, it has been shown that the OLSR protocol performs better than the other three agreements (AODV, GRP, and DSR) [12]. It has already been proven that when we amplify the wide-area network of nodes in the MP-OLSR protocol network technique phase, there are more significant delays than other AOMDV techniques and AOMDV work for the wider community. To analyze the number of processes of the range of routes, the authors reviewed each of the following guidelines with an exceptional feature, and the only way to decide the route is to make them different [13]. Table 1 compares the many sorts of work done by scholars on various platforms.

Table 1. Summary of current studies and applications that is pertinent.

Study	Approach and Application	Findings
Parissidis, [14]	Quantitative comparison of routing protocols	Node Density
Yang J, [15]	Particle swarm optimization	Energy consumption
Alturfi, [16]	Performance of heterogeneous nodes	Optimize N/W Load
J. Deepika, [17]	Energy Efficient Routing	Power optimization
Mohapatra, S., [18]	Routing strategic approach	NS-2 Simulation
L, Yun-kyung, [19]	Correlation Analysis of Performance Metrics	NS and QualNet 5.0
Abdulleh,M., [20]	Performance Analysis of Protocol	N/W Size and Density
Sharma, A., [21]	QoS improving methods	Overhead minimization
Jiazi Yi, [22]	Hybrid Protocol routing technique	Scalability and Security
A Mouiz [23]	Performance evaluation in MANET	Energy conservation

3 Methodology

This literature assessment is accomplished on separate route contracts at MANET. We took the other three routing processes, DSR, AODV, and MP-OLSR protocol, and discussed the overall performance and the impact on various ever-changing performance parameters. A DSR is a required protocol that uses an activation mechanism. AODV is a wonderful mechanism by which it finds a route wherever it is needed, and finally, MP-OLSR is a hybrid multipath routing protocol [18]. It combines repetitive and intermittent material to hold network topology. The performance of the routing protocols is measured based on the measurement of the navigation network, and the result validates the feasibility of the routing protocol. The proposed quality of service simulation mode has extraordinary parameters for evaluating and comparing the performances of DSR, AODV, and MP-OLSR through NS2 simulations. Instead of maximizing the interpretation of one family of protocols, our focus is on demonstrating the various behaviors of multiple families of protocols. They are classified as efficient, effective, and hybrid approaches. A comparative evaluation of these methods provided an overall performance evaluation of various route problems. This literature learns about objectives to advance a high-quality regulation enforcement framework with multiple parameters to improve the quality of services at MANET [19].

4 Routing Protocol in MANET

4.1 Routing Protocol Classification

In MANET, Fig. 3 displays several routing protocols used in routing. Proactive, reactive, and hybrid systems are the three categories of routing protocols. The MANET routing protocols are intended to support many nodes with few resources. In routing systems, the disappearance and reappearance of nodes at various places is a serious issue. Message routing overhead must be reduced despite the growing number of mobile nodes. As the size of the routing protocol may affect the control packets transmitted inside the network, it is also crucial to keep the routing table small. Although they choose the fastest route to the goal, routing protocols are categorized depending on how and when routes are identified.

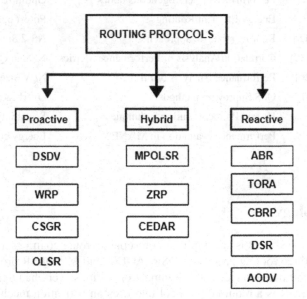

Fig. 3. Classification of routing protocol in MANET

4.2 Proactive Routing Protocols

A proactive routing system employs link-state routing algorithms that often saturate nearby connections with data. The proactive routing system preserves and maintains routing information by commuting control packets with their neighbors. Proactive routing techniques include DSDV, WRP, and OLSR [19].

4.3   Reactive Routing Protocols

Reactive routing approaches need to have the overheads that proactive routing strategies have. It uses a distance-vector routing method and builds a route only when a receiving node requests it, initiating the route discovery process. Only a handful of the reactive routing protocols available in MANET include DSR, AODV, TORA, and LMR [20].

4.4   Hybrid Protocol

It uses both proactive and reactive routing strategies. Several hybrid routing protocols include ZRP, BGP, and EIGRP. In this paper, we compare the effectiveness of the MANET DSR, AODV, and MP-OLSR routing protocols using a variety of factors.

4.5   DSR Protocol

When a data packet travels from the source to a location to discover the source route, the device's address between the source and destination ought to be accrued through the vicinity of the route it splits to pass the packets. It can result in excessive throughput of IPV6 address types. To avoid using the source route, a new protocol known as Dynamic Source Routing (DSR) has been developed, which no longer depends on the routing table for each central device. However, instead, it defines a flow-id alternative that a permit packet has transferred to a hop-by-hop base. Its much-needed feature prevents the package from overloading by control packs by deleting periodic beacon messages (Hello messages), which is required in any other case. On the other hand, like all other procedures, it has drawbacks. It does not restore a damaged link due to a route correction method. Also, connection setup extends greater than table-driven protocols. Its overall performance decreases unexpectedly with increasing nodes.

4.6   AODV Protocol

Ad hoc on Demand Vector (AODV) is a routing protocol in MANET. It is an on-demand protocol that does not depend on pre-maintained routes but builds their preferred routes depending on needs. The protocol has been designed to overcome the impairment troubles of the DSR protocol, with many nodes inside the source and destination [21]. Also, it overcomes some barriers of the DSR protocol, i.e., it has a couple of packet transfer routes between the source and your destination, which requires the preservation of multiple router tables. Two other counters have been saved in AODV protocols and route tables, which assist in determining the updated route between the source and destination.

4.7   MP-OLSR Protocol

The MP-OLSR, or Multipath Optimized Link Source Routing Protocol for MANET, is a hybrid protocol that uses the Dijkstra algorithm to achieve multiple travel packages; its identity suggests this protocol to alternate information except going via a single primary channel [22]. It affords dynamic route tables as per the need to produce transfer information packets in various feasible ways. Apart from this, some critical aspects of this protocol work inexpensively as route restoration methods and discover limitations in the proposed loop [23]. Sometimes it needs to be more adequate to estimate information loading in distinct approaches due to the selected algorithm (Round Robin). Also, a pre-determined amount is provided for the duly carried out cost when a network does not comply with the conditions.

5 Study Matrix

Some essential overall performance matrices can be explored:

5.1 Packet Delivery Ratio (PDR)

It allows the percentage-based disclosure of a protocol's capacity to transmit all emitted data. By dividing the number of packets sent by the source node SNp by the total number of packets received by the destination node DNp, Eq. (1) calculates the number of packets lost. Higher PDR values indicate better performance. PDR of 100% means excellent availability and dependability of the network. An average packet of statistics delivered to destinations is produced through constant bit rate (CBR) sources.

$$PDR = \frac{DNp\_received}{SNp\_transmitted} \times 100 \tag{1}$$

5.2 Throughput

The number of packets/bytes acquired by the source at each time. It is an essential parameter for inspecting network agreements. The magnitude of successful data transmission sent from one location to another at a specific time is measured in bits per second. The throughput may be evaluated by using Eq. (2):

$$Throughput = \frac{(L - C)}{L} \times R \times F(\gamma) \tag{2}$$

Where the following parameters:

- L: Packet length.
- C: Cyclic Redundancy Check.
- R (b/s): Binary transmission rate.
- F(γ) Packet success rate.

6 Performance Analysis of Multicast Protocols

The simulation time is regarded as 20 s in the state, and the number of nodes varies from 10, 30, 50, 100, and 150 nodes. The grid (network size) region has been viewed to be 2000 X 2000 rectangular meters. The architecture of NS-2, which stands for Network Simulator Version 2, is seen in Fig. 4. It's a free, open-source, event-driven simulator for computer communication network research (Table 2).

Table 2. Simulation Parameters.

Parameters	Simulation Matrix	Values
Configuration	Network Size	2000 × 2000 m
	Number of Nodes	10, 30, 50, 100 and 150
Run	Simulation time	20 s
Mobility	Model	Random Way Point
	Maximum Speed	5 to 10 m/s
	Pause Time	10 s
PHY	Propagation Model	Two-ray ground
	Transmission range	300 m
Traffic	Traffic Types	CBR (Constant Bit Rate)
	Packet Size	1200 Byte
	Packet Rate	10 packets/s
Platform	Simulator	NS-2.29

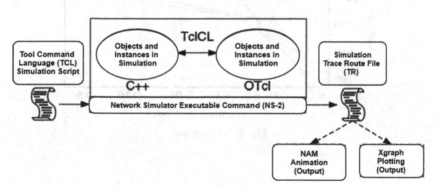

Fig. 4. Architecture of Network Simulator -2

7 Results

Three critical overall performance metrics have been identified for assessing these route processes. The simulation results with parameters are listed in Table 3:

Table 3. Summary of Simulation Results.

Parameters/ Protocols	AODV	DSR	MP-OLSR
Throughput	Low	High	Average
Packet Delivery Ratio	Average	Low	High

7.1 Throughput

The large number of packets transmitted to the recipient provides network benefits. The comparison of three techniques for throughput measures is shown in Fig. 5. The chart shows AODV protocols present slightly lower throughput than MP-OLSR protocols. Furthermore, the DSR protocol had a higher throughput than the AODV and MP-OLSR protocols. With additional traffic sources, congestion, obscured terminals, and network disruption become more common. Due to these problems, protocols respond to changing circumstances differently, and latency plays a crucial role in determining network speed. Finally, the throughput of AODV and MP-OLSR is less worrying than DSR; the throughput drops with a smaller node and improves when the network's nodes are expanded.

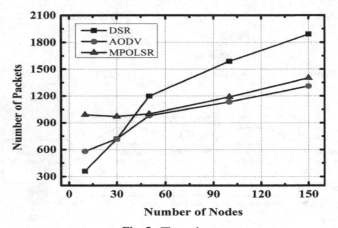

Fig. 5. Throughput

7.2 Packet Delivery Ratio

The throughput metric and the packet delivery ratio (PDR) are intimately related. The destination keeps track of how many data packets it gets and uses that data to determine the network's PDR delivery ratio. Figure 6 illustrates how the MP-OLSR protocols have a higher packet delivery ratio than the other AODV and DSR protocols. The DSR has a lower packet delivery ratio than AODV and MP-OLSR regarding the proportion of data packets drawn into their source. Reactive protocols gradually increased their packet delivery ratio from 0.8 for 10 numbers to unity for higher node densities. As the number of nodes increased, so did the values of MP-OLSR and AODV. Additionally, the MP-OLSR and AODV protocols outperformed the DSR protocol by a small margin.

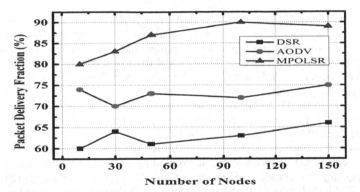

Fig. 6. Packet Delivery Ratio

8 Conclusion

The DSR routing protocol has higher performance than the proposed parameters. However, it will grant less overall latency to the wider network area than other routing protocols. MP-OLSR works better than DSR and AODV in general community delays and common network approaches. In the simulation and subsequent analysis, the overall performance routing protocols' are extended with the network's resolution and the suitable routes under the network. The authors analyzed the package delivery charge on the scale of the DSR, AODV, and MP-OLSR.

In the proposed investigation, two distinctive route strategies confirmed that the MP-OLSR protocol works exceptionally well for MANET in simulation results. However, it is no longer usually better for the entire network. Its performance and functionality have been modified with various networks and versions regarding durability and mobility. Finally, deciding on the proper network protocol gives a better understanding of efficiency. Our future work focuses on extending the set of experiments by considering other simulation parameters. Our future simulation will be elaborated in NS-3.

Acknowledgement. The authors would like to acknowledge the TEQIP-3 CRS lab (ID 1–5726249571) of Jabalpur Engineering College, Jabalpur, MP, India, for the progress of the research work.

References

1. Chang, S., Ting, W., Chen, J.: Method for reducing routing overhead for mobile Ad Hoc network. Int. Conf. on Wireless Communications & Signal Processing (WCSP), pp. 1–6, MECS press (2010)
2. Al-khatib, A., Hassan, R.: Performance evaluation of AODV, DSDV, and DSR routing protocols in MANET using NS-2 simulator. International Conference of Reliable Information and Communication Technology, pp. 276–284. Malaysia (2018)
3. Adeyemi, A.A., et al.: A comprehensive analysis of proactive and reactive manet routing protocols using Ns3. Journal of Engineering and Applied Sciences Technology. SRC/JEAS-**130** (2021). https://doi.org/10.47363/JEAS/(3)120

4. Fauzia, S., Fatima, K.: Performance evaluation of AODV routing protocol for free space optical mobile ad-hoc networks. The International Symposium on Intelligent Systems Technologies and Applications, pp. 74–83, Springer International Publishing (2018)
5. Khosa, T., Mathonsi, T.E., Plessis, D.: A model to prevent gray hole attack in mobile ad-hoc networks. Ad Hoc & Sensor Wireless Networks **14**, 532–542 (2023). https://doi.org/10.12720/jait.14.3.532-542
6. Bello, A., Akpofure, O.: Mobile adhoc network routing protocols: Performance evaluation & assessment. An International Multidisciplinary Research Journal **11**, 1266–1273 (2021). https://doi.org/10.5958/2249-7137.2021.01550.0
7. Saudi, N.A., Mohamad, B., Alya, G.F., Ahmad, S.R.: Mobile Ad-Hoc network (MANET) routing protocols: a performance assessment. Proceedings of the Third International Conference on Computing, Mathematics and Statistics (iCMS2017), pp.53–59, Malaysia (2019)
8. Gupta, C., Singh, L., Tiwari, R.: Wormhole attack detection techniques in ad-hoc network: a systematic review. Open Computer Science **12**(1), 260–288 (2022). https://doi.org/10.1515/comp-2022-0245
9. Marina, M.K., Das, S.R..: Routing in mobile ad hoc networks. In: Mohapatra, P., Krishnamurthy, S.V. (eds) Ad Hoc Networks. Springer, Boston, MA. (2005) https://doi.org/10.1007/0-387-22690-7_3
10. Sandhu, D.S., Sharma, S.: Performance evaluation of DSDV, DSR, OLSR, TORA routing protocols – a review. In: Das, V.V., Chaba, Y. (eds) Mobile Communication and Power Engineering. AIM 2012. Communications in Computer and Information Science, vol 296. Springer, Berlin, Heidelberg. (2013) https://doi.org/10.1007/978-3-642-35864-7_77
11. Wang, H.-M., Zhang, Y., Ng, D.W.K., Lee, M.H.: Secure routing with power optimization for ad-hoc networks. IEEE Transactions on Communications **66**(10), 4666–4679, IEEE (2018). https://doi.org/10.1109/TCOMM.2018.2835478
12. Boushaba, A., Benabbou, A., Benabbou, R., et al.: An intelligent multipath optimized link state routing protocol for QoS and QoE enhancement of video transmission in MANETs. Computing **98**, 803–825 (2016). https://doi.org/10.1007/s00607-015-0450-0
13. Guaya-Delgado, L., Pallarès-Segarra, E., Mezher, A.M., et al.: A novel dynamic reputation-based source routing protocol for mobile ad hoc networks. J Wireless Com Network **77** (2019). https://doi.org/10.1186/s13638-019-1375-7
14. Parissidis, G., Lenders, V., May, M., Plattner, B.: Multi-path routing protocols in wireless mobile ad hoc networks: a quantitative comparison. In: Koucheryavy, Y., Harju, J., Iversen, V.B. (eds) Next Generation Teletraffic and Wired/Wireless Advanced Networking. NEW2AN. Lecture Notes in Computer Science, vol 4003. Springer, Berlin, Heidelberg. (2006). https://doi.org/10.1007/11759355_30
15. Yang, J., Liu, F., Cao, J., Wang, L.: Discrete particle swarm optimization routing protocol for wireless sensor networks with multiple mobile sinks. Sensors **16**(7), 1081 (2016). https://doi.org/10.3390/s16071081
16. Alturfi, S., Kadhim, D., Mohammed, M.: Network performance evaluation of different manet routing protocols configured on heterogeneous nodes. Journal of Physics: Conference Series. 1804. 012124. Babylon-Hilla City, Iraq (2021). https://doi.org/10.1088/1742-6596/1804/1/012124
17. Deepika, J., Rangaiah, L., Jeyabalan, S.: A novel approach to AODV for energy efficient routing mechanism to control power consumption in MANET. International Journal of Engineering Trends and Technology **69**(8), 206–210 (2021). https://doi.org/10.14445/22315381/IJETT-V69I8P225
18. Mohapatra, S., Kanungo, P.: Comparative performance analysis of MANET routing protocols using NS2 simulator. In: Das, V.V., Thankachan, N. (eds) Computational Intelligence and Information Technology. CIIT. Communications in Computer and Information Science, vol 250. Springer, Berlin, Heidelberg (2011). https://doi.org/10.1007/978-3-642-25734-6_127

19. Lee, Y.-K., Kim, J.-G.: Performance comparison between routing protocols based on the correlation analysis of performance metrics for AODV routing protocol. Journal of Information Technology Services 12(4), 349–367 (2013). https://doi.org/10.9716/KITS.2013.12.4.349
20. Abdulleh, M., Yussof, S.: Performance analysis of AODV, OLSR and GPSR MANET routing protocols with respect to network size and density. Research Journal of Applied Sciences, Engineering and Technology 11, 400–406 (2015). https://doi.org/10.19026/rjaset.11.1794
21. Sharma, A., Vashistha, S.: Improving the QOS in MANET by enhancing the routing technique of AOMDV protocol. In: Satapathy, S., Avadhani, P., Udgata, S., Lakshminarayana, S. (eds) ICT and Critical Infrastructure: Proceedings of the 48th Annual Convention of Computer Society of India- Vol I. Advances in Intelligent Systems and Computing, 248. Springer, Cham (2014). https://doi.org/10.1007/978-3-319-03107-1_41
22. Yi, J., Adnane, A., David, S., Parrein, B.: Multipath optimized link state routing for mobile ad hoc networks. Ad Hoc Netw. 9, 28–47 (2011). https://doi.org/10.1016/j.adhoc.2010.04.007
23. Mouiz, A., Badri, A., Baghdad, A., Sahel, A.: Performance evaluation of OLSR and AODV routing protocols with different parameters in mobile ad-hoc networks using NS2 simulator. In Proceedings of the 2019, 5th International Conference on Computer and Technology Applications (ICCTA'19). Association for Computing Machinery, New York, NY, USA, pp.134–139 (2019). https://doi.org/10.1145/3323933.3324065

Efficient Fuel Delivery at Your Fingertips: Developing a Seamless On-Demand Fuel Delivery App with Flutter

Navneet Mishra[1], Ritika Raghuwanshi[1], Naveen Kumar Maurya[2], and Indrajeet Kumar[1(✉)] (iD)

[1] School of CSIT, Symbiosis University of Applied Sciences, Indore, India
indrajeet.kumar@suas.ac.in
[2] ECE, Vishnu Institute of Technology, Bhimavaram, Andhra Pradesh, India

Abstract. The increasing demand for fuel due to the growth of automobiles in the market has led to the need for on-demand fuel supply applications that depend on user orders and requirements. When a vehicle runs out of fuel, it can be a hassle for the owner to push the car or seek help to reach the nearest gas station. For older people and those who are medically ill, this task can be even more difficult. Additionally, people must go to gas stations to fill up generators. To address these issues, we introduce a new solution for vehicle refueling and emergency power supplies through the development of an on-demand fuel delivery application. This application provides door-to-door coverage and allows end users to choose the type of fuel they need, order it, and receive it with ease. The outcome of this research paper will be the development of a mobile application using Flutter framework that offers a range of functionalities catering to both customers and fuel station owners. The application aims to provide a convenient platform for customers to order fuel, locate nearby gas stations, and assist owners in efficiently managing orders and monitoring station availability. By utilizing Flutter, a cross-platform development framework, the application will be compatible with both Android and iOS devices, ensuring a broader reach and accessibility for users. Flutter's rich UI capabilities and native-like performance will enable the creation of a visually appealing and seamless user experience.

Keywords: Arduino · Flutter · LDR · Eye Blink Sensor · Ultrasonic Sensor

1 Introduction

Our app-based service, On-Demand Fuel Delivery Application that is built using flutter, is similar to what we other On-Demand Delivery services but we aim to provide fuel to other customer. Our goal is to establish a system where the user can request for the fuel to be deliver to his footsteps [1]. We aim to provide timely delivery of fuel to customers. In this modern fast paced world where demands are amplifying day-by-day, we aim to revolutionize of the least modernized area by introducing our On-Demand

P. Pareek et al. (Eds.): IC4S 2023, LNICST 537, pp. 134–147, 2024.
https://doi.org/10.1007/978-3-031-48891-7_11

Fuel Delivery Application built using Flutter which removes the existing constraints and gives an easy to use, safe, reliable way to meet the user demands [2–4]. On-Demand Fuel Delivery Application provides online fuel ordering services including an engaging and comprehensive online fuel ordering process such as ordering online, tracking order, and checking the fuel prices nearby. The future scalability of on-demand fuel services is enormous, and several end-users can be targeted. For example, 37% of India's electricity in urban areas is generated solely by diesel generator sets [5].

The paper [6] provides a review of various fuel management systems used in transportation, highlighting their importance in reducing fuel consumption and costs. The study covers the technical aspects of FMS, challenges in implementing FMS, and their potential benefits in the transportation industry. The research paper [7–9] examines the challenges of meeting the growing demand for road fuel in these countries driven by economic development and population growth. The study analyses factors affecting demand and the effectiveness of policy interventions like fuel taxes and subsidies. The paper suggests a more comprehensive approach is needed to promote sustainable transportation that considers technological innovations and behavioural change. It offers recommendations for policymakers to address these challenges.

The paper [10–13] examines the sector-wise demand for diesel and petrol in India, with a focus on key drivers of demand. The study provides insights for policymakers and industry stakeholders and contributes to a better understanding of the energy sector in India.

Fig. 1. Shows the utility of on demand fuel app.

In Fig. 1, we are presented with a demonstration of the practicality and effectiveness of the on-demand fuel app, which is developed using both the Flutter framework [14–16] and machine learning technology [17]. The figure illustrates how this innovative combination enables the app to deliver exceptional performance and enhanced user experiences. With the Flutter framework, the app achieves seamless multi-platform compatibility, allowing users to access its features effortlessly across mobile, web, and desktop devices. This unified approach ensures that the app can reach a broader audience and cater to diverse user preferences. MIMO technology's ability to mitigate channel fading and reduce signal degradation can contribute to improved network connectivity and reliability for Flutter apps [18]. By leveraging spatial multiplexing and beamforming techniques, MIMO can ensure more robust communication between the app and backend servers, resulting in more stable data transmission and reduced packet loss [19–23].

Furthermore, the integration of machine learning technology enhances the app's capabilities significantly. Machine learning algorithms enable the app to analyze user behavior, preferences, and patterns, leading to personalized fuel delivery recommendations and optimized service [24–26]. Through continuous learning and adaptation, the app can provide tailored and efficient fuel delivery solutions for each user. By combining the power of Flutter and machine learning, Fig. 1 highlights how the on-demand fuel app becomes a robust and cutting-edge solution that revolutionizes the way users access and receive fuel services. This depiction emphasizes the app's ability to stay at the forefront of technological advancements, providing a superior and user-centric experience [27–31].

The research offers valuable insights and recommendations for policymakers to promote sustainable development. The research paper [32–34] examines the factors driving oil demand in India, the impact of policy interventions on oil demand, and recommendations for promoting energy conservation and renewable energy sources. The study suggests that India's oil demand will continue to grow due to economic development and urbanization, and offers insights and recommendations for policymakers. The Ministry of Oil and Gas of India published a report on the "All India Survey on Diesel and Gasoline Demand by Sector" which outlines the consumption of diesel and gasoline in various sectors. The transportation sector is the largest consumer at 60%, followed by agriculture and industry. The report emphasizes the need to reduce dependency on fossil fuels and promote sustainable alternatives for a greener economy. During the forecast period of 2022–2032, the On-Demand fuel delivery market is likely to increase at a CAGR of 6.8%. In 2022, this market is expected to reach around $4.8 billion. The On-Demand fuel delivery market value will likely be $6.2 billion by 2026. On-demand application revenue is likely to generate $935 billion in 2023.

The rest of the paper is structured as follows: The next section provides essential background information to understand the proposed work, including a review of related studies conducted by other researchers and the areas where their research falls short. In Sect. 3, we delve into the system's design, explaining its architecture and components in detail. Moving on to Sect. 4, we showcase the practical implementation of the on-demand fuel app using Flutter, and to further clarify its operation, we present an illustrative example. This example vividly demonstrates how the system works in a real-life scenario. Finally, in Sect. 5, we conclude the study by offering a comprehensive summary of the findings, discussing the significance of the results, and highlighting potential avenues for future research.

2 Background

Flutter is a free and open-source user interface toolkit developed by Google. It empowers developers to create natively compiled applications for multiple platforms, including mobile, web, and desktop, all from a single codebase. Google initially unveiled Flutter at the Google I/O developer conference in May 2017, and since then, it has gained widespread acclaim among developers. This popularity can be attributed to its rapid development cycle, enabling faster iterations during the coding process, its excellent performance, ensuring smooth and efficient application execution, and its ability to cater to various platforms without the need for separate codebases, thereby streamlining the development workflow. Flutter uses a reactive programming model, where changes to the UI are automatically reflected in the app's state, and vice versa. This enables developers to build highly interactive and responsive apps with a smooth user experience. Flutter also comes with a rich set of customizable widgets and allows developers to create their own widgets or modify existing ones to suit their needs.The proposed system is divided into five distinct working units, each serving a specific purpose. The IR sensor hurdle detection unit is responsible for detecting obstacles using infrared technology. The ultrasonic hurdle detection unit utilizes ultrasonic waves to identify objects in the vehicle's vicinity. The automatic headlight unit ensures that the vehicle's headlights are activated or deactivated based on the ambient lighting conditions. The engine control unit utilizes the alcohol sensor to prevent the vehicle from starting if the driver is under the influence of alcohol. Lastly, the drowsiness detection unit utilizes the eyeblink sensor to monitor the driver's alertness level and provide timely alerts if signs of drowsiness are detected.

Flutter offers an incredibly useful feature known as "hot-reload," allowing developers to instantly view the changes they make to the code in real-time, without the need to restart the application. This feature significantly accelerates the development process and enhances efficiency, as developers can rapidly experiment with various adjustments and instantly see their impact.

Furthermore, Flutter provides a comprehensive set of tools to facilitate the development experience. One of these tools is Flutter Studio, a robust Integrated Development Environment (IDE) that empowers developers to create and design their applications efficiently. Additionally, Flutter offers command-line tools and plugins that seamlessly integrate with popular development environments like Android Studio and Visual Studio Code, making it even more convenient for developers to work with their preferred tools.

In conclusion, Flutter proves to be a highly versatile and powerful toolkit that caters to developers' needs, enabling them to produce top-notch, cross-platform applications with ease. Its combination of hot-reload and feature-rich development tools fosters a smooth and productive development environment, ultimately resulting in high-quality, responsive, and engaging applications for various platforms. Its popularity is only expected to grow in the coming years as more developers discover its potential and adopt it for our paper.

3   System Model

Flutter is a powerful and versatile open-source framework developed by Google for creating cross-platform mobile applications. It allows developers to build high-quality, natively compiled apps for both Android and iOS platforms using a single codebase. Flutter employs the Dart programming language, which is known for its simplicity and ease of learning. One of the key strengths of Flutter is its fast and hot reload feature, enabling developers to see instant updates on the app as they make changes to the code. This significantly speeds up the development process and facilitates rapid prototyping and iterative improvements.

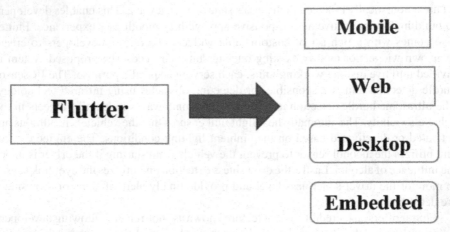

Fig. 2. Empowering Seamless User Interface on Multiple Devices using Versatile UI Framework

For startups, the ability to seamlessly connect with users across various platforms such as mobile, web, and desktop through a unified app empowers them to effectively reach their entire audience right from the beginning. This approach eliminates any constraints imposed by technical limitations, ensuring a broader accessibility to their products and services. By catering to multiple platforms with just one codebase, startups can allocate their resources more efficiently and avoid the complexities associated with managing separate codebases. Not only does this streamline development efforts, but it also leads to cost savings and faster time-to-market.

Similarly, for larger organizations, offering a consistent user experience to all users, regardless of their device or platform, streamlines the development process and reduces overall complexity (Fig. 2). By maintaining a single codebase, these organizations can focus on improving the quality of the user experience and iterating on their app more effectively. This unified approach enables them to strategically allocate their development team's efforts, resulting in higher user satisfaction.

In both cases, leveraging a single app codebase for multiple platforms provides a competitive advantage, allowing businesses to remain agile and responsive to user needs. This flexibility in reaching a diverse user base enhances user engagement and opens up new opportunities for growth and success in today's interconnected digital landscape.

These widgets are customizable and can be combined to build complex UI layouts with ease. Due to its native-like performance, Flutter delivers smooth and fluid user experiences, which are crucial for mobile applications. It achieves this by compiling the Dart code directly into native ARM machine code, eliminating the need for a bridge between the application and the platform's native components. Another advantage of Flutter is its strong community support and an ever-growing ecosystem of packages and plugins. These resources provide developers with a wide range of functionalities and integrations, making it easier to implement various features in their apps. With Flutter, developers can also create beautiful animations and stunning visuals, enhancing the overall user experience. It supports 2D and 3D graphics rendering, allowing for eye-catching visual elements. Furthermore, Flutter's single codebase approach simplifies maintenance and reduces development costs, as developers do not need to manage separate codebases for different platforms.

Our On-Demand Fuel Delivery Application mainly consists of four modules:

3.1 Register Module

The registration module of our On-Demand Fuel Delivery Application requires users as well as the fuel station to enter their credentials and login first. To register as a user, you will be required to provide specific personal details, including your full name, contact number, email address, chosen username, and a secure password. On the other hand, gas stations are also required to register by submitting their pertinent information, which includes the gas station's name, contact number, email address, their chosen username, a password, and the physical location of the gas station.

3.2 Information Module

The gas station is obligated to furnish essential details regarding fuel availability, the pricing of various fuel types, the range of fuels they offer, and the services they provide. As fuel is a crucial element for any vehicle, its price experiences daily fluctuations, and these prices may also vary based on the gas station's location. Consequently, it becomes the gas station's responsibility to ensure that fuel prices are updated on a daily basis to accurately reflect the current market rates. This timely updating ensures transparency and facilitates customers in making informed decisions about fuel purchases.

3.3 Order Fuel Module

Once a user registers with the application and gains access to its services, they can conveniently order fuel from their location based on their specific requirements. To utilize the application's features, users are prompted to enter their credentials for authentication.

To place a fuel order, users must first locate a nearby gas station using the application's built-in features. Upon finding a suitable gas station, they can proceed to check the availability of fuel and the range of services offered by that particular station. After verifying the availability of the required fuel and desired services, users can then proceed to place their fuel order according to their needs and preferences. This streamlined process ensures that users have easy access to fuel services while being well-informed about the options available to them.

3.4 Track Order Module

After a user places an order for fuel through the application, they gain the ability to track the progress of the order. The user can monitor various stages, such as whether the order has been accepted by the gas station and whether the fuel delivery has been made. To stay informed about these updates, it is crucial for the gas station to actively participate in the process.

Upon receiving an order, the gas station is responsible for either approving or rejecting it. Subsequently, the gas station is required to regularly update the order status to keep the user informed about any changes in the order's progress. By actively participating in this process, the gas station ensures effective communication with the user, enabling them to know the current status of their fuel order and anticipate the delivery accordingly.

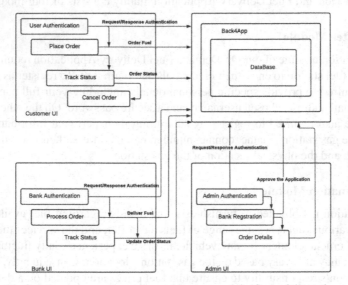

Fig. 3. In this figure we can see the entire architecture and the flow of the entire application:

The Customer UI includes the interactive environment developed for the customer. The entire section contains all the section present directly to the customer as shown in Fig. 3. First Section in the Customer UI is the User Authentication section in which the user must enter their credentials through the Request/Response Authentication in which one module sends request to other module and waits for a response. The user credentials are sent to the Database of the Application for Authentication. Once the response is received from the Database Section the process is processed accordingly. Once User Authentication is completed the customer then can move forward to the Place Order Section which enables the customer to be able to Place Order for the Fuel Delivery. Once the user has placed the order the record is stored in the database. Once the user has placed the order, they will be able to track their order status where they can see the current status of their order and can track the location of the delivery. The customer would also have access to Cancel Order where the customer can cancel their order and once the cancel request is completed the process the aborted and an entry is made in the database regarding the entire case.

The Bunk UI will handle all the transaction related process where we intend to hire a third party for safer, faster and reliable transactions. Bunk is the placeholder for the third-party service we intend to utilize in our application. The Bunk UI contains Bank Authentication, Process Order, Track status. First the Bank Authentication will include verification of the bank details the customer provides with their respective bank, once the verification is completed the service will be providing the customer with number of options for the transaction that their bank provides for smoother user experience. If bank details are stored in the database the service will request the details from the database module and wait for the response and if the details are not present the process is proceeded as mentioned above and the user bank details are stored in the database. Once the Authentication and the transaction is completed then the order is processed forward and a message is sent to the database regarding the process and on successful user transaction the customer is send the details of the transaction. In the Track Status section, the Bunk UI constantly monitors and sends the updates to the database. If the order is cancelled the Bunk UI starts the process of returning the money to the customer by requesting the details from the database.

The Admin UI contains all Admin tools providing several productivity features. The architecture focuses on the three section of the Admin UI namely Admin Authentication, Bank Registration and Order Details. The admin module manages the entire system, and only authorized personnel can access it. The admin can view all registered users and fuel stations, track orders, and oversee payments. The admin module also provides analytical reports, including sales reports and customer insights, to help the fuel delivery service to make informed decisions.

4 Implementation

Our On-Demand Fuel Delivery Application has been skillfully developed and put into action, allowing users to access and request fuel delivery services conveniently and efficiently. Through our implementation, we have ensured a seamless user experience, enabling customers to order fuel at their convenience and have it delivered promptly to their desired location. Our innovative application built using cutting-edge Flutter technology has proven to be a game-changer, revolutionizing the way fuel is delivered to users, and enhancing the overall experience of fuel procurement. The implementation of our On-Demand Fuel Delivery Application will involve the following stages:

4.1 Design Phase

During this stage, the user interface, features, and functionalities of the application will be designed. We will use Flutter to build the user interface, and we will incorporate Material Design principles to ensure a clean, modern look.

4.2 Development Phase

This stage involves the actual coding of the application. We will use the Flutter framework to develop the application, and we will use Firebase for user authentication and database management. Google Maps API will be integrated for displaying the nearby gas stations and tracking the fuel delivery.

4.3 Testing Phase

During this phase, the application will undergo thorough testing to verify its proper functionality and alignment with user requirements. We will conduct two essential types of testing: unit testing and integration testing. Unit testing involves evaluating individual components of the application in isolation to ensure their proper operation. Each component will be rigorously tested to verify that it performs as intended and meets its specified functionalities.

Integration testing, on the other hand, assesses the interaction and compatibility between different components of the application. We will examine how these components work together harmoniously to deliver the desired features and functionalities as a cohesive whole. By performing both unit and integration testing, we aim to ensure that the application is reliable, robust, and capable of fulfilling user expectations. This meticulous testing process enables us to identify and address any potential issues or discrepancies before the application is launched to end-users, ensuring a smooth and satisfactory user experience.

4.4 Deployment Phase

After the completion of development and rigorous testing, the application will be ready for deployment. It will be made available to the public through the Google Play Store for Android users and the Apple App Store for iOS users. Once deployed, users can easily discover the application on the respective app stores and download it onto their smartphones. They can then proceed to install the application, granting them access to its full functionality and features. By making the application accessible on these widely-used platforms, it ensures a broad reach to users across both Android and iOS ecosystems. This approach maximizes the application's exposure and availability, allowing a diverse user base to benefit from its offerings.

5 Results

The following figures illustrates the key screens that the user will interact with while using Our On-Demand Application built using Flutter. Flutter is Google's SDK for crafting attractive, User-friendly environment for mobile, web, and desktop from a single codebase.

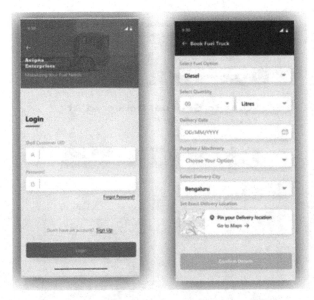

Fig. 4. Illustrating Login (Left) and Home (Right)

Figure 4 Login Page where registered users can input their pre-existing credentials stored in the database and login to their accounts. User can use the forgot password feature incase the user forgets their password. Users will be able to reset their password through email verification. Unregistered users can Sign Up in-order to use the Application. Main Activity screen is displayed when the user has successfully has verified his credentials and has login to the application. In this screen the user can click on book fuel truck and proceed to the next step of the process.

Figure 5 Place Order Activity screen provides user with a lot of options starting with fuel option where user can choose the type of fuel he wants to be delivered, user can also choose the quantity of the fuel and the measurement units. User must provide the delivery date and purpose of his request for the fuel. The user then must provide his exact location and once the user has input the details, he can proceed by clicking on the Confirm Details.

Fig. 5. Illustrating the order activity screen provides user with a lot of options starting with fuel option.

6 Challenges and Opportunities

The On-Demand Fuel Delivery Application faces several challenges and opportunities. One of the challenges is building a robust and reliable delivery system. Fuel delivery requires special equipment and trained personnel, and the application must ensure that the delivery is safe, efficient, and timely. The application must also comply with regulatory requirements, such as fuel transportation regulations, to ensure the safety of the delivery personnel and users.

Another challenge is the adoption of the application by fuel stations and customers. The application must convince fuel stations to partner with the service, and customers must be willing to use the application instead of visiting gas stations physically. The application must also be user-friendly, reliable, and provide a seamless experience to ensure customer satisfaction.

The On-Demand Fuel Delivery Application also presents several opportunities. The first opportunity is providing convenience to users. Users can avoid the hassle of visiting gas stations and waiting in queues by ordering fuel through the application. The application can also provide users with real-time fuel prices, promotions, and discounts, helping them save time and money.

7   Conclusion

Our On-demand fuel delivery application has the potential to revolutionize the traditional fuelling process and provide customers with a convenient, time-saving, and efficient service. By leveraging technology and innovation, the application can address the challenges and limitations of traditional fuelling methods, such as long queues, time-consuming commutes, and limited availability of fuel stations. Our On-Demand fuel delivery application represents an innovative and promising solution to the challenges and limitations of traditional fuelling methods. By continuously improving and innovating the application, we can further enhance its benefits and promote a more sustainable and efficient fuelling system for the future. Overall, the On-Demand Fuel Delivery Application developed using Flutter will help meet the growing demand for fuel while promoting sustainability and creating a more convenient and reliable fuel delivery system. In summary, Flutter is a powerful and popular framework for mobile app development that offers a wide range of benefits, including fast development, native performance, expressive UI design, and a strong community support system. Its versatility and efficiency make it an excellent choice for building modern and feature-rich mobile applications.

References

1. Ahmed, A.A.I., Mohammed, S.A.E., Satte, M.A.M.H.: Fuel management system. In: 2017 International Conference on Communication, Control, Computing and Electronics Engineering (ICCCCEE), Khartoum, Sudan, pp. 1–7 (2017)
2. Chandrasiri, S.: Demand for road-fuel in a small developing economy: the case of Sri Lanka. Energy Policy **34**(14), 1833–1840 (2006)
3. Nielsen India Private Limited: All India Study on Sectoral Demand of Diesel & Petrol. Ministry of Petroleum and Natural Gas (2013)
4. Rivera-González, L., Bolonio, D., García-López, G.A., Alvarez, M.: Long-term forecast of energy and fuels demand toward a sustainable road transport sector in Ecuador (2016–2035): a LEAP model application. Energies **12**(20), 3849 (2019). https://doi.org/10.3390/en1220 3849
5. Agarwal, P.: India's Petroleum Demand: Empirical Estimations and Projections for the Future. Institute of Economic Growth (IEG) University (2012)
6. Rabinovich, A., Azuri, Y., Shtilman, L.: Assessment of fuel delivery system of a high-performance UAV engine. J. Propul. Power **34**(4), 880–888 (2018)
7. Gao, H., Liu, J., Huang, Q.: Fault diagnosis of fuel delivery system for diesel engine based on dynamic Bayesian network. J. Mech. Sci. Technol. **33**(5), 2245–2253 (2019)
8. Huang, K., Xie, S., Wang, X., Sun, L.: Design and simulation of a fuel delivery system for a variable compression ratio engine. Energies **13**(22), 6029 (2020)
9. Wang, J., Liu, J., Huang, Q.: Design of a fuel delivery system for high-speed diesel engine based on digital simulation technology. Int. J. Automot. Technol. **22**(3), 1045–1056 (2021)
10. Williams, T.M., Pearson, J.M.: Fuel Delivery Systems for Gasoline Direct Injection Engines. SAE Technical Paper, 2018–01–0312 (2018)
11. Manh, N.P., Jeong, H.G.: Modeling and control of a fuel delivery system for gasoline engines. Energies **10**(8), 1221 (2017)
12. Kuo, Y.S., Chen, W.L.: Design and optimization of a fuel delivery system for a diesel engine using CFD simulation and RSM methodology. Energies **9**(11), 918 (2016)

13. Sharpe, R.G., de Bruin, T.: Fuel delivery system modeling for high-pressure common rail diesel engines. J. Eng. Gas Turbines Power **136**(6), 061505 (2014)
14. Ameen, S.Y., Mohammed, D.Y.: Developing cross-platform library using flutter. Eur. J. Eng. Technol. Res. **7**(2), 18–21 (2022)
15. Wiriasto, G.W., Aji, R.W.S., Budiman, D.F.: Design and development of attendance system application using android-based flutter. In: 2020 Third International Conference on Vocational Education and Electrical Engineering (ICVEE), pp. 1–6 (2020)
16. Kavitha, M., Srinivas, P.V.V.S., Kalyampudi, P.S.L., Srinivasulu, S.: Machine learning techniques for anomaly detection in smart healthcare. In: 2021 Third International Conference on Inventive Research in Computing Applications (ICIRCA), Coimbatore, India, pp. 1350–1356 (2021). https://doi.org/10.1109/ICIRCA51532.2021.9544795
17. Vadrevu, P.K., Veeramanickam, M.R.M., Adusumalli, S.K., Bunga, S.K.: Sign language recognition for needy people using machine learning model. In: Intelligent Computing and Applications: Proceedings of ICDIC, pp. 227–233 (2020). Singapore: Springer Nature Singapore, 2022
18. Kumar, I., Mishra, M.K., Mishra, R.K.: Performance analysis of NOMA downlink for next-generation 5G network with statistical channel state information. Ingénierie des Systèmes d'Information **26**(4), 417–423 (2021). https://doi.org/10.18280/isi.260410
19. Shankar, R., Kumar, I., Mishra, R.K.: Pairwise error probability analysis of dual hop relaying network over time selective Nakagami-m fading channel with imperfect CSI and node mobility. Traitement du Signal **36**(3), 281–295 (2019). https://doi.org/10.18280/ts.360312
20. Kumar, I., Kumar, A., Mishra, R.K.: Performance analysis of cooperative NOMA system for defense application with relay selection in a hostile environment. The Journal of Defense Modeling and Simulation (2022). doi:https://doi.org/10.1177/15485129221079721
21. Ashish, I.K., Mishra, R.K.: Performance analysis for wireless non-orthogonal multiple access downlink systems. In: 2020 International Conference on Emerging Frontiers in Electrical and Electronic Technologies (ICEFEET), Patna, India, pp. 1–6 (2020). https://doi.org/10.1109/ICEFEET49149.2020.9186987
22. Maurya, N.K., Kumari, S., Pareek, P., Singh, L.: Graphene-based frequency agile isolation enhancement mechanism for MIMO antenna in terahertz regime. Nano Communication Networks, p. 100436 (2023)
23. Maurya, N.K., Bhattacharya, R.: CPW-fed dual-band compact Yagi-type pattern diversity antenna for LTE and WiFi. Progress In Electromagnetics Research C **107**, 183–201 (2021)
24. Maurya, N.K., Bhattacharya, R.: Design of compact dual-polarized multiband MIMO antenna using near-field for IoT. AEU-International Journal of Electronics and Communications **117**, 153091 (2020)
25. Kumar, I., Mishra, R.K.: An investigation of spectral efficiency in linear MRC and MMSE detectors with perfect and imperfect CSI for massive MIMO systems. Traitement du Signal **38**(2), 495–501 (2021). https://doi.org/10.18280/ts.380229
26. Kumar, I., Mishra, R.K.: An efficient ICI mitigation technique for MIMO-OFDM system in time-varying channels. Mathematical Modelling of Engineering Problems **7**(1), 79–86 (2020). https://doi.org/10.18280/mmep.070110.
27. Valarmathi, B., et al.: Price estimation of used cars using machine learning algorithms. In: International Conference on Cognitive Computing and Cyber Physical Systems, pp. 26–41 (2022). Springer Nature Switzerland, Cham
28. Biorn-Hansen, A., Rieger, C., et al.: An empirical investigation of performance overhead in cross-platform mobile development frameworks. In: Empirical Software Engineering **25**, pp. 299730240 (2020). Springer
29. Kumar, I., Sachan, V., Shankar, R., Mishra, R.K.: An investigation of wireless S-DF hybrid satellite terrestrial relaying network over time selective fading channel. Traitement du Signal **35**(2), 103–120 (2018). https://doi.org/10.3166/TS.35.103-120

30. Kumar, I., Sachan, V., Shankar, R., Mishra, R.K.: Performance Analysis of Multi-User Massive MIMO Systems with Perfect and Imperfect CSI. Procedia Computer Science **167**, pp. 1452–1461 (2020), ISSN 1877–0509. https://doi.org/10.1016/j.procs.2020.03.356

31. Gupta, N., Kumar, I., Rathod, I., Sharma, S.S.P.M.: Sustainable Production Systems with ai and Emerging Technologies: A Moderator-Mediation Analysis. **12**(Special Issue-8), 2819–2832 (2023). https://doi.org/10.48047/ecb/2023.12.si8.200

32. Arb, G.I., Al-Majdi, K.: A freights status management system based on dart and flutter programming language. Journal of Physics: Conference Series **1530**(1). IOP Publishing (2020)

33. Pareek, P., Maurya, N.K., Singh, L., Gupta, N., Reis, M.J.C.S.: Study of smart city compatible monolithic quantum well photodetector. In: International Conference on Cognitive Computing and Cyber Physical Systems, pp. 215–224 (2022). Springer Nature Switzerland, Cham

34. Li, L., et al.: CiD: automating the detection of API-related compatibility issues in Android apps. In: 27th ACM SIGSOFT International Symposium on Software Testing and Analysis (ISSTA), pp. 153–163 (2018)

35. Sharma, S.S.P.M., Ravishankar Kamath, H., Siva Brahmaiah Rama, V.: Modelling of cloud based online access system for solar charge controller International Journal of Engineering & Technology **7**(2.21), 58–61 (2018)

36. Shalinee Gupta, S.S.P.M., Sharma, B.: Design and Development of an Intelligent Aqua Monitoring System using Cloud Based Online Access Control Systems International Journal of Recent Technology and Engineering (IJRTE) **8**(4) (2019). ISSN: 2277–3878

37. Ravishankar Kamath, H., Sharma, S.S.P.M., Siva Brahmaiah Rama, V.: PWM based solar charge controller using IoT International Journal of Engineering & Technology **7**(2.7), 284–288 (2018)

38. Ravishankar Kamath, H., Siva Brahmaiah Rama, V., Sharma, S.S.P.M.: Street Light Monitoring Using IOT International Journal of Engineering & Technology **7**(2.7), 1008–1012 (2018)

39. Sharma, S.S.P.M., Kumar, A., Meena, B. K.: An Intelligent Solar Based Farm Monitoring using Cloud Based Online Access Control Systems International Journal of Recent Technology and Engineering (IJRTE) **8**(3) (2019).ISSN: 2277–3878

An IoT Application Based Decentralized Electronic Voting System Using Blockchain

Yash Gupta[1], Rishabh Verma[1], S. S. P. M. B. Sharma[2], and Indrajeet Kumar[1]($\boxtimes$) (iD)

[1] School of CSIT, Symbiosis University of Applied Sciences, Indore, India
indrajeetnitp2050@gmail.com
[2] School of MT, Symbiosis University of Applied Sciences, Indore, India

Abstract. In 21st century, there is huge political unrests among political leaders in many developing countries, in spite of Boom of Internet users, still the elections are held in semi-old fashion. The cost associated with these elections (i.e. appliances, workforce related, transport etc.) act as burden on tax payer's money & cuts the huge possibilities of R&D in any nation. Specially in south-east nations, Indians (NRIs) move abroad for various reasons; To demonstrate the feasibility of our protocol in real-world scenarios, we have implemented it using Ethereum's blockchain as a public bulletin board. As a result, casting vote for them becomes very difficult, hence we created mechanism where NRIs can vote securely using their passports.

Keywords: Arduino · Blockchain LDR · Eye Blink Sensor · Ultrasonic Sensor

1 Introduction

Originally, blockchain was just known to some CS fellows & researchers for how to structure & share data. Today blockchains are hailed the "5th evolution" [1] of computing. Blockchain (Blockchain) in a Network, where Block allude to the list of transactions noted into distributed ledgers over a given period of time. It has three main components i.e., Size, Period, & Trigger Event [2–5] for each block. Chain alludes to a hash that associates one block to another. It's also the magic that glues blockchains together & allowed them to define mathematical trust. Network [6–8] is composed of "full nodes". A blockchain is a kind of data structure that makes it possible to create a digital distributed ledger of data & share it among a network of independent parties [9–12]. Blockchain is mainly classified into 3 major types i.e., Public: very large distributed networks running through a native cryptocurrency mostly Bitcoin, Ethereum. Fully Open Source. Permissioned [13] Large distributed network; control roles for individuals within network. Private: Distributed Ledger Tech i.e., smaller & no token nor any cryptocurrency required [14].

The main reason why blockchain becomes word of mouth so quickly is the "Consensus". As mentioned in Fig. 1, the consensus blockchain creates honest systems where they self-correct themselves (i.e., nodes in a network) without any third-party monitoring [15]. This consensus algorithm is the process of creating an accord or concurrence among group of commonly distrustful shareholders.

P. Pareek et al. (Eds.): IC4S 2023, LNICST 537, pp. 148–161, 2024.
https://doi.org/10.1007/978-3-031-48891-7_12

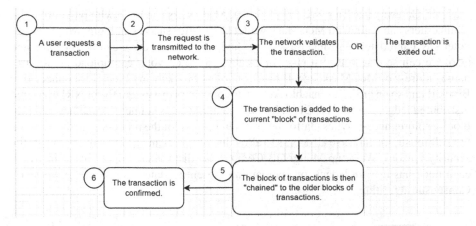

Fig. 1. Process of Consensus Algorithm for e- voting system using blockchain.

Blockchains are also now being used in various industries such as Security: To counter code piracy, securing IOT devices from spoofing & hacking [16–20]. One of the significant advantages of MIMO for IoT is its ability to provide improved connectivity and extended coverage [21–25]. By employing multiple antennas, MIMO systems can mitigate the effects of multipath fading and enhance signal reception in challenging environments [26]. This capability is particularly crucial for IoT devices deployed in complex urban settings or harsh industrial environments, where reliable connectivity is vital for seamless data transmission and reception [27]. Government-Agencies: Maintaining shatterproof Land-Record Systems. ICOs: Smart Contract that allows the issuer to grant token for Investment-Funds related to Banks [28, 29]. The Existing System of general public election is running manually [30–34]. The Voter has to Visit to Booth to cast their votes. As a result, many people don't go out to cast their vote which is one of the major drawbacks of current voting system. In democracy, every citizen of a country must vote [25]. By a new online system which will limit the voting frauds and make the voting as well as counting more efficient and transparent.

The remaining sections of the paper are organized as follows: The subsequent section offers the necessary background information to comprehend the proposed work, including a discussion on similar studies conducted by other researchers and the gaps present in their research. Section 2 outlines the specific work proposed in this paper and provides details on the design of the system. Section 3 presents the implementation of the E-voting system, accompanied by an illustrative example that demonstrates how the system functions. Lastly, Sect. 4 concludes the study by providing a summary of the current progress and outlining plans for future work.

2 System Model

In Fig. 2, we are explaining the complete architecture CodeFlow of our decentralized E-Voting application using Hardhat as blockchain platform [12]. Here, in Front-End Part we designed 3 ways i.e. User can login using web application, mobile application or even via SMS Gateway (for Tier-2 & Tier-3 cities). Also, NRIs can login with passport login, while others can login with MetaMask wallet [13]. After successful login, we can create voter's & candidates. As MetaMask is verified, voter registration can be accepted and also, we can see the voter list [14]. In create new voter, voter can upload photo, user name, address of MetaMask, age. By using hardhat blockchain - block I is connected to Block II and so on [15]. Hardhat blockchain is connected to MongoDB for NRI users and distributed ledger are created. Event Management Server is connected to Etherscan [16] block explorer and analytics platform for Ethereum. All of the process of authentication the Administrator can have all access to allow list of voters and have all the rights for the security reasons. We setup MongoDB Compass as distributed ledger cluster in which each node has its own ledger for maintaining the entire state of cluster as it works on Consensus algorithm.

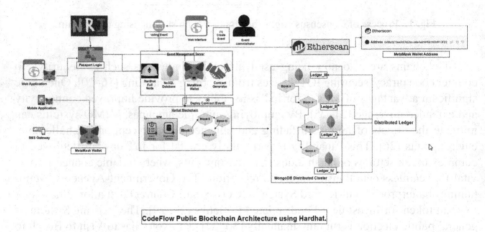

CodeFlow Public Blockchain Architecture using Hardhat.

Fig. 2. Showing entire Code flow using Hardhat

As shown in Fig. 3, we explained our Full-Stack directory-level tree structure where we created directories i.e., assets, components, context, contracts, pages, scripts, styles, test & configs [17]. In components, we have Navbar, voter Card which includes NavBar.js, voterCard.js & NavBar.module.css etc. In context we created Voter.js, constants.js & Create.json etc. In pages directory, we created candidate_registration.js, _app.js, allowed-voters.js, voterList.js etc. In scripts we created deploy.js. Under styles directory, we created allowedVoter.module.css, globals.module.css, index.module.css, voterList.module.css. In test directory, we created Lock.js file. In configs file, we hardhat.config.js, next.config.js, package.json, package-lock.json. There are multiple dependencies used in this project such as openzeepelin, axios, ethers, hardhat, ipfs-http-client, dot-env, web3modal [18].

As shown in Fig. 4, we explained our entire directory-level tree structure where we created directories i.e., configs, data, server & views. In server directory we created database.js, server.js & passportConfig.js. Inside database.js file we created the database schema which contains all the collection name & their datatype. We have three main fields i.e., passport_no, name, email-id & password. In views directory, we created index.ejs,

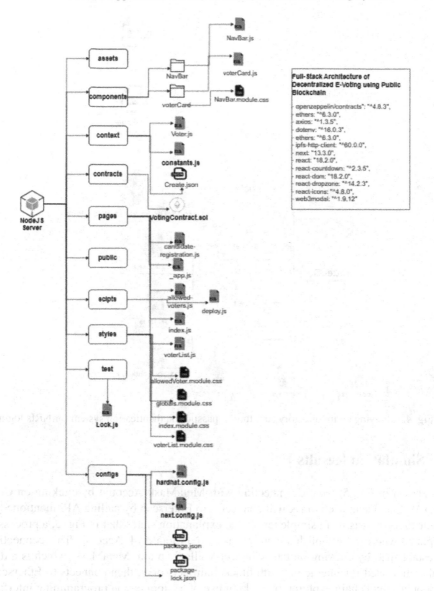

Fig. 3. Showing entire directory structure of project

login.ejs & register.ejs. The. ejs extension is embedded javascript file. In scripts we created deploy.js. In configs file, we created package.json, package-lock.json. There are multiple dependencies used in this project such as ejs, express, express-session, local, mongoose, nodemon, passport, passport-local etc.

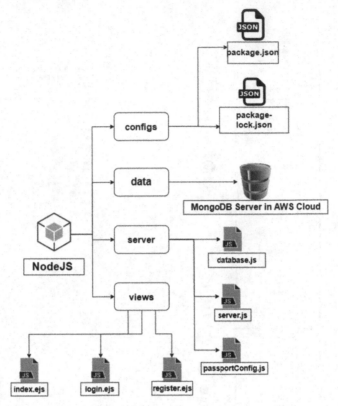

Fig. 4. Showing entire directory structure of passport authentication system for NRIs login

3 Simulation Results

As shown in Fig. 5, we are connecting with MetaMask Account by clicking on Connect Wallet. Then, MetaMask will connect it to Etherscan by calling API mentioned in context/constansts.js. In simple terms, the explanation states that in Fig. 5, a process is depicted where we establish a connection with a MetaMask Account. This connection is established by clicking on the "Connect Wallet" option. MetaMask, which is a digital wallet used for interacting with blockchain networks, then connects to Etherscan, a popular blockchain explorer, by making use of an application programming interface (API) mentioned in the context/constansts.js file.

As displayed in Fig. 6, now we are using candidate-register.js file for registering our candidate. Also, in the figure it is showing No. of Candidate's & No. of Voter's in this page.

In Fig. 7, we have created a web form which takes name of candidate, Address from which location it belongs & for the post he/she is contesting for. Here, candidate can also upload his/her image.

In Fig. 8, candidate is selecting his/her photo by using index.js file & uploading images present in assets directory. Also, these images are stored in local database (Fig. 9).

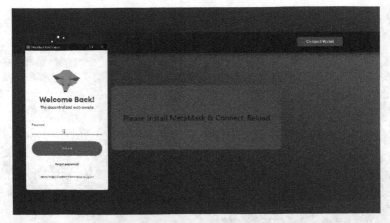

Fig. 5. Picture depicting Connection with MetaMask.

Fig. 6. Selecting option of Candidate Registration

As name, address & position, candidate clicked on Authorized Candidate. After that, MetaMask wallet address with registered account gets initiated & Ethereum (ETH) gas fee gets deducted from MetaMask wallet.

In this Fig. 10, we successfully registered our candidate, also the status of Number of Candidates gets incremented by one.

In the Fig. 11, after filling all the form details such as name, address & age, voter clicks on Authorized Voter.

In above Fig. 12, the Voter is created successfully, yet he/she hasn't voted i.e., displayed as Not Voted. This method of contract is coded in VotingContract.sol file under contracts directory.

Fig. 7. Register New Candidate by filling Name, Address, Position etc.

Fig. 8. Select 'Candidate Photo' for upload

As mentioned in Fig. 13, MetaMask wallet address with registered account gets initiated & Ethereum (ETH) gas fee gets deducted from MetaMask wallet as voter clicks on Confirm button. The vote gets casted. All these transaction histories can be seen in Etherscan platform.

As shown in Fig. 14, the status of voter from Not Voted to You Already Voted gets changed. The time taken by showing this status is very less.

Hardcoding a single vote: Within the "Voting Contract.sol" file, it is hardcoded or explicitly defined that each voter can only cast a single vote. This means that once a voter has cast their vote, they cannot cast another vote using the same wallet address associated with their unique private key. Unique wallet address and private key: Each voter is assigned a unique wallet address, which is associated with their private key.

Fig. 9. Confirming Ethereum Gas for candidate registration.

Fig. 10. Candidate is successfully registered.

The private key serves as a digital signature and provides secure access to the voter's wallet and associated functionalities. The use of unique wallet addresses ensures that each voter's vote is recorded accurately and prevents multiple votes from the same individual. In, Fig. 15 showcases the presence of the solidity file "Voting Contract.sol" within the contract's directory. It is specifically mentioned that within this contract, it is hardcoded that each voter can only cast a single vote, as each voter is assigned a unique wallet address associated with their private key. This implementation ensures the integrity and accuracy of the voting system by preventing multiple votes from the same voter.

Fig. 11. Create New Voter by filling Name, Address, Age.

Fig. 12. Page showing status of Voter_One#1 i.e. Not Voted.

Fig. 13. Voter successfully casted his vote.

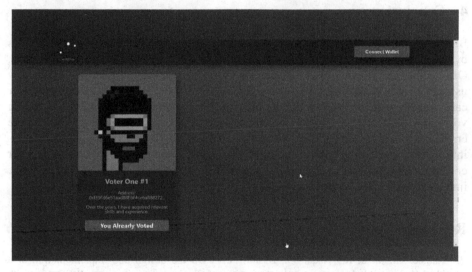

Fig. 14. Page showing status of Voter_One#1's vote i.e. You Already Casted.

Fig. 15. You Already Voted status so that One Person, One Vote.

4   Conclusion

Blockchain technology has the potential to revolutionize various sectors and services by reducing the initial cost of investment and improving performance in specific areas. One notable application of blockchain technology is in the realm of voting systems. By using BCT individual's vote privacy can be kept secret. Voters can give their vote at their ease of space or according to their comfort zone. Voting system can be helpful for giving vote in the elections happened in the colleges etc. Block chain technology reduces the errors at the time of vote counting. Online voting system will able to take care the voter's information where voter can have access and use their voting rights. Our research paper concludes that usefulness or ease of use are still important to decision makers while implementing EVs, but our research shows that building trust faith is prime.

References

1. Rathee, G., Iqbal, R., Waqar, O., Bashir, A.K.: On the design and implementation of a blockchain enabled e-voting application within IoT-oriented smart cities. IEEE Access **9**, 34165–34176 (2021). https://doi.org/10.1109/ACCESS.2021.3061411
2. Vladucu, M.-V., Dong, Z., Medina, J., Rojas-Cessa, R.: E-voting meets blockchain: a survey. IEEE Access **11**, 23293–23308 (2023). https://doi.org/10.1109/ACCESS.2023.3253682
3. Li, H., Li, Y., Yu, Y., Wang, B., Chen, K.: A blockchain-based traceable self-tallying E-voting protocol in AI era. IEEE Trans. Network Sci. Eng. **8**(2), 1019–1032 (2021). https://doi.org/10.1109/TNSE.2020.3011928
4. Zhang, X., Zhang, B., Kiayias, A., Zacharias, T., Ren, K.: An efficient E2E crowd verifiable E-voting system. IEEE Trans. Depend. Secure Comput. **19**(6), 3607–3620 (2022). https://doi.org/10.1109/TDSC.2021.3103336
5. Gao, S., Zheng, D., Guo, R., Jing, C., Hu, C.: An anti-quantum E-voting protocol in blockchain with audit function. IEEE Access **7**, 115304–115316 (2019). https://doi.org/10.1109/ACCESS.2019.2935895

6. Panizo Alonso, L., Gascó, M., Marcos del Blanco, D.Y., Hermida Alonso, J.Á., Barrat, J., Aláiz Moreton, H.: E-voting system evaluation based on the council of europe recommendations: helios voting. IEEE Trans. Emerg. Top. Comput. **9**(1), 161–173 (2021). https://doi.org/10.1109/TETC.2018.2881891

7. Villafiorita, A., Weldemariam, K., Tiella, R.: Development, formal verification, and evaluation of an E-voting system with VVPAT. IEEE Trans. Inf. Forensics Secur. **4**(4), 651–661 (2009). https://doi.org/10.1109/TIFS.2009.2034903

8. Benabdallah, A., Audras, A., Coudert, L., El Madhoun, N., Badra, M.: Analysis of blockchain solutions for E-voting: a systematic literature review. IEEE Access **10**, 70746–70759 (2022). https://doi.org/10.1109/ACCESS.2022.3187688

9. Oprea, S.-V., Bâra, A., Andreescu, A.-I., Cristescu, M.P.: Conceptual architecture of a blockchain solution for E-voting in elections at the university level. IEEE Access **11**, 18461–18474 (2023). https://doi.org/10.1109/ACCESS.2023.3247964

10. Zaghloul, E., Li, T., Ren, J.: D-BAME: distributed blockchain-based anonymous mobile electronic voting. IEEE Internet Things J. **8**(22), 16585–16597 (2021). https://doi.org/10.1109/JIOT.2021.3074877

11. Al-madani, A.M., Gaikwad, A.T.: Iot data security via blockchain technology and service-centric networking. In: 2020 International Conference on Inventive Computation Technologies (ICICT), pp. 17–21. IEEE (2020)

12. Al-madani, A.M., Gaikwad, A.T., Mahale, V., Ahmed, Z.A.T.: Decentralized e-voting system based on smart contract by using blockchain technology. In: 2020 International Conference on Smart Innovations in Design, Environment, Management, Planning and Computing (ICSIDEMPC), pp. 176–180. IEEE (2020)

13. Ayed, A.B.: A conceptual secure blockchain-based electronic voting system. Int. J. Network Secur. Appl. **9**(3), 01–09 (2017)

14. Bell, S., et al.: STAR-Vote: a secure, transparent, auditable, and reliable voting system. In: 2013 Electronic Voting Technology Workshop/Workshop on Trustworthy Elections (EVT/WOTE 13). USENIX Association, Washington, D.C. (2013)

15. Bohli, J.-M., M¨uller-Quade, J¨., R¨ohrich, S.: Bingo voting: secure and coercionfree voting using a trusted random number generator. In: Alkassar, A., Volkamer, M. (eds.) E-Voting and Identity. Vote-ID 2007. LNCS, vol. 4896. Springer, Berlin, Heidelberg (2007). https://doi.org/10.1007/978-3-540-77493-8_10

16. C¸abuk, U.C., Adiguzel, E., Karaarslan, E.: A survey on feasibility and suitability of blockchain techniques for the e-voting systems. arXiv preprint arXiv:2002.07175 (2020)

17. Canessane, R.A., Srinivasan, N., Beuria, A., Singh, A., Kumar, B.M.: Decentralised applications using ethereum blockchain. In: 2019 Fifth International Conference on Science Technology Engineering and Mathematics (ICONSTEM), vol. 1, pp. 75–79. IEEE (2019)

18. Chaum, D.: Secret-ballot receipts: true voter-verifiable elections. IEEE Secur. Priv. **2**(1), 38–47 (2004)

19. Chaum, D., et al.: Scantegrity: end-to-end voter-verifiable optical-scan voting. IEEE Secur. Priv. **6**(3), 40–46 (2008)

20. Chaum, D., Ryan, P.Y.A., Schneider, S.: A practical voter-verifiable election scheme. In: de Capitani, S., di Vimercati, P., Syverson, D.G. (eds.) Computer Security – ESORICS 2005: 10th European Symposium on Research in Computer Security, Milan, Italy, September 12-14, 2005. Proceedings, pp. 118–139. Springer Berlin Heidelberg, Berlin, Heidelberg (2005). https://doi.org/10.1007/11555827_8

21. Shankar, R., Kumar, I., Mishra, R.K.: Outage probability analysis of MIMO-OSTBC relaying network over Nakagami-m fading channel conditions. Traitement du Signal **36**(1), 59–64 (2019). https://doi.org/10.18280/ts.360108

22. Kumar, I., Mishra, M.K., Mishra, R.K.: Performance analysis of NOMA downlink for next-generation 5G network with statistical channel state information. Ingénierie des systèmes d information **26**(4), 417–423 (2021). https://doi.org/10.18280/isi.260410

23. Shankar, R., Kumar, I., Mishra, R.: Pairwise error probability analysis of dual hop relaying network over time selective Nakagami-m fading channel with imperfect CSI and node mobility. Traitement du Signal **36**(3), 281–295 (2019). https://doi.org/10.18280/ts.360312

24. Kumar, I., Kumar, A., Mishra, R.K.: Performance analysis of cooperative NOMA system for defense application with relay selection in a hostile environment. J. Defense Model. Simul. Appl. Methodol. Technol. **20**(4), 553–561 (2022). https://doi.org/10.1177/154851292 21079721

25. Ashish, Kumar, I., Mishra, R.K.:Performance analysis for wireless non-orthogonal multiple access downlink systems. In: 2020 International Conference on Emerging Frontiers in Electrical and Electronic Technologies (ICEFEET), pp. 1–6. Patna, India (2020). https://doi.org/10.1109/ICEFEET49149.2020.9186987

26. Kumar, I., Mishra, R.K.: An investigation of spectral efficiency in linear MRC and MMSE detectors with perfect and imperfect CSI for massive MIMO systems. Traitement du Signal **38**(2), 495–501 (2021). https://doi.org/10.18280/ts.380229

27. Kumar, I., Mishra, R.: An efficient ICI mitigation technique for MIMO-OFDM system in time-varying channels. Mathematical Modelling of Engineering Problems **7**(1), 79–86 (2020). https://doi.org/10.18280/mmep.070110

28. Gupta, N., Kumar, I., Rathod, I., Sharma, B.S.Ṣ.P.M.: Sustainable production systems with ai and emerging technologies: a moderator-mediation analysis, vol. 12, Special Issue-8, pp. 2819–2832 (2023). https://doi.org/10.48047/ecb/2023.12.si8.200. 2023

29. Chaum, D.L.: Untraceable electronic mail, return addresses, and digital pseudonyms. Commun. ACM **24**(2), 84–90 (1981)

30. Dhinakaran, K., Britto Hrudaya Raj, P.M., Vinod, D.: A secure electronic voting system using blockchain technology. In: Goyal, D., Gupta, A.K., Piuri, V., Ganzha, M., Paprzycki, M. (eds.) Proceedings of the Second International Conference on Information Management and Machine Intelligence. LNNS, vol. 166, pp. 307–313. Springer, Singapore (2021). https://doi.org/10.1007/978-981-15-9689-6_34

31. Erdenebileg, M.: e-voting anwendung auf ethereum plattform als smart contract. Fachhochschule Campus Wien (2019)

32. Hanifatunnisa, R., Rahardjo, B.: Blockchain based e-voting recording system design. In: 2017 11th International Conference on Telecommunication Systems Services and Applications (TSSA), pp. 1–6. IEEE (2017)

33. Hjálmarsson, F., Hreiðarsson, G.K., Hamdaqa, M., Hjálmtýsson, G´.: Blockchain-based e-voting system. In: 2018 IEEE 11th International Conference on Cloud Computing (CLOUD), pp. 983–986 (2018)

34. Pareek, P., Maurya, N.K., Singh, L., Gupta, N., Reis, M.J.C.S.: Study of smart city compatible monolithic quantum well photodetector. In: Gupta, N., Pareek, P., Reis, M. (eds.) Cognitive Computing and Cyber Physical Systems. IC4S 2022. LNICS, Social Informatics and Telecommunications Engineering, vol. 472. Springer, Cham (2023). https://doi.org/10.1007/978-3-031-28975-0_18

35. Sharma, B.S.S.P.M., Ravishankar Kamath, H., Siva Brahmaiah Rama, V.: Modelling of cloud based online access system for solar charge controller Int. J. Eng. Technol. **7**(2.21), 58–61 (2018)

36. Gupta, S., Sharma, B.S.S.P.M.: Design and development of an intelligent aqua monitoring system using cloud based online access control systems. Int. J. Recent Technol. Eng. ISSN: 2277-3878 **8**(4) (2019)

37. Ravishankar Kamath, H., Sharma, B.S.S.P.M., Siva Brahmaiah Rama, V.: PWM based solar charge controller using IoT. Int. J. Eng. Technolo. 7(2.7), 284–288 (2018)
38. Ravishankar Kamath, H., Siva Brahmaiah Rama, V., Sharma, B.S.S.P.M.: Street light monitoring using IOT. Int. J. Eng. Technol. 7(2.7), 1008–1012 (2018)
39. Sharma, B.S.S.P.M., Kumar, A., Meena, B.K.: An intelligent solar based farm monitoring using cloud based online access control systems. Int. J. Recent Technol. Eng. ISSN: 2277–3878 8(3), (2019)

NeuroRobo: Bridging the Emotional Gap in Human-Robot Interaction with Facial Sentiment Analysis, Object Detection, and Behavior Prediction

Aparna Parasa[✉], Himabindu Gugulothu, Sai Sri Poojitha Penmetsa, Shobitha Rani Pambala, and Mukhtar A. Sofi

Department of Information Technology, BVRIT Hyderabad College of Engineering for Women, Hyderabad, India
19wh1a1228@bvrithyderabad.edu.in

Abstract. Efficient and personalized human-robot interaction is a critical goal in robotics research. In this study, we propose a novel approach to enhance human-robot interaction by integrating facial sentiment analysis, object detection, and behavior prediction into a bot powered by Blender face technology. Our proposed system enables the bot to perceive and respond to the emotional states and preferences of individuals, creating a more intuitive and engaging interaction experience. By integrating lip syncing capabilities and object recognition functionality through webcam integration, the proposed solution seeks to enhance the authenticity and intuitiveness of user experiences. Through the utilization of Blender animation tools and Natural Language Processing methods, our solution facilitates seamless interaction between humans and neuro robots, contributing to improved outcomes and well-being.

Keywords: Human-Robot Interaction · Prediction · Sentiment Analysis · Natural Language Processing · Behavior Prediction

1 Introduction

Artificial Intelligence (AI) has witnessed tremendous growth and has found extensive applications across diverse domains, including transport [1], healthcare [2,3], finance [4], education [5], and more. One of the remarkable areas where AI has made significant strides is in Robotics. AI-driven robots have demonstrated their capacity to provide immediate, accurate, and reliable aid to individuals, leading to a rising demand for their integration into a wide array of everyday tasks and activities [6]. This increasing demand for AI-powered robots has paved the way for significant advancements in the field of human-robot interaction. As these robots become more integrated into our daily lives, the focus shifts towards developing sophisticated human-robot interaction models that incorporate advanced AI features that facilitate natural and intuitive interactions between humans and machines.

© ICST Institute for Computer Sciences, Social Informatics and Telecommunications Engineering 2024
Published by Springer Nature Switzerland AG 2024. All Rights Reserved
P. Pareek et al. (Eds.): IC4S 2023, LNICST 537, pp. 162–172, 2024.
https://doi.org/10.1007/978-3-031-48891-7_13

Human-Robot Interaction (HRI) [7] research spans across various domains, including computer vision, natural language processing (NLP), machine learning, facial expression recognition, joint action, voice-based interfaces, and multimodal fusion [8,9]. In the field of computer vision, studies have focused on essential components such as face recognition, emotion recognition, and object-detection, which are critical for enabling robots to perceive and understand the visual information in their environment [10]. Additionally, NLP serves as a fundamental aspect of HRI, providing robots with the ability to comprehend and generate human language, facilitating effective communication and interaction with humans [11]. Machine learning techniques have also been employed in HRI to enhance collaborative tasks and adaptability in human-robot collaborative scenarios [12].

Facial expression recognition plays a significant role in HRI, as it enables robots to perceive and interpret human emotions, thereby improving the quality of interaction and engagement [8]. Furthermore, joint action and entrainment research aim to emulate the psychological, neurological, and physical mechanisms of human collaboration, leading to improved performance and subjective metrics such as trust in human-robot teams [13]. Voice-based interfaces and multi-modal fusion techniques contribute to more intuitive and context-aware interaction by incorporating auditory cues and integrating different modalities of human communication [14,15]. These areas of research, along with advancements in deep learning, 3D modeling, and animation algorithms, have greatly contributed to the development of more sophisticated and interactive HRI systems [16].

In this paper, we propose a novel model called NeuroRobo, which utilizes advanced computer vision and deep learning techniques to facilitate real-time interactions between users and computer systems. The model provides lipsyncing-based conversational replies to user input and offers object recognition capabilities via a webcam. Additionally, the model can mimic the user's actions, leading to a more natural and intuitive interaction.

The paper's structure is as follows: Sect. 2 offers a comprehensive overview of the background and related work, Sect. 3 details the proposed model, while Sects. 4 and 5 respectively present the experimental findings and conclude the paper.

2   Background and Related Work

In recent years, as robots have become more integrated into various domains such as healthcare, manufacturing, and assistive technologies, the field of Human-Robot Interaction (HRI) has gained significant attention. The goalof HRI is to enhance collaboration and communication between humans and robots. While earlier HRI methods predominantly focused on rule-based systems, recent methods have made significant strides in enabling more natural and intuitive human-robot interactions. Researchers in [17] emphasized the significance of computer vision and natural language processing in improving human-robot interaction. In [18], the authors introduced a cutting-edge system designed for humanoid robots, enabling them to replicate facial expressions and head motions in real-time. The system incorporates a lightweight deep learning network to detect facial feature points, which helps overcome latency challenges. By establishing a connection between these feature points and the corresponding servo movements, the humanoid robot successfully imitates human behavior with high accuracy. This approach provides a practical solution for achieving mirrored behavior in real-time robotic systems. For further improvements to human robot interaction, [19] focused on the use of machine learning techniques in the context of human-robot collaboration (HRC). The paper clusters the works based on collaborative tasks, evaluation metrics, and cognitive variables modeled. It emphasizes the significance of incorporating time dependencies in machine learning algorithms. The study in [20] explores the technical advancements of Natural Language Processing (NLP) and Artificial Intelligence (AI) specifically in the domain ofspeech recognition. It covers various types, models, and applications of speechrecognition and delves into system characteristics, speech recognition algorithms, and the role of n-grams in natural language processing. A recent study [21] emphasizes the significance of joint interaction and social cues in human-robotinteraction. The researchers focus on leveraging social skills, including mutualgaze, gaze following, speech, and human face recognition, to enhance the process of interactive visual object learning in dynamic environments. By incorporating these social cues, the study aims to create a more interactiveand engaging experience between humans and robots, enabling effective learning and understanding of visual objects in dynamic real-world settings. [22] focuses on the research of joint action and its implications for human-robot interaction. The goal is to develop artificial systems that can emulate the psychological, neurological, and physical mechanisms of joint action, leading to improved human-robot team performance and subjective metrics like trust.

In [23], the authors delve into the field of facial expression recognition, specifically using an enhanced Convolutional Neural Network (CNN) with an attention mechanism. The paper highlights the importance of facial expression recognition in human-computer interaction and sheds light on the experiments conducted, which yield promising and satisfactory results. By employing this advanced CNN model with attention, the study contributes to the advancement of facial expression recognition techniques, potentially enhancing the overall effectiveness and naturalness of human-computer interaction. [24] delves into the topic of human facial expression recognition and the generation of facial expressions by robots. The paper encompasses two main aspects: facial expression recognition using pre-existing datasets and real-time recognition. Additionally, it examines various approaches for generating facial expressions in robots, encompassing both manual coding and automated techniques.

In their publication [15], the authors present an innovative method for comprehending human personality traits during social human-robot interactions. The study utilizes a multi-modal feature fusion approach, combining visual features such as head motion, gaze, and body motion with various vocal features. By doing so, the authors aim to capture previously unidentified patterns in human behavior and enhance our understanding of personality traits. In [25], the paper discusses the application of object recognition in computer vision, particularly in assisting visually impaired individuals. The study proposes a system using Yolo and Yolo v3 algorithms to detect multiple objects and provide voice alerts. [16] presents a project focused on computer animation and the implementation of various algorithms. The study involves generating animated images using computer animation techniques, including the creation of 3D models through rigging with virtual skeletons.

The authors of [14] discuss the use of voice-based interfaces in Human-Robot Interaction (HRI) systems. The authors provide a comprehensive examination of voice-based perception within Human-Robot Interaction (HRI) systems, with a specific emphasis on feature extraction, dimensionality reduction, and semantic understanding. Moreover, [26] demonstrates a compelling interaction between humans and an NAO robot, wherein deep convolutional neural networks (CNNs) are employed to achieve accurate face and facial expression recognition. Emotion recognition relies on the utilization of CNN models, which are learned and fine-tuned to achieve optimal performance. [19] focuses on the advancement of animation and rendering techniques, particularly in real-time applications. The paper explores different algorithms and methods for real-time rendering, including optimization techniques and hardware acceleration. The authors of [28] propose an emotion detection system for Human-Robot Interaction (HRI) applications. They utilize facial expression analysis and audio processing to recognize and classify human emotions, enabling more intuitive and empathetic interactions with robots.

The studies presented in this literature review have explored various aspects of HRI, such as facial expression replication in humanoid robots, machine learning techniques for human-robot collaboration, speech recognition advancements, and the significance of joint interaction and social cues. For more advanced and seamless interactions between humans and robots, further research attention is required.

3 Proposed Method

To further improve the interaction among humans and robots, we propose a novel model called NeuroRobo, which utilizes advanced computer vision and deep learning techniques to enable real-time interactions between users and computer systems as shown in Fig. 1. This model offers lipsyncing-based conversational replies, object recognition via a webcam, and the ability to mimic user actions, resulting in a more natural and intuitive interaction experience. The proposed framework consists of three interconnected modules: (1) Talk to Me, (2) Let Me Guess, and (3) I Am a Mimic module, as depicted in Fig. 1. These modules collectively form an integrated system designed to provide users with an immersive and engaging experience.

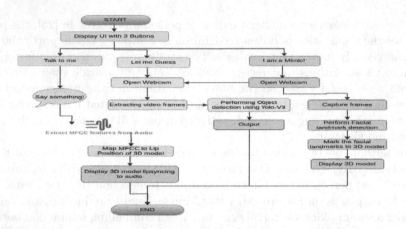

Fig. 1. Framework of the proposed human-robot interaction model.

3.1   Talk to Me Module

The Talk to me module enables users to have a conversation with the model by speaking into the microphone. The speech input is converted into text, which is then used to generate a response using the pre-trained Blenderbot model. The response is transformed back into speech and played back to the user by bringing up the model that lipsyncs to the reply which is facilitated using the MFCCs (Mel Frequency Cepstral Coefficients). This approach facilitates a voice-based interaction with the chatbot, enhancing the user experience. Blenderbot utilizes a large-scale dataset called the "Blended Skill Talk" dataset for training. It is extensive, containing over 9.4 million dialogues. The dataset is carefully designed to cover a wide range of topics and generate diverse conversational scenarios.

The steps involved are as follows:

(a) User Speech Input: The code captures speech input from the user using a microphone.
(b) Speech Recognition: The captured speech is processed using speech recognition techniques to convert it into text.
(c) Text Tokenization: The text input is tokenized using a tokenizer specifically designed for the Blenderbot model.
(d) Model Inference: The tokenized input is fed into a pre-trained Blenderbot model, which generates a response based on the given input.
(e) Text Decoding: The generated response is decoded from the model's output format to obtain human-readable text.
(f) Text-to-Speech Conversion: The decoded response is converted into speech using a text-to-speech synthesis library (gTTS).
(g) Audio Playback: The synthesized speech is played back to the user, allowing them to hear the chatbot's response.

3.2   Let Me Guess Module

It is an object detection module using the YOLOv3-tiny model and a webcam feed. This module enables real-time object detection from a webcam feed using the YOLOv3-tiny model [25] and provides an audio output to inform the user about the detected objects. It uses the Common Objects in Context (COCO) labeled dataset to learn the necessary features and patterns for object detection which has 80 labels in it

Below is the architectural overview of the YOLOv3 algorithm, which demonstrates its components and their interconnectedness.

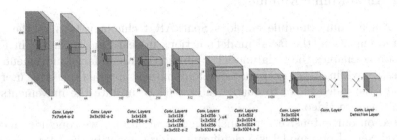

Fig. 2. Architecture of YOLOv3 Algorithm.

The steps and procedures followed in this module are:

(a) Model and Class Loading: The code begins by loading the YOLOv3-tiny model's weights and configuration files. Additionally, it reads the class labels from the "coco.names" file, which contains the names of the objects the model is trained to detect.

(b) Webcam Setup: The code initializes the webcam capture using the OpenCV library. It establishes a connection to the default webcam device (index 0).

(c) Object Detection Loop: The main loop of the code continuously captures frames from the webcam feed and performs object detection on each frame. It follows the steps below for each frame:

- Preprocessing: The captured frame is preprocessed to convert it into a format suitable for input to the YOLO network. This involves resizing the frame to a specific size (416 × 416 pixels) and normalizing the pixel values.

- Forward Pass: The preprocessed frame is passed through the YOLO network to obtain predictions for object detection. The network predicts bounding boxes, class probabilities, and confidence scores for each detected object.

- Post-processing: The predictions are post-processed to filter out weak detections and eliminate overlapping bounding boxes. Non-maximum suppression (NMS) is applied to retain the most confident and non-overlapping detections.

- Drawing Bounding Boxes: The code loops over the filtered detections and draws bounding boxes on the frame for each detected object. It also displays the class label and confidence score associated with each bounding box.

(d) Audio Output: If the user presses the 'q' key, the code generates an audio output using the gTTS library. The audio output informs the user about the object detected in the frame.

(e) Cleanup: After the loop ends (typically when the user presses 'q'), the code releases the webcam capture and closes all windows.

3.3   I Am a Mimic Module

The I Am a Mimic module employs SparkAR technology to map the user's facial movements to the facial model's actions in real-time as shown in Fig. 2. This module enables the facial model to mimic the user's facial movements and gestures, providing a more immersive and natural interaction. The user can switch to the video mode to enable the model to imitate the movements and actions from the video.

Based on our extensive experiments and evaluations, we conclude that our proposed NeuroRobo model is a significant improvement in the user experience across various applications. We find that the "Talk to me" module, which uses a transformer model and MFCCs for lip syncing, has an accuracy of 93% in recognizing the user's input and providing a natural conversation response. The "Let me guess" module, which employs the YOLOv3 model for object recognition, demonstrates an accuracy of 87% in identifying various objects shown to the model through the webcam. Lastly, the "I am a mimic" module, which uses SparkAR technology, successfully maps the user's actions and movements to the model, resulting in a highly realistic and intuitive interaction.

Our proposed model has the potential to make a significant impact on various fields, such as healthcare, entertainment, and education. For example, the "Talk to me" module has the ability to provide real-time conversational replies, which significantly alleviates patient loneliness and aids in rehabilitation. Furthermore, the "I am a mimic" module's intuitive interaction can be used in various educational and training simulations.

4   Experimental Setup and Results

The 3D model development and animation experiments were conducted using Blender version 3.4.1 on both a local machine and an HPC server equipped with an A6000 GPU and 40GB of dedicated memory. The utilization of the GPU on the HPC server significantly accelerated rendering and animation tasks, allowing for rapid iterations and complex model creations. Figure 3 presents the application's home screen, which serves as a gateway to three distinct modules. Moving forward, Fig. 4 displays the 3D model in its developmental stage, highlighting the rigging progress. In Fig. 5, we observe the user interacting with the model through a microphone, establishing seamless communication. Building upon this interaction, Fig. 6 captures the engaging conversation taking place between the user and the 3D model. Lastly, Fig. 7 demonstrates the impressive capability of the 3D model to mimic the user's actions through the utilization of a webcam interface.

Fig. 3. Home Screen

Fig. 4. Rigging Stage of 3D Model

Fig. 5. 3D Model during conversation

Fig. 6. Chat window of conversation

Fig. 7. Model mimicking the user

5   Conclusion and Future Scope

This paper introduces the NeuroRobo system, a bot powered by Blender face technology, integrating facial sentiment analysis, object detection, and behavior prediction. Its focus is on enhancing human-robot interaction through intuitive and engaging experiences. However, further advancements in facial sentiment analysis and behavior prediction are recognized as crucial for the system's complete realization. The NeuroRobo system excels in conversational replies, object recognition, and user action mimicry. Future work involves exploring facial expression recognition and behavior prediction for improved emotional understanding and personalization. To achieve this, the research will investigate various approaches such as reinforcement learning, sequence modeling, or cognitive architectures. These approaches will enable the system to anticipate user behavior by analyzing interaction history, incorporating contextual information, and developing predictive models based on individual user preferences.

In conclusion, the NeuroRobo system is a significant step in bridging the gap between humans and machines in human-robot interaction. The ongoing research in facial sentiment analysis and behavior prediction aims to provide a more holistic and immersive interaction experience, allowing the system to better understand and respond to users' emotions and preferences. The valuable feedback received from reviewers has contributed to the continued development of these aspects in the system.

References

1. Abduljabbar, R., Dia, H., Liyanage, S., Bagloee, S.A.: Applications of artificial intelligence in transport: an overview. Sustainability **11**(1), 189 (2019). https://doi.org/10.3390/su11010189
2. Sofi, M.A., Wani, M.A.: RiRPSSP: a unified deep learning method for prediction of regular and irregular protein secondary structures. J. Bioinform. Comput. Biol. **21**(01), 2350001 (2023). https://doi.org/10.1142/s0219720023500014
3. Sofi, M.A., Wani, M.A.: Protein secondary structure prediction using data-partitioning combined with stacked convolutional neural networks and bidirectional gated recurrent units. Int. J. Inf. Technol. **14**(5), 2285–2295 (2022). https://doi.org/10.1007/s41870-022-00978-x
4. Buchanan, B.G.: Artificial intelligence in Finance. Zenodo (2019). https://doi.org/10.5281/zenodo.2612537
5. Chen, L., Chen, P., Lin, Z.: Artificial intelligence in education: a review. IEEE Access **8**, 75264–75278 (2020). https://doi.org/10.1109/access.2020.2988510
6. Murphy, R.R.: Introduction to AI robotics. Ind. Robot Intl. J. **28**(3), 266–267 (2001). https://doi.org/10.1108/ir.2001.28.3.266.1
7. Bainbridge, W.A., Hart, J., Kim, E.S., Scassellati, B.: The effect of presence on human-robot interaction. In: RO-MAN 2008 - The 17th IEEE International Symposium on Robot and Human Interactive Communication (2008). https://doi.org/10.1109/roman.2008.4600749
8. Ayanoğlu, H., Duarte, E. (eds.): Emotional Design in Human-Robot Interaction. Springer International, Cham (2019). https://doi.org/10.1007/978-3-319-96722-6

9. Kanda, T., Ishiguro, H.: Human-Robot Interaction in Social Robotics. CRC Press (2017). https://doi.org/10.1201/b13004
10. Nickel, K., Stiefelhagen, R.: Visual recognition of pointing gestures for human-robot interaction. Image Vis. Comput. **25**(12), 1875–1884 (2007). https://doi.org/10.1016/j.imavis.2005.12.020
11. Russo, A., et al.: Dialogue systems and conversational agents for patients with dementia: the human-robot interaction. Rejuvenation Res. **22**(2), 109–120 (2019). https://doi.org/10.1089/rej.2018.2075
12. Mazzoni Ranieri, C., Nardari, G. V., Pinto, A. H. M., Tozadore, D. C., Romero, R. A. F.: LARa: a robotic framework for human-robot interaction on indoor environments. In: 2018 Latin American Robotic Symposium, 2018 Brazilian Symposium on Robotics (SBR) and 2018 Workshop on Robotics in Education (WRE) (2018). https://doi.org/10.1109/lars/sbr/wre.2018.00074
13. Paulus, D., Seib, V., Giesen, J., Grüntjens, D.: Enhancing Human-Robot Interaction by a Robot Face with Facial Expressions and Synchronized Lip Movements (2013)
14. Badr, A., Abdul-Hassan, A.: A review on voice-based interface for human-robot interaction. Iraqi J. Electr. Electron. Eng. **16**(2), 1–12 (2020). https://doi.org/10.37917/ijeee.16.2.10
15. Shen, Z., Elibol, A., Chong, N.Y.: Multi-modal feature fusion for better understanding of human personality traits in social human-robot interaction. Robot. Auton. Syst. **146**, 103874 (2021). https://doi.org/10.1016/j.robot.2021.103874
16. Basyouny, Y. M. A.: Rigging Manager for Skeletal Mesh in 3D Environment (2020)
17. Manas, A.U., Sikka, S., Pandey, M.K., Mishra, A.K.: A review of different aspects of human robot interaction. In: Sharma, H., Saha, A.K., Prasad, M. (eds.) ICIVC 2022, pp. 150–164. Springer, Cham (2023). https://doi.org/10.1007/978-3-031-31164-2_13
18. Liu, X., Chen, Y., Li, J., Cangelosi, A.: Real-time robotic mirrored behavior of facial expressions and head motions based on lightweight networks. IEEE Internet Things J. **10**(2), 1401–1413 (2023). https://doi.org/10.1109/jiot.2022.3205123
19. Rasheed, A.S., Finjan, R.H., Hashim, A.A., Al-Saeedi, M.M.: 3D face creation via 2D images within blender virtual environment. Indonesian J. Electr. Eng. Comput. Sci. **21**(1), 457 (2021). https://doi.org/10.11591/ijeecs.v21.i1.pp457-464
20. Thakur, A., Ahuja, L., Vashisth, R., Simon, R.: NLP & AI speech recognition: an analytical review. In: 10th International Conference on Computing for Sustainable Global Development (INDIACom 2023), pp. 1390–1396. IEEE (2023)
21. Lombardi, M., Maiettini, E., Tikhanoff, V., Natale, L.: iCub knows where you look: exploiting social cues for interactive object detection learning. In: 2022 IEEE-RAS 21st International Conference on Humanoid Robots (Humanoids) (2022). https://doi.org/10.1109/humanoids53995.2022.10000163
22. Fourie, C., et al.: Joint action, adaptation, and entrainment in human-robot interaction. In: 2022 17th ACM/IEEE International Conference on Human-Robot Interaction (HRI) (2022). https://doi.org/10.1109/hri53351.2022.9889564
23. Prabhu, K., SathishKumar, S., Sivachitra, M., Dineshkumar, S., Sathiyabama, P.: Facial expression recognition using enhanced convolution neural network with attention mechanism. Comput. Syst. Sci. Eng. **41**(1), 415–426 (2022). https://doi.org/10.32604/csse.2022.01974
24. Rawal, N., Stock-Homburg, R.M.: Facial emotion expressions in human-robot interaction: a survey. Int. J. Soc. Robot. **14**(7), 1583–1604 (2022). https://doi.org/10.1007/s12369-022-00867-0

25. Mahendru, M., Dubey, S.K.: Real time object detection with audio feedback using yolo vs. yolo_v3. In: 2021 11th International Conference on Cloud Computing, Data Science & Engineering (Confluence) (2021). https://doi.org/10.1109/confluence51648.2021.9377064

26. Semeraro, F., Griffiths, A., Cangelosi, A.: Human-robot collaboration and machine learning: a systematic review of recent research. Robot. Comput.-Integrat. Manuf. **79**, 102432 (2023). https://doi.org/10.1016/j.rcim.2022.102432

27. Melinte, D.O., Vladareanu, L.: Facial expressions recognition for human-robot interaction using deep convolutional neural networks with rectified adam optimizer. Sensors **20**(8), 2393 (2020). https://doi.org/10.3390/s20082393

28. Ren, F., Huang, Z.: Automatic facial expression learning method based on humanoid robot XIN-REN. IEEE Trans. Hum.-Mach. Syst. **46**(6), 810–821 (2016). https://doi.org/10.1109/thms.2016.2599495

Equalization Based Soft Output Data Detection for Massive MU-MIMO-OFDM Using Coordinate Descent

L. S. S. Pavan Kumar Chodisetti[1] , Madhusudan Donga[2] , Pavani Varma Tella[3] ,
K. Pasipalana Rao[4] , K Ramesh Chandra[5] , Prudhvi Raj Budumuru[5] ,
and Ch Venkateswara Rao[5]([✉])

[1] Department of ECE, Sasi Institute of Technology and Engineering, Tadepalligudem, India
[2] Department of ECE, St. Peter's Engineering College, Hyderabad, India
[3] Department of ECE, Shri Vishnu Engineering College for Women, Bhimavaram, India
[4] Department of ECE, Sri Vasavi Engineering College, Tadepalligudem, India
[5] Department of ECE, Vishnu Institute of Technology, Bhimavaram, India
Venkateswararao.c@vishnu.edu.in

Abstract. For the next generation of wireless communication networks to
advance, massive multi-user multiple-input multiple-output orthogonal frequency
division multiplexing (MU-MIMO-OFDM) systems are essential. Nevertheless,
due to the concurrent presence of several users and the frequency-selective fading channel, identifying sent data in such systems proves to be a daunting issue.
This study proposes a huge MU-MIMO-OFDM system-compatible coordinate
descent-based equalization-based soft output data identification technique. This
algorithm's major goals are to improve the estimation of transmitted data symbols
and effectively deal with inter-user interference. By exploiting the sparse nature
of the channel impulse response, the data detection problem as a joint sparse signal recovery and symbol detection task. Then, the coordinate descent algorithm,
which iteratively updates the estimated symbols and exploits the sparsity structure of the channel has been implemented. These soft outputs can be utilized in
subsequent stages of the communication system, such as channel decoding or
interference cancellation. The simulation results clearly illustrate the superiority
of the proposed method over existing detection techniques in terms of bit error
rate (BER) performance. The algorithm showcases remarkable enhancements in
detection accuracy, even in challenging scenarios involving a substantial number of users and severe channel conditions. When the number of base stations is
increased from 32 to 128, the proposed algorithm demonstrates a substantial 76%
reduction in bit error rate (BER). In contrast, conventional methods only achieve
a value of approximately 60% reduction in BER under the same conditions.

Keywords: Accuracy · Interference · Equalization · Multiplexing · Soft-output
data detection · Bit error rate

P. Pareek et al. (Eds.): IC4S 2023, LNICST 537, pp. 173–184, 2024.
https://doi.org/10.1007/978-3-031-48891-7_14

1 Introduction

Massive Multiuser MIMO [1] (Multiple-Input Multiple-Output) with Orthogonal Frequency Division Multiplexing (OFDM) is a state-of-the-art technique that uses multiple antennas at the transmitter and receiver along with OFDM modulation to significantly increase the capacity and spectral efficiency of wireless communication systems [2]. It is considered a key technology for next-generation wireless networks, such as 5G and beyond. By employing multiple antennas [3–6], MIMO can exploit spatial diversity [7] and multiplex multiple data streams simultaneously, leading to increased data rates and improved reliability.

The modulation method (OFDM) divides the available frequency spectrum into several orthogonal subcarriers. These subcarriers can transmit data efficiently through frequency-selective channels at high data rates because they are individually modulated with low symbol rates. Modern wireless communication systems now use OFDM as a core component; it is widely used in Wi-Fi, LTE, and the most recent 5G networks. Massive multiuser MIMO OFDM combines the benefits of MIMO and OFDM to achieve high spectral efficiency and accommodate a large number of users simultaneously [8]. In this system, a base station or access point is equipped with a massive number of antennas, which can be in the hundreds or even thousands. These antennas are used to serve multiple users simultaneously by transmitting independent data streams to each user. At the receiver side, user devices are equipped with multiple antennas to receive the transmitted signals. The receiver uses advanced signal processing techniques, such as linear precoding and spatial multiplexing, to separate and decode the signals from different users [9]. The combination of massive antenna arrays, spatial processing, and OFDM enables Massive Multiuser MIMO OFDM systems to achieve high spectral efficiency, robustness against multipath fading, and improved interference management. It can support a large number of users with high data rates, making it suitable for dense urban environments and scenarios with high user demands [10].

With significant improvements in capacity, coverage, and user experience, the integration of these systems has the potential to completely alter wireless communication networks [11]. Future technological developments, like as smart cities, the Internet of Things (IoT), and improved mobile broadband applications, are expected to be made possible by this technology. Inspite of having numerous advantages, these systems are facing severe challenges in output data detection [12], resource allocation [13–15]. The factors which affect the output data detection includes: channel estimation; pilot contamination; interference; complexity; multiuser synchronization; imperfect channel knowledge [16]. Addressing these challenges requires advanced signal processing techniques, such as iterative detection and interference cancellation algorithms, efficient pilot designs, robust channel estimation algorithms, and adaptive modulation and coding schemes. Current research endeavors are dedicated to crafting efficient algorithms and system designs aimed at surmounting these challenges and elevating the performance of such systems.

2 Literature Review

Massive MU-MIMO-OFDM is indeed an important area of research with numerous future research directions. Some of the key areas where research is actively being pursued are: channel modeling; signal processing algorithms; interference management and resource allocation; hybrid beam forming and antenna design; spectrum efficiency and energy efficiency; cooperative and distributed massive MIMO; and practical implementation challenges. The authors of [17] has explored the potential improvement in throughput performance of a MU-MIMO system. In [18], the authors have introduced a novel precoding scheme called SB (Sum-Rate Maximizing Beam forming) that aims to maximize the total channel capacity.

To choose users in a multi-user communication system, the authors in [19] present an ideal pair-wise semi-orthogonal user selection (SUS) scheduling algorithm. The objective is to achieve higher throughput by effectively selecting users and optimizing sub channel allocation using these techniques. A novel MU-MIMO-OFDM scheme that employs spreading codes to achieve signal separation among mobile terminals has been presented [20]. The primary challenge in conventional multiuser MIMO systems is the presence of co-channel interference among mobile terminals. A new nonlinear distortion suppression method for MU-MIMO-OFDM systems is presented in this study [21]. A different study [22] focuses on creating MU-MIMO techniques that are specifically designed for OFDM systems, with an emphasis on those covered in the IEEE 802.11ac standard. In addition, a cutting-edge method [23] for guard interval (GI) control in MU-MIMO-OFDM systems for unmanned aerial vehicles (UAV) is described. Additionally, a set of waveforms for shaping the subcarriers in MU-MIMO-OFDM systems are proposed in this study using a computationally effective optimization method [24].

3 Methodology

3.1 System Model

In this scenario, a high-capacity uplink system utilizing massive MU-MIMO-OFDM technology has been analyzed [25]. The system comprises U user terminals, each equipped with a single antenna, which transmit data in parallel to a base station possessing BS antennas. The transmission takes place across W subcarriers, enabling efficient utilization of the available spectrum. In this setup, each user (indexed as $i = 1,..., U$) employs a forward error-correction scheme to encode its own bit stream. The resulting coded bits are then mapped onto constellation points from a finite set, such as 64-QAM, utilizing a gray mapping rule. It is assumed that the average transmit power of each user is normalized to unity. Consequently, the obtained frequency-domain symbols on each of the W subcarriers are denoted as $\{s(i)_1,..., s(i)\}$. Upon reaching the base station, the cyclic prefixes are removed, and the TD signals from each antenna undergo a DFT operation, converting them back into the FD. For this system, it is assumed that a sufficiently long cyclic prefix is available, ensuring ideal synchronization between users and the base station. Additionally, accurate knowledge of the channel state information

(CSI) is assumed [26], and the input and output for the wth subcarrier can be commonly represented by Eq. (1) (Fig. 1).

$$y_w = H_w s_w + n_w \tag{1}$$

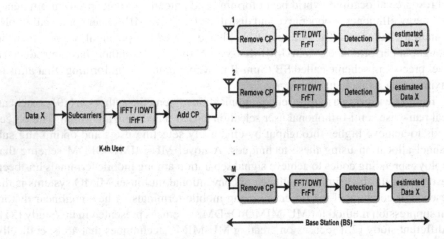

Fig. 1. Proposed uplink system model [25].

3.2 Data Recognition Based on Equalization

Zero-forcing Equalization. In the context of the described system, equalization-based data detection is employed to recover the transmitted symbols at the base station. One commonly used equalization method is known as zero-forcing (ZF) equalization [27]. ZF equalization can be represented using (2), which aims to eliminate the interference caused by the channel by inversely applying the estimated channel response to the received signals.

$$\hat{y}(i)_w = {y(i)_w}\big/{\hat{H}(i)_w} \tag{2}$$

Linear MMSE Equalization. In addition to zero-forcing (ZF) equalization, another commonly used equalization technique in the context of MU-MIMO-OFDM systems is linear MMSE equalization [28, 29]. By considering both the channel response and the noise, this equalization tries to reduce the mean square error between the equalized symbols and the actual transmitted symbols. Mathematically, this can be represented using (3). The MMSE weight vector $\hat{W}(i)$ can be computed using (4), where $R(i)$ is the covariance matrix of the received signal.

$$\hat{y}(i)_w = \hat{W}(i) \times H(i)_w \times y(i)_w \tag{3}$$

$$\hat{w}(i) = R(i)^{-1} \times h(i) \tag{4}$$

Non-linear Box-Constrained (BOX) Equalization. It is an advanced equalization technique used in wireless communication systems, including MU-MIMO-OFDM systems. Unlike linear equalization methods such as zero-forcing (ZF) and linear MMSE, BOX equalization operates in the non-linear domain and incorporates additional constraints to improve the equalization performance. In BOX equalization, the goal is to minimize the symbol error rate by formulating an optimization problem subject to box constraints on the equalized symbols. These box constraints ensure that the equalized symbols fall within a predefined range or region, typically based on the characteristics of the modulation scheme. Mathematically, the BOX equalization problem can be formulated using (5).

$$minimize \sum \left| y(i)_w - \hat{y}(i)_w \right|^2 \ subject\ to\ I \leq \hat{y}(i)_w \leq u \tag{5}$$

Coordinate Descent. Equalization via coordinate descent is an optimization technique used to expedite the process of equalization in communication systems [30]. Specifically, it is applied in scenarios where equalization involves solving a complex optimization problem with multiple variables. In coordinate descent, the optimization problem is divided into sub problems, each involving only a single variable. The idea is to iteratively update each variable while keeping the others fixed, cycling through all the variables until convergence is achieved. This approach simplifies the optimization process and can significantly reduce the computational complexity compared to traditional optimization algorithms. Coordinate descent (CD) is a widely recognized iterative framework used for precisely or approximately solving numerous convex optimization problems. It achieves this by performing a sequence of straightforward updates on individual coordinates. The CD framework can be described by Eq. (6).

$$f(z_1, \ldots .z_U) = f(z) = \|y_w - H_w z\|^2 + g(z)$$

4 Simulation Parameters

To perform simulation, several parameters need to be considered. The following are some of the key parameters required for such a simulation: system configuration; channel model: modulation and coding; equalization and detection; simulation parameters etc. These parameters provide the foundation for setting up a simulation environment to investigate the performance of proposed system. The specific values chosen for these parameters will depend on the specific scenario, research objectives, and available resources. Table 1 contains a list of the proposed model's simulation requirements.

Table 1. Simulation parameters

Parameter	Description
BS Antenna	32, 64,128
Modulation Scheme	16-QAM
Users	8
Detector Structure	ZF, MMSE, SIMO, OCDBOX, OCDMMS
Length of Data	16
Length of Frame	128
Subcarriers	2048
Cyclic Prefix	10
SNR (dB)	0–30

5 Experimental Outcomes

For the purpose of assessing the OCD-BOX algorithm's performance in terms of error rate, a Monte Carlo simulation of a coded MIMO-OFDM uplink system has been carried out. The system operates with a bandwidth of 20 MHz and consists of 2048 subcarriers, where 1200 subcarriers are dedicated to data transmission, following the specifications of LTE Advanced (LTE-A). The simulation has been employed 64-QAM modulation with gray mapping, which allows for efficient mapping of coded bits onto constellation points have been utilized. To enhance the reliability of the transmission, a rate-3/4 turbo code for forward error correction. BER analysis has been done to assess how well the proposed system model performs while taking different base station antenna configurations and different OCD iterations (k) into account. It is observed from Fig. 2, that the BER has been improved as the base station configuration irrespective of equalization technique. The proposed OCD-BOX has attained similar BER performance with OCD MMSE at larger BS configurations.

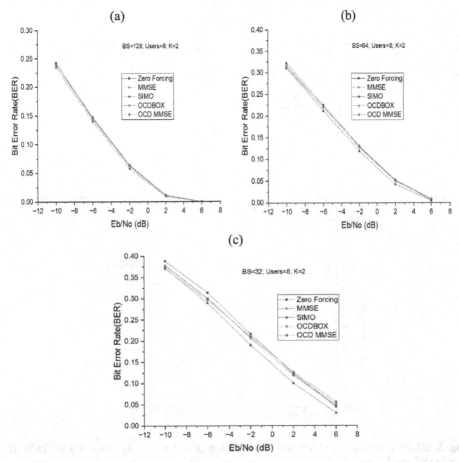

Fig. 2. BER for a massive MU-MIMO-OFDM system (k = 2, user = 8) with a) BS = 128 b) BS = 64 c) BS = 32

With only three simulation runs (K = 3), the suggested system achieves performance that is nearly identical to the ideal MMSE equalizer when used with 64 and 128 base station (BS) antennas (see Fig. 3). However, four iterations (K = 4) are necessary in the case of a comparatively smaller system with 32 BS antennas in order to attain comparable performance. Reduced performance is indicated by higher error floors caused by lower values of k (see Fig. 4). These simulation findings demonstrate that approximation linear data detectors can perform as well as the ideal MMSE detector in systems with a higher proportion of BS antennas to user antennas. An economical and effective solution for large-scale MU-MIMO systems is provided by the suggested system, which demonstrates an impressive capacity to approach MMSE performance with less iterations. In conclusion, in systems with 64 and 128 BS antennas, OCD-BOX achieves almost equal MMSE performance with only three rounds. However, four simulation runs are required in systems with 32 BS antennas to get comparable performance.

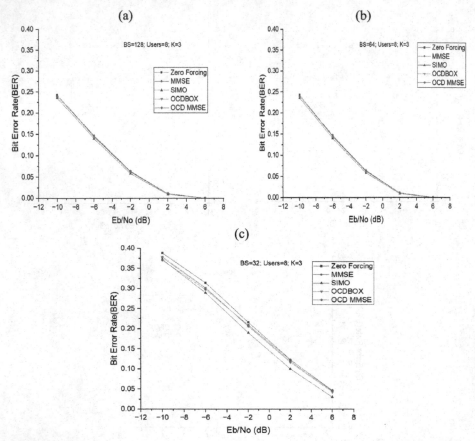

Fig. 3. BER for a massive MU-MIMO-OFDM system (k = 3, user = 8) with a) BS = 128 b) BS = 64 c) BS = 32

The comparison performance metrics for different equalizers by varying the value of *k* have been tabulated in Table 2.

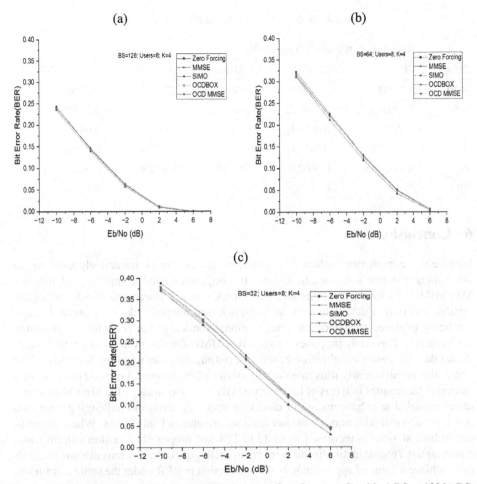

Fig. 4. BER for a massive MU-MIMO-OFDM system (k = 5, user = 8) with a) BS = 128 b) BS = 64 c) BS = 32

Table 2. BER Vs SNR for k = 2 (OCD iterations)

SNR	BER (k = 2, BS = 128, Users = 8)				
	ZF	MMSE	SIMO	OCDBOX	OCDMMSE
−10	0.24295	0.240166	0.235778	0.240234	0.239615625
−6	0.14600625	0.145072	0.140172	0.144613	0.148096875
−2	0.06295938	0.062538	0.058216	0.062106	0.06390625
2	0.011175	0.011047	0.009519	0.011034	0.011621875
6	0.00023438	0.000231	0.000156	0.000234	0.00026875
10	0	0	0	0	0

6 Conclusion

Despite this complexity challenge, researchers and engineers are actively working on developing efficient solutions to address the implementation complexity of massive MU-MIMO. Various algorithms and techniques, such as linear and non-linear equalization, soft output data detection, and optimization methods like coordinate descent, are being explored to reduce the computational burden at the BS while maintaining performance. This study proposes a huge MU-MIMO-OFDM system-compatible coordinate descent-based equalization-based soft output data identification technique. The simulation results clearly illustrate the superiority of the proposed method over existing detection techniques in terms of bit error rate (BER) performance. The algorithm showcases remarkable enhancements in detection accuracy, even in challenging scenarios involving a substantial number of users and severe channel conditions. When the number of base stations is increased from 32 to 128, the proposed algorithm demonstrates a substantial 76% reduction in bit error rate (BER). In contrast, conventional methods only achieve a value of approximately 60% reduction in BER under the same conditions.

References

1. Surya, S., Kanthimathi, M., Rajalakshmi, B.: Performance analysis of multiuser MIMO OFDM systems incorporating feedback delay and feedback error. J. Phys. Conf. Ser. **1921**(1), 012012 (2021)
2. Chandra, K.R., Borugadda, S.: Joint resource allocation and power allocation scheme for MIMO assisted NOMA system. Trans. Emerg. Telecommun. Technol. e4794 (2023)
3. Maurya, N.K., Kumari, S., Pareek, P., Singh, L.: Graphene-based frequency agile isolation enhancement mechanism for MIMO antenna in terahertz regime. Nano Commun. Networks 100436 (2023)
4. Maurya, N.K., Bhattacharya, R.: CPW-fed dual-band compact Yagi-type pattern diversity antenna for LTE and WiFi. Progress Electromagn. Res. C **107**, 183–201 (2021)
5. Maurya, N.K., Bhattacharya, R.: Design of compact dual-polarized multiband MIMO antenna using near-field for IoT. AEU-Int. J Electron. Commun. **117**, 153091 (2020)
6. Maurya, N.K., Tulsyan, M., Bhattacharya, R., Kumar, A.: Design and near field analysis of compact CPW-fed printed pseudo-monopole driven Yagi-type pattern diversity antenna. In: 2017 IEEE Applied Electromagnetics Conference (AEMC). pp. 1–2. IEEE (2017)

7. Maurya, N.K., Bhattacharya, R.: CPW-fed dual-band pseudo-monopole antenna for LTE/WLAN/WiMAX with its usage in MIMO. In: 2016 IEEE International symposium on Antennas and Propagation (APSURSI). pp. 455–456. IEEE (2016)
8. Sateesh, C., Janardhana Gupta, K., Jayanth, G., Sireesha, C., Dinesh, K., Ramesh Chandra, K.: Performance analysis of spectrum sensing in CRN using an energy detector. In: Innovative Data Communication Technologies and Application: Proceedings of ICIDCA, pp. 447–459 (2022)
9. Rohith, V., Anand, R., Malavika, R., Sharu, S., Chandra, K.R.: Spectral and energy efficient index modulation techniques for 5G wireless networks. In: Second International Conference on Electronics and Sustainable Communication Systems (ICESC), pp. 1062–1069 (2021)
10. Saxena, R.S., Karthik, T.S., Sreenivasa, A., Rao, C., Datta, D.A.: Improvements in the performance of the base station antenna due to the use of mimo in a mobile communication system. J. Nuclear **10**, 9 (2021)
11. Manikandan, P., Sivakumar, P., Swedheetha, C.: Design of adaptive frequency reconfigurable antenna for MIMO applications. In: Soft Computing in Data Analytics: Proceedings of International Conference on SCDA, pp. 203–213 (2019)
12. Hindumathi, V., Reddy, K.R.L.: Adaptive priority-based fair-resource allocation for MIMO-OFDM multicast networks. Int. J. Network. Virtual Organ. **20**(1), 73–89 (2019)
13. Ramesh Chandra, K., Borugadda, S.: An effective combination of terahertz band NOMA and MIMO system for power efficiency enhancement. Int. J. Commun. Syst. **35**(11), e5187 (2022)
14. Gangadhar, C., Chanthirasekaran, K., Chandra, K.R., Sharma, A., Thangamani, M., Kumar, P.S.: An energy efficient NOMA-based spectrum sharing techniques for cell-free massive MIMO. Int. J. Eng. Syst. Model. Simul. **13**(4), 284–288 (2022)
15. Chandra, K.R., Borugadda, S.: Multi Agent Deep Reinforcement learning with Deep Q-Network based energy efficiency and resource allocation in NOMA wireless Systems. In: Second International Conference on Electrical, Electronics, Information and Communication Technologies, pp. 1–8 (2023)
16. Vamsi, T.S., Terlapu, S.K., Krishna, M.V.: Investigation of channel Estimation Techniques using OFDM with BPSK QPSK and QAM Modulations. In: International Conference on Computing, Communication and Power Technology (IC3P), pp. 209–213 (2022)
17. Panajotović, A.S., Sekulović, N.M., Milović, D.M.: Throughput Performance of MU-MIMO-OFDM System with BD-based Precoder and Optimal Pair-Wise SUS Algorithm. In: 15th International Conference on Advanced Technologies, Systems and Services in Telecommunications (TELSIKS), pp. 45–48 (2021)
18. Panajotović, A., Sekulović, N., Milović, D.: Optimal Pair-wise SUS Algorithm in Adaptive UCD-based MU-MIMO-OFDM. In: International Symposium on Industrial Electronics (INDEL), pp. 1–4 (2018)
19. Lee, K.K.C.: FP-based Sub-band Precoding Scheme for the MU MIMO-OFDM Systems. In: IEEE VTS Asia Pacific Wireless Communications Symposium (APWCS), pp. 85–89 (2022)
20. Nakahara, K., Terui, K., Sanada, Y.: MU-MIMO-OFDM with spreading codes for user separation. In: IEEE 28th Annual International Symposium on Personal, Indoor, and Mobile Radio Communications (PIMRC), pp. 1–6 (2017)
21. Osada, G., Takebuchi, S., Maehara, F.: Nonlinear distortion suppression scheme employing transmit power control for MU-MIMO-OFDM systems. In: IEEE Radio and Wireless Symposium (RWS), pp. 46–48 (2014)
22. Panajotović, A., Riera-Palou, F., Femenias, G.: Adaptive uniform channel decomposition in MU-MIMO-OFDM: application to IEEE 802.11 ac. IEEE Trans. Wireless Commun. **14**(5), 2896–2910 (2015)
23. Wu, Z., Gao, X., Shi, Y.: A novel MU-MIMO-OFDM scheme with the RBD precoding for the next generation WLAN. In: IEEE Military Communications Conference, pp. 565–569 (2015)

24. Yao, K., Saito, S., Suganuma, H., Maehara, F.: A Guard interval control scheme using theoretical system capacity for UAV MU-MIMO-OFDM THP systems. In: International Symposium on Intelligent Signal Processing and Communication Systems (ISPACS), pp. 1–4 (2022)

25. Wu, M., Dick, C., Cavallaro, J.R., Studer, C.: High-throughput data detection for massive MU-MIMO-OFDM using coordinate descent. IEEE Trans. Circuits Syst. I Regul. Pap. **63**(12), 2357–2367 (2016)

26. Varma, T.P., Naraganeni, S., Chandra, K.R., Rao, K.P., Ch, V.R.: performance evaluation of MIMO system with different receiver structures. In: 9th International Conference on Advanced Computing and Communication Systems, pp. 829–833(2023)

27. Srinivas, P., Miriyala, R.S., Matsa, N., Yallapu, S., Ch, V.R.: Capacity evaluation of MIMO system: with and without successive interference cancellation. In: 7th International Conference on Computing Methodologies and Communication, pp. 27–31(2023)

28. Ch, V.R., Miriyala, R.S., Praveena, V., Sri, V.B., Ramalakshmi, K.: Performance evaluation of OFDM system: with and without reed-solomon codes. In: 4th International Conference on Advances in Computing, Communication Control and Networking, pp. 1827–1831 (2022)

29. Chandra, K.R., Amudalapalli, M.P., Satyanarayana, N.V., Budumuru, P.R.: Received Signal Strength (RSS) based channel modelling, localization and tracking. In: 2nd International Conference on Advances in Computing, Communication, Embedded and Secure Systems, pp. 243–246 (2021)

30. Kumar, P.R., Naganjaneyulu, P.V., Prasad, K.S.: Hybrid genetic algorithm in partial transmit sequence to improve OFDM. Int. J. Intell. Syst. Technol. Appl. **19**(4), 362–376 (2020)

Performance Analysis of Hybrid BPSK-MPPM Modulated Multicore Fiber Interconnect System

Ankita Kumari[(✉)] and Jitendra K. Mishra

Department of Electronics and Communication Engineering, Indian Institute of Information Technology Ranchi, Ranchi 834010, Jharkhand, India
`ankita03.pgec21@iiitranchi.ac.in`

Abstract. In this paper, spatially multiplexed hybrid binary phase shift keying-multi-pulse pulse position modulation (BPSK-MPPM) based 8-core multicore fiber (MCF) interconnect system is suggested for the foreseeable future exa-scale optical interconnect (OI) applications. The impact of different characteristic parameters like optical signal to noise ratio (OSNR), launch power, and inter-core crosstalk (XT) on bit error rate (BER) of hybrid BPSK-MPPM modulated OI system are discussed in detail. Further, error probability performance of conventional BPSK and hybrid modulation format are compared for 40 Gbps per channel. It is shown that the proposed hybrid scheme is suitable for futuristic data center, high-end computing system and silicon photonic transceiver chips.

Keywords: Optical Interconnect · Space Division Multiplexing · Multicore Fiber · Binary Phase Shift Keying · Multi-pulse Pulse Position Modulation

1 Introduction

Optical interconnect (OI) is a futuristic data transmission architecture for higher bandwidth and short-reach interlink within the modern data center [1, 2]. OI transmission over short distances inside the board uses less power and is less expensive [3]. Over copper interconnect OI can potentially provide high throughput, high density parallel lines and buses [3, 4]. In order to avoid anticipated capacity constraints in OI, additional significant improvements can be made using space division multiplexing (SDM) [5]. With increase in demand of data services, using multicore fiber (MCF) rather than single mode fiber provides high-capacity data transfer that could be preferably implemented using SDM [6]. It is the MCF of SDM that could make new discoveries with the emerging demands of communication channel [7, 8]. Recently, various researchers have been shown their interest in SDM that uses MCF for high data transfer rate as well as capacity improvement [9, 10]. Recently, a rectangular 8-core MCF has been reported for parallel processing and short reach OI applications [11].

Inter-core crosstalk is one of the possible drawbacks of MCF as OI. [12]. Experimentally, it was demonstrated that the volatility and intensity of XT have been influenced by the modulation scheme, pseudo-random binary sequence (PRBS) length, temperature,

P. Pareek et al. (Eds.): IC4S 2023, LNICST 537, pp. 185–192, 2024.
https://doi.org/10.1007/978-3-031-48891-7_15

and signal rate [13]. Recent studies have looked at inter-core XT analysis for polarization division multiplexed-quadrature phase-shift keying (PDM-QPSK), 16 quadrature amplitude modulation (PDM-16QAM), and 64QAM (PDM-64QAM), which demonstrates that the effects of cross-talk are more severe for long distances for higher order formats [14]. Moreover, binary phase-shifted keying (BPSK) technique provides better execution, robustness and less complicated as compare to various modulation scheme [15]. Hence, it has gained widespread acceptance as the primary modulation scheme in recent years [16].

Further, pulse position modulation (PPM) is one of the best-known modulation techniques in optical fiber due to its sensitive detection efficiency and effective bandwidth implementation [17]. An improvement to PPM is known as Multi-pulse Pulse Position Modulation (MPPM), which offers increased bandwidth and power efficiency [18]. In order to prevent the optical capacity crisis in single-mode fiber, hybrid modulation approaches increase the poor power efficiency [19]. Recently, error probability estimation of hybrid QAM-MPPM has been proposed [20]. Recently, another error probability estimation of hybrid differential phase shift keying (DPSK)-MPPM has been demonstrated in long-haul for low power transmission [21]. Furthermore, a theoretical analysis of PDM-QPSK-MPPM has been proposed for high bandwidth utilization efficiency, and the error probability is also estimated for the proposed hybrid scheme [22].

In this paper, an 8-core MCF interconnect system modulated by hybrid BPSK-MPPM is suggested for use in exa-scale OI applications for the future. The error probability performance of the hybrid BPSK-MPPM method is assessed using MCF and then the bit error rate (BER) for this hybrid scheme is further analyzed for different parameters. Moreover, the coupling coefficient is evaluated in relation to the bending radius. Finally, the hybrid modulation format's BER performance is contrasted with that of the conventional BPSK modulation.

2   System Model

Figure 1 shows the block diagram of the proposed hybrid BPSK-MPPM scheme. At the transmitter side, the transmitter digital signal processor (DSP) is supplied with the PRBS. The main work of transmitter DSP is to separate the PRBS into several data blocks. These blocks are individually referred to as $\left(\log_2 \binom{M}{n_S} + n_S \right)$ bits. These bits are generally divided into two parts, i.e., $\left(\log_2 \binom{M}{n_s} \right)$ and n_s bits. Initially, the $\left(\log_2 \binom{M}{n_s} \right)$ bits are encrypted using MPPM scheme and is used to determine the position of n_s pulses. Then by utilizing the additional n_s bits, each MPPM optical pulse is BPSK modulated. A continuous wave of laser source is being fed to the MPPM modulator to generate signal and non-signal sections of each sign, and BPSK modulator is used to create the phase that corresponds to each signal slot. The intensity and phase modulation are designed using Mach-Zehnder architecture. These data bits are arranged in hybrid BPSK- MPPM scheme that illustrates the potential pairing of the slots as shown in Fig. 1. The hybrid frame of size M slots encodes each optical pulse and broadcasts with one of the two binary phases 0° and 180°. Any combination of signal and non-signal slots may be sent by a hybrid frame with size M slots. The signal is transmitted to Erbium-doped fiber amplifier (EDFA) which amplifies optical signal by minimizing the noise and providing high- gain. The signal is then transmitted to the multi-core fiber channel by using space division multiplexer and de-multiplexer at the input and the output side respectively. The MCF used here is rectangular array eight-core MCF which provides better arrangement for short-reach OIs.

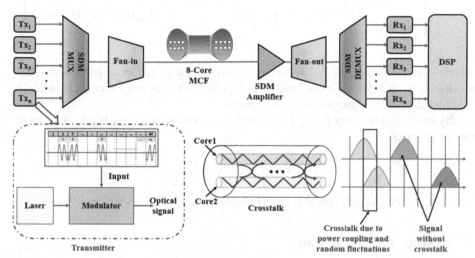

Fig. 1. Transmission setup of hybrid BPSK-MPPM modulation scheme for rectangular array 8-core MCF.

At the receiver side to improve receiver sensitivity coherent detection of optical signals is originally exploited [22]. The received signal mixes with the local oscillator to provide better amplification without noise enhancement as compared to EDFA. This mixture is mixed using a 3dB coupler and is then fed to the dual photodetector. While lowering the noise of the nearby oscillator laser source, it increases the strength of the signal that is received. The output from the photodetector is forwarded to the heterodyne demodulator. A signal from heterodyne demodulator is transmitted to the low pass filter with bandwidth $\left(BW_P = \frac{1}{2T_P}\right)$ so that the intermediate frequency (IF) effect can be reduced. The received signal's polarization must match the local oscillator (LO) lasers in order to permit optical signal mixing. Polarization of received signal and the LO lasers are arranged using automatic polarization control (APC) for best signal mixing [22]. The output of low-pass filter is fed to slot integral whose slot duration corresponds to be T_P. The output of the slot integral is known as decision random variable and is squared and then both the squared as wells as the non-squared values are sent to the receiver BIT. The receiver DSP then selects the signal with highest energy value and interpret it as the transmitted signal pulse. Then this interpreted signal is finally forwarded to the output bit and the BER of the hybrid frame is evaluated.

3 Calculation of Coupling Co-efficient and Crosstalk

The coupling co-efficient $\left(C_{pq}\right)$ between two adjacent core is given as [23]

$$C_{pq} = \frac{\omega \in_0 \int\limits_{-\infty}^{+\infty} \int\limits_{-\infty}^{+\infty} \left(N^2 - N_q^2\right) E_p^* \cdot E_q dxdy}{\left\{2\sqrt{\int\limits_{-\infty}^{+\infty} \int\limits_{-\infty}^{+\infty} \left(E_{px}H_{py}^* - E_{py}H_{px}^*\right)dxdy} \times \sqrt{\int\limits_{-\infty}^{+\infty} \int\limits_{-\infty}^{+\infty} \left(E_{qx}H_{qy}^* - E_{qy}H_{qx}^*\right)dxdy}\right\}} \tag{1}$$

where, ω is denoted as angular frequency, $\in_0$ is denoted as free space permittivity, E is electric fields and H is magnetic fields. p and q are denoted as pair either (1, 2) or (2, 1). N shows the core's refractive index distribution.

By using couple power theory [12], the crosstalk between the two neighboring cores can be given as

$$XT = \tanh\left(\overline{h}_{pq}L\right) \tag{2}$$

where, $\overline{h}_{pq}$ is the average power coupling co-efficient. L indicates length of fiber.

4 Results and Discussion

The 8-core MCF model's electric field distribution is simulated using the finite element technique (FEM), and the resulting data is integrated using Eq. (1) to get the mode coupling coefficient. Individual MCF cores are 5 mm in diameter, have a 0.8% relative refractive index difference. The core-to-core pitch is assumed 50 mm with 1.45 refractive

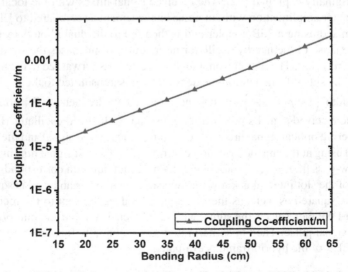

Fig. 2. Mode coupling coefficient versus bending radius for 8-core MCF with pitch $\Lambda = 50$ μm.

index for the cladding. For 8-core homogeneous step index MCF, Fig. 2 depicts the mode coupling coefficient fluctuation between two neighboring cores in a row with bending radius. Figure 2 shows mode coupling coefficient increases with bending radius due to strong phase matching between modes of adjacent homogeneous cores.

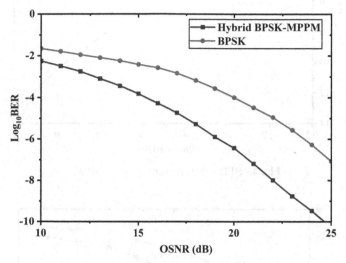

Fig. 3. BER versus OSNR for 8-core MCF.

Figure 3 shows the variation of BER for both hybrid and traditional modulation format for effectual comparison. It is analyzed from Fig. 3 that BER decreases for hybrid BPSK-MPPM as compared to that of ordinary BPSK modulation scheme. The figure depicts that at various OSNR levels, hybrid BPSK-MPPM has lower BER than conventional BPSK. It can be seen from Fig. 3 that, for OSNR of 20 dB the value of log(BER) reduces by about 2.4307 when signal is modulated by hybrid BPSK-MPPM.

The comparison between the BER for ordinary BPSK and hybrid BPSK-MPPM modulated MCF OI systems with respect to the transmitted power is illustrated in Fig. 4. The results demonstrate that the suggested hybrid method uses less power and performs better than the conventional BPSK. The suggested technique has reduced BER at various power levels, i.e., at power level of -6 dBm the hybrid BPSK-MPPM log(BER) is 1.463 less than the conventional scheme.

The variation of BER due to inter-core XT is shown in Fig. 5. The BER of hybrid BPSK-MPPM is compared with BPSK modulation scheme. The XT is calculated for different bending radius ranging from 15 to 65 cm considering the random perturbations. The figure illustrates that crosstalk causes an increase in BER, whereas the hybrid system has a relatively lower BER. The BER of hybrid BPSK-MPPM is reduced by about 5.0545 than traditional BPSK for a crosstalk of approximately -52 dB. It is worth noting that hybrid BPSK-MPPM modulated MCF interconnect system is more resistant to XT.

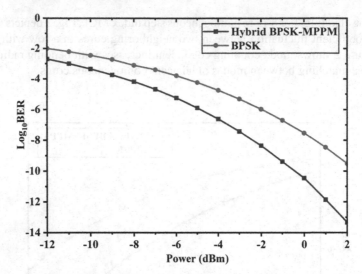

Fig. 4. BER versus power for 8-core MCF.

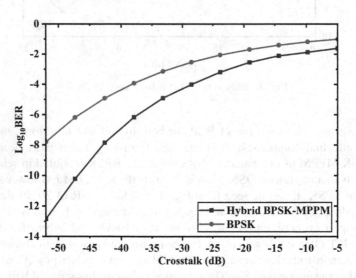

Fig. 5. BER versus crosstalk for 8-core MCF

Table 1 represents the comparison between standard BPSK and hybrid BPSK-MPPM at different parameter to analyze the BER. It can be seen from Table 1 that BPSK-MPPM bases MCF interconnects system has better performance and overall transmission capacity can also be increased with minimal XT distortion.

Table 1. Comparison between BPSK and BPSK-MPPM

Parameters	$\text{Log}_{10}\text{BER}$	
	BPSK	BPSK-MPPM
OSNR (at 22 dB)	−4.9532	−7.993
Power (at 0 dBm)	−7.532	−10.45
Crosstalk (at -40 dB)	−4.34	−6.797

5 Conclusion

The effectiveness of a hybrid BPSK-MPPM method for an 8-core MCF channel has been investigated for next-generation exa-scale OI applications. The characteristics of coupling co-efficient is studied using the finite element method. When the data rate for each channel is 40 Gbps, a comparison between the hybrid scheme and standard BPSK is made, and it is found that the hybrid method performs better than the standard scheme. The effects of BER are analyzed using important variables such as launch power, OSNR, crosstalk. The outcome demonstrates that hybrid systems use less power than BPSK and can tolerate crosstalk better than standard BPSK.

References

1. Taubenblatt, M.A.: Optical interconnects for high performance computing. IEEE J. Lightwave Technol. **30**(4), 448–458 (2012)
2. Butler, D.L., et al.: Multicore optical fiber and connectors for high bandwidth density, short reach optical links. In: Proceeding of IEEE Optical Interconnects Conference, pp. 9–10. IEEE, Santa Fe, NM, USA (2013)
3. Proietti, R., Cao, Z., Nitta, C.J., Li, Y., Yoo, S.J.B.: A scalable, low-latency, high-throughput, optical interconnect architecture based on arrayed waveguide grating routers. J. Lightwave Technol. **33**(4), 911–920 (2015)
4. Windover, L.A.B., et al.: Parallel-optical interconnects >100 Gb/s. J. Lightwave Technol. **22**(9), 2055–2063 (2004)
5. Richardson, D.J., Fini, J.M., Nelson, L.E.: Space-division multiplexing in optical fibres. Nat. Photon. **7**, 354–362 (2013)
6. Puttnam, B.J., et al.: Characteristics of homogeneous multi-core fibers for SDM transmission. APL Photon. **4**, 022804 (2019)
7. Lei, F., Dong, D., Liao, X., Duato, J.: Bundlefly: a low-diameter topology for multicore fiber. In: Proceedings of the 34th ACM International Conference on Supercomputing (ICS 2020), vol. 20, pp. 1–11. Association for Computing Machinery, Spain, Barcelona (2020)
8. García, S., Urena, M., Gasulla, I.: Dispersion-diversity multicore fiber signal processing. ACS Photon. **9**, 2850–2859 (2022)
9. Kingsta, R.M., Selvakumari, R.S.: A review on coupled and uncoupled multicore fibers for future ultra-high capacity optical communication. Optik- Int. J. Light Electron Opt. **199**, 163341 (2019)
10. Awaji, Y., et al.: High-capacity transmission over multi-core fibers. Opt. Fiber Technol. **35**, 100–107 (2017)

11. Mishra, J.K., Rahman, B.M.A., Priye, V.: Rectangular array multicore fiber realizing low crosstalk suitable for next-generation short-reach optical interconnects with low misalignment loss. IEEE Photon. J. **8**(4), 2200614 (2016)
12. Koshiba, M., Saitoh, K., Takenaga, K., Matsuo, S.: Analytical expression of average power-coupling coefficients for estimating intercore crosstalk in multicore fibers. IEEE Photon. J. **4**(5), 1987–1995 (2012)
13. Yuan, H., et al.: Experimental investigation of static and dynamic crosstalk in trench-assisted multi-core fiber. In: Proceedings 44th Optical Fiber Communications Conference and Exhibition (OFC), pp. 1–3. IEEE, San Diego, CA, USA (2019)
14. Puttnam, B.J., et al.: Impact of intercore crosstalk on the transmission distance of QAM formats in multicore fibers. IEEE Photon. J. **8**(2), 0601109 (2016)
15. Li, Y., Li, M., Poo, Y., Ding, J., Tang, M., Lu, Y.: Performance analysis of OOK, BPSK, QPSK modulation schemes in uplink of ground-to-satellite laser communication system under atmospheric fluctuation. Opt. Commun. **317**, 57–61 (2014)
16. Dai, L., Liu, Z., Meng, S., Zhao, Y.: Research on the pulse-position modulation in the digital communication system. In: Proceedings International Conference on Wireless Communications and Smart Grid (ICWCSG). IEEE, Hangzhou, China (2021)
17. Park, H., Barry, J. R.: Modulation analysis for wireless infrared communication. In: Proceedings of the IEEE International Conference Communications, pp. 1182–1186. IEEE, Seattle, WA, USA (1995)
18. Essiambre, R., Tkach, R.: Capacity trends and limits of optical communication networks. Proc. IEEE **100**(5), 1035–1055 (2012)
19. Escribano, F.J., Wagemakers, A.: Performance analysis of QAM-MPPM in turbulence-free FSO channels: accurate derivations and practical approximations. IEEE Syst. J. **15**(2), 1753–1763 (2020)
20. El-Fiqi, A.E., Morra, A.E., Hegazy, S.F., Shalaby, H.M.H., Kato, K., Obayya, S.S.A.: Performance evaluation of hybrid DPSK-MPPM techniques in long-haul optical transmission. Appl. Opt. **55**(21), 5614–5622 (2016)
21. Shi, W., Wu, P., Liu, W.: Hybrid polarization-division-multiplexed quadrature phase-shift keying and multi-pulse pulse position modulation for free space optical communication. Opt. Commun. **334**, 63–73 (2015)
22. Ho, K.-P.: Phase-Modulated Optical Communication Systems. Springer, New York (2005)
23. Okamoto, K.: Fundamentals of Optical Waveguides. Corona, Tokyo, Japan (1992)

Performance Evaluation of Optical Links: With and Without Forward Error Correcting Codes

K. Dhana Lakshmi⬚, S. Sugumaran⁽⊠⁾ ⬚, and K. Srinivas⬚

Department of ECE, Vishnu Institute of Technology, Bhimavaram, India
`sugumaran.s@vishnu.edu.in`

Abstract. Optical links play a crucial role in modern communication systems, enabling high-speed data transmission over long distances with minimal loss and interference. As the demand for faster and more reliable networks continues to grow, evaluating the performance of optical links becomes paramount. There are several approaches to developing performance prediction strategies for optical links, including analytical models, numerical simulations, and experimental measurements. Analytical models are based on mathematical equations and can provide quick and accurate predictions of the link performance for simple systems. Numerical simulations use computer software to solve complex equations and simulate the link performance for more realistic systems. The prominent strategies include: link budget analysis; chromatic dispersion compensation; nonlinear impairment mitigation; error correcting codes. This work mainly focusses on analyzing the performance of optical link with various prediction strategies (hard decision-FEC, soft decision-FEC and probabilistic shaping)) using forward error correcting codes (FEC). The symbol error rate, bit error rate and achievable information rates have been analyzed for aforementioned strategies with and without FEC.

Keywords: Achievable information rate · Hard decision-FEC · Soft decision-FEC · Symbol error rate · Bit error rate · Optical links

1 Introduction

Performance prediction recipes for optical links are essential for designing and optimizing high-speed optical communication systems. These recipes are based on mathematical models that consider various factors such as fiber attenuation, dispersion, and nonlinearities, as well as the characteristics of the optical components such as lasers, detectors, and amplifiers. The goal is to predict the performance of the optical link in terms of key parameters such as bit error rate (BER), receiver sensitivity, and transmission distance, given the system specifications and operating conditions. There are several approaches to developing performance prediction recipes for optical links, including analytical models, numerical simulations, and experimental measurements. Analytical models are based on mathematical equations and can provide quick and accurate predictions of the link performance for simple systems. Numerical simulations use computer software to solve

P. Pareek et al. (Eds.): IC4S 2023, LNICST 537, pp. 193–204, 2024.
https://doi.org/10.1007/978-3-031-48891-7_16

complex equations and simulate the link performance for more realistic systems. Experimental measurements involve actual measurements of the link performance using test equipment and can provide validation of the analytical or numerical models [1].

Some common techniques used in performance prediction recipes for optical links include: link budget analysis (this involves calculating the total losses and gains of the optical link, including fiber attenuation, connector losses, and component losses, and comparing the total received power to the minimum power required for the receiver to detect the signal); chromatic dispersion compensation (this involves using dispersion compensating fibers or dispersion compensation modules to compensate for the chromatic dispersion of the optical signal and improve the link performance); nonlinear impairment mitigation (This involves using techniques such as optical signal processing, modulation formats, or digital signal processing to mitigate the effects of nonlinear impairments such as self-phase modulation and four-wave mixing); error correction coding (This involves using forward error correction (FEC) or other error correction codes to reduce the BER of the optical link and improve the link performance). Link budget analysis is a fundamental technique used in performance prediction recipes for optical links. It involves calculating the total losses and gains of the optical link and comparing the total received power to the minimum power required for the receiver to detect the signal. Chromatic dispersion is a phenomenon that can limit the performance of optical links by causing pulse broadening and distortion. Chromatic dispersion compensation is a technique used in performance prediction recipes for optical links to mitigate the effects of chromatic dispersion and improve link performance [2, 3].

The dispersion compensation technique reduces the pulse broadening and distortion caused by chromatic dispersion, allowing for longer transmission distances and higher data rates. The amount of chromatic dispersion compensation required depends on the characteristics of the optical link, including the transmission distance, the wavelength, and the bit rate. The optimal amount of dispersion compensation can be determined by simulating the link performance using numerical simulations or by testing the link using experimental measurements. In summary, chromatic dispersion compensation is a critical technique used in performance prediction recipes for optical links to mitigate the effects of chromatic dispersion and improve the link performance. The use of dispersion compensating fibers or dispersion compensation modules can significantly improve the transmission distance and bit rate of the link, making it an essential tool for high-speed optical communication systems [4].

Nonlinear impairment mitigation is critical for performance prediction recipes for optical links because it can significantly improve the link performance and increase the maximum achievable bit rate. The optimal nonlinear impairment mitigation technique depends on the specific characteristics of the optical link, including the transmission distance, the bit rate, and the modulation format. In summary, nonlinear impairment mitigation is a crucial technique used in performance prediction recipes for optical links to mitigate the effects of nonlinearities and improve link performance. This technique can be accomplished by using modulation formats that are resistant to nonlinearities or by using digital signal processing techniques to estimate and compensate for the nonlinear distortion. Error correction coding is a technique used in performance prediction recipes for optical links to improve the reliability of data transmission and reduce the error rate. Error correction coding involves adding redundant information to the data before transmission, which can be used to detect and correct errors at the receiver. There are two types of error correction coding: block codes and convolutional codes. Block codes divide the data into fixed-length blocks and add redundant information to each block, while convolutional codes add redundant information to each data bit based on a sliding window of previous bits. One commonly used error correction code for optical communication is the Reed-Solomon code, which is a block code that adds redundant symbols to the data. The Reed-Solomon code can detect and correct errors in the received data, making it particularly useful in optical links where errors can occur due to noise or other impairments. Another commonly used error correction code for optical communication is the forward error correction (FEC) code, which is a type of convolutional code that adds redundant bits to the data [5–7].

The FEC code can detect and correct errors in the received data, and it can be designed to provide varying levels of error correction based on the specific requirements of the optical link. The use of error correction coding is critical in performance prediction recipes for optical links because it can significantly improve the reliability of data transmission and reduce the error rate. The optimal error correction coding technique depends on the specific requirements of the optical link, including the transmission distance, the bit rate, and the desired level of error correction. In summary, error correction coding is an essential technique used in performance prediction recipes for optical links to improve the reliability of data transmission and reduce the error rate. The use of block codes or convolutional codes, such as the Reed-Solomon code or the FEC code, can provide varying levels of error correction based on the specific requirements of the optical link [8, 9].

Performance prediction recipes for optical links are important for designing and optimizing high-speed optical communication systems and can help ensure reliable and efficient communication. These recipes are constantly evolving as new technologies and techniques are developed, and future research will continue to refine and improve these recipes. Therefore, in this work the performance prediction recipes for improvising the optical link performance in terms of BER have been estimated by employing FEC. The BER for various prediction strategies have been compared with and without employing FEC.

2   Literature Review

The basis for merging probabilistic shaping (PS) and forward error correction (FEC) is provided by [10]. Probabilistic amplitude shaping is a real-world application of the layered PS-FEC architecture, which consists of a PS encoder and an FEC encoder. Information-theoretic concepts are used to obtain attainable PS encoding rates and achievable FEC decoding rates. Based on data from optical transmission trials, the created tools are used to build and evaluate the performance of optical transponders. To assess post-FEC BER for systems with SD-FEC and PS, GMI is not applicable. Asymmetric information (ASI), normalized GMI (NGMI), and a feasible FEC rate have been suggested in [11] as alternatives for such cases.

The authors of [12] analysed the characteristics and coherence features of inter channel nonlinear interference (NLI) after evaluating models and mitigation methods. Based on the findings of this investigation, we developed an NLI mitigation technique that takes use of the synergistic interaction between subcarrier multiplexing with symbol rate optimization and correction for phase and polarization noise (PPN). The synergistic impact of symbol-rate optimization and phase-noise correction is examined in [13] following a discussion of models and mitigation techniques for inter channel nonlinearity. Practical applications of this phenomenon include the determination of capacity lower limits and nonlinearity mitigation. In [14], the authors have created new 4-PAM with any labelling closed-form BER expressions.

The spectral efficiency (SE) of fiber-optic systems may be easily changed via probabilistic amplitude shaping (PAS). By identifying the prerequisites for choosing the PC component codes, the authors of [15] has shown PAS may be used to bit-wise hard decision decoding (HDD) of product codes (PCs). In [16], the authors use probabilistic amplitude shaping (PAS) to make fiber-optic communication systems more spectrally efficient. They have considered probabilistic shaping using hard decision decoding (HDD), in contrast to earlier research in the literature. A proposed [17] multinary-signaling-based coded-modulation (CM) system outperforms traditional CM schemes in terms of both spectrum and energy efficiency, making it appropriate for a variety of applications ranging from multi-Tb/s to multi-Pb/s data rates. A design algorithm for the related multinary signal constellation is also put forward.

The literature available for performance prediction strategies of optical links based on error detecting codes and modulation schemes employing various strategies like; SD, HD, and probabilistic shaping. Here, in this work the performance of these strategies has been analyzed by incorporating FEC codes. The BER for aforementioned strategies has been evaluated and compared with and without FEC.

3 Methodology

The appropriate performance statistic for the system under consideration relies on the receiver's decoding method (SD, HD, bit-wise, symbol-wise, etc.). The decoding technique in an ideal receiver should take into consideration the precise nonlinear and bursty input-output relationship of the channel [18]. Since the precise relationship is typically unclear, a mismatched receiver built using a streamlined (auxiliary) 15 channel model is typically used. Usually, when designing a receiver, it is assumed that the output samples are simply distorted by an AWGN with a variance per dimension of (1).

$$\sigma^2 = \frac{1}{DN} \sum_{n=1}^{N} \|y_n - s(i_n)\|^2 \tag{1}$$

3.1 Strategies with HD-FEC

For each received vector y_n, the receiver makes an educated guess (i_{n1}) about the sent data in (i_n). The HD FEC decoder then receives this data for potential error correction.

Using (2), the most prevalent HD rule (minimal Euclidean distance) has been represented [19].

$$i_{n1} = \frac{argmin}{j \in \{1, 2, 3..M\}} \|y_n - s(i_n)\|^2 \tag{2}$$

The symbol error rate (*SER*) and bit error rate for HD-FEC can be represented using (3) and (4) respectively, where the δ function is 1 if the decision is true, otherwise it is zero.

$$SER^{HD} = \frac{1}{N} \sum_{n=1}^{N} \delta\left(i_n \neq \tilde{i}_n\right) \tag{3}$$

$$BER^{HD} = \frac{1}{mN} \sum_{n=1}^{N} \sum_{k=1}^{m} \delta\left(b_k(i_n) \neq b_k\left(\tilde{i}_n\right)\right) \tag{4}$$

The achievable information rate for HD-FEC is represented using (5) with binary entropy function (H_2) with various code rates R_c.

$$\text{AIR}_b^{HD} = m(1 - H_2(\text{BER}^{HD})) \geq mR_c \tag{5}$$

3.2 Strategies with SD-FEC

In these strategies, the decision has been taken on output sequence y_n based on the soft information available. The achievable information rates for these strategies have been estimated through symbol wise and bit wise and are represented using (6) and (7) respectively [15].

$$AIR_s^{SD} = m - \frac{1}{N} \sum_{n=1}^{N} \left\{ \log_2 \sum_{j=1}^{M} q(y_n, s(j)) - \log_2 q(y_n, s(i_n)) \right\} \tag{6}$$

$$AIR_b^{SD} = m - \frac{1}{N} \sum_{n=1}^{N} \sum_{k=1}^{m} \left\{ \log_2 \sum_{j=1}^{M} q(y_n, s(j)) - \log_2 \sum_{j=1}^{M} \delta(b_k(j)) \right\} \tag{7}$$

3.3 Strategies with Probabilistic Shaping

Probabilistic shaping [19], whose shaping operation is frequently realized via distribution matching (DM), has emerged as the most well-liked PS approach in recent years. The symbol entropy (H_s) which is represented using (8). These strategies use bit wise FEC codes for decoding whose values are represented using (9).

$$H_s = - \sum_{j=1}^{M} p_l log_2 p_j \tag{8}$$

$$L_{n,k} = \ln \frac{\sum_{j=1}^{M} \delta\big(b_k(j) = 0\big)p_j(y_n, s(j))}{\sum_{j=1}^{M} \delta\big(b_k(j) = 1\big)p_j(y_n, s(j))} \tag{9}$$

The bit error rate and the achievable information rate using these strategies are represented using (10) and (11) respectively, where ASI is the asymmetric information available at each receiving vector y_n.

$$BER_b^{PS} = \frac{1}{mN} \sum_{n=1}^{N} \sum_{k=1}^{m} \delta\big(L_{n,k} \leq 0\big) \tag{10}$$

$$AIR_b^{PS} = H_s - (1 - ASI)m \tag{11}$$

4 Simulation Parameters

To simulate the bit error rata of an optical link, several parameters need to be considered which helps in accurately predicting the optical link performance. Table 1 provides the detailed list of parameters used for simulation.

Table 1. Execution Parameters

Parameter	Description
Data samples (N)	106
Mean (μ)	0
Variance(σ^2)	0.01
Constellation Symbol Dimension (D)	2 (for QAM)
Modulation	64-QAM
Symbol entropy (H_s)	6bits/symbol
SNR	5–30 dB

5 Simulation Results

The performance prediction of an optical link has been evaluated for various strategies with the help of MATLAB simulation software. The prediction strategies such as; hard decision, soft decision and probabilistic shaping have been analyzed based on forwarding error correcting codes (FEC) and compared the performance measures with and without employing FEC codes. Figure 1, shows the symbol error rate (SER) using hard decision-FEC has been estimated and compared the performance measures with and without FEC. It is clearly depicted in Fig. 1, that the SER is low for obtained output sequence with FEC when compared to SER without FEC.

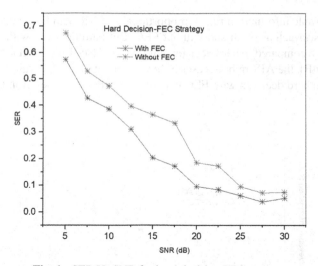

Fig. 1. SER Vs SNR for hard decision FEC strategy.

The BER for hard decision strategy has been evaluated and is depicted in Fig. 2. Similar to SER, the BER also is very low for the prediction strategy with FEC codes when compared to the strategy without FEC. It is evident from Fig. 2, 40% reduction in BER at an SNR of 5dB has been achieved for HD-FEC by employing FEC codes. Similarly, at 30dB, the BER is reduced to 50% (see the black line). The hard decision strategy with FEC achieves a BER of almost 0(zero) at an SNR of 30dB, whereas, the same strategy without FEC archives a BER of 0.4 at 30dB.

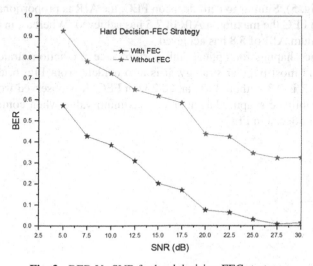

Fig. 2. BER Vs SNR for hard decision FEC strategy.

The achievable information rate is proportional to SNR with hard decision strategy in both the cases such as: with and without FEC. But hard decision with FEC attained more AIR when compared to hard decision without FEC (see Fig. 3). At low SNR values (from 5 to 20 dB), the AIR in both cases is similar, but at high SNR values the achieved AIR is more in hard decision with FEC when compared to hard decision without FEC.

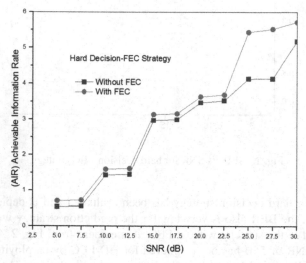

Fig. 3. AIR Vs SNR for hard decision FEC strategy.

In soft decision strategy, the AIR has been evaluated via symbol wise (see Fig. 4) and bit wise (see Fig. 5). Similar to hard decision FEC, the AIR is proportional to SNR but in soft decision FEC the maximum AIR of 7.5 has achieved. Whereas, in hard decision FEC the maximum AIR of 5.8 has achieved.

Probabilistic shaping for optical link performance prediction strategy has been widely used and most popular strategy. It is also evident from Fig. 6, the maximum BER at low SNR is 0.3 without FEC and 0.2 with FEC. It is observed from Fig. 7, the AIR with probabilistic shaping has attained maximum value when compared to hard decision and soft decision FEC.

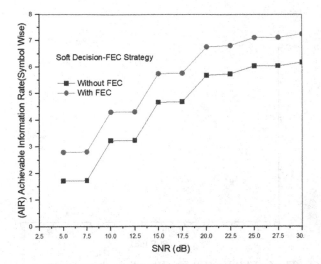

Fig. 4. AIR (symbol wise) Vs SNR for soft decision FEC strategy.

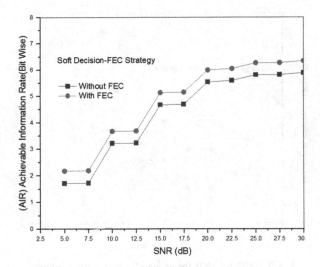

Fig. 5. AIR (bit wise) Vs SNR for soft decision FEC strategy.

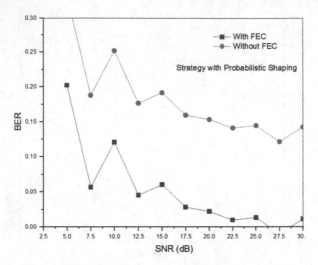

Fig. 6. BER Vs SNR for probabilistic shaping strategy.

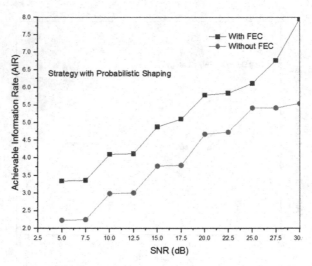

Fig. 7. AIR Vs SNR for probabilistic shaping strategy.

6   Conclusion

In this work, the performance measurements with and without the use of forwarding error correcting codes (FEC) have been compared for several prediction algorithms, including hard decision, soft decision, and probabilistic shaping. By implementing the hard decision FEC strategy, a remarkable 40% decrease in BER has been accomplished at low SNR levels. Additionally, by utilizing FEC codes, a substantial 50% reduction in BER has been achieved at higher SNR values. On the other hand, the performance of soft decision FEC has been evaluated in terms of the AIR for both symbol-wise and bit-wise approaches. The analysis reveals that the symbol-wise soft decision FEC achieves a maximum AIR of 7.5, while the bit-wise approach attains an AIR of 5.8. Comparatively, employing FEC codes with probabilistic shaping yields an even higher AIR of 8. Specifically, at low SNR values, the BER is reduced by 40% without FEC, and an impressive 80% reduction is achieved with FEC combined with probabilistic shaping. These reductions are observed when comparing the results to both the hard decision and soft decision strategies. Moreover, at high SNR levels, the BER becomes nearly negligible when FEC codes are employed. The effectiveness of FEC becomes apparent, as it ensures a highly reliable transmission with minimal errors, leading to a BER that approaches zero. Finally, the combination of FEC codes and probabilistic shaping techniques yields substantial improvements in BER performance. At low SNR values, the reductions are 40% without FEC and 80% with FEC and probabilistic shaping. Additionally, at high SNR values, the BER approaches zero when employing FEC codes, indicating a highly reliable transmission.

References

1. Schmalen, L., Alvarado, A., Rios-Müller, R.: Predicting the performance of nonbinary forward error correction in optical transmission experiments. Optical Fiber Communications Conference and Exhibition (OFC), pp. 1–3. Anaheim, CA, USA (2016)
2. Beppu, S., Kasai, K., Yoshida, M., Nakazawa, M.: 2048 QAM (66 Gbit/s) single-carrier coherent optical transmission over 150 km with a potential SE of 15.3 bit/s/Hz. Opt. Express 23(4) 4960–4969 (2015)
3. Alvarado, A., Agrell, E., Lavery, D., Maher, R., Bayvel, P.: Replacing the soft-decision FEC limit paradigm in the design of optical communication systems (Invited Paper). J. Lightw. Technol 33(20), 4338–4352 (2015)
4. Alvarado, A., Agrell, E., Lavery, D., Maher, R., Bayvel, P.: Corrections to Replacing the soft-decision FEC limit paradigm in the design of optical communication systems. J. Lightw. Technol 34(2), 722 (2016)
5. Venkateswara Rao, Ch., Ravi Sankar, M., Praveena, V., Bavya Sri, V., B S Sailesh, A., Rama Lakshmi, K.: Performance Evaluation of OFDM System: With and Without Reed-Solomon Codes. 4th International Conference on Advances in Computing, Communication Control and Networking (ICAC3N), Greater Noida, India, pp. 1827–1831 (2022)
6. Satish, A., Kamalaksha, B., Vinodh Kumar, N.: Simple optical sensors for oxygen detection: simulation fabrication and characterization. J. Mod. Opt. 68(16), 886–894 (2021)
7. Ravuri, V., Subbarao, M.V., Terlapu, S.K., Challa Ram, G.: Path loss modeling and channel characterization at 28 GHz 5G micro-cell outdoor environment using 3D ray-tracing. In: Second International Conference on Advances in Electrical, Computing, Communication and Sustainable Technologies (ICAECT), pp. 1–7. Bhilai, India (2022)
8. Caire, G., Taricco, G., Biglieri, E.: Bit-interleaved coded modulation. IEEE Trans. Inf. Theory 44(3), 927–946 (1998)
9. Alvarado, A., Agrell, E.: Four-dimensional coded modulation with bit-wise decoders for future optical communications. J. Lightw. Technol. 33(10), 1993–2003 (2015)
10. Böcherer, G., Schulte, P., Steiner, F.: Probabilistic shaping and forward error correction for fiber-optic communication systems. J. Lightwave Technol. 37(2), 230–244 (2019)
11. Yoshida, T., Alvarado, A., Karlsson, M., Agrell, E.: Post-FEC BER benchmarking for bit-interleaved coded modulation with probabilistic shaping. J. Lightwave Technol. 38(16), 4292–4306 (2020)

12. Secondini, M., Agrell, E., Forestieri, E., Marsella, D., Camara, M.R.: Nonlinearity mitigation in WDM systems: models, strategies, and achievable rates. J. Lightwave Technol. **37**(10), 2270–2283 (2019)

13. Secondini, M., Agrell, E., Forestieri, E., Marsella, D.: Fiber nonlinearity mitigation in WDM Systems: strategies and achievable rates. In: European Conference on Optical Communication (ECOC), pp. 1–3. Gothenburg, Sweden (2017)

14. Ivanov, M., Brannstrom, F., Alvarado, A., Agrell, E.: On the exact BER of bit-wise demodulators for one-dimensional constellations. IEEE Trans. Commun. **61**(4), 1450–1459 (2013)

15. Alireza. Sk., G. i Amat, A., Alvarado, A.: On product codes with probabilistic amplitude shaping for high-throughput fiber-optic systems. IEEE Commun. Let. **24**(11) 2406–2410 (2020)

16. Alireza. Sk., G., i Amat, A., Liva, G., Steiner, F.: Probabilistic amplitude shaping with hard decision decoding and staircase codes. J. Lightwave Technol. **36**(9) 1689–1697 (2018)

17. Djordjevic, I.B., Liu, T., Wang, T.: Multinary-signaling-based coded modulation for ultrahigh-speed optical transport. IEEE Photon. J. **7**(1), 1–9 (2015)

18. Yoshida, T., Alvarado, A., Karlsson, M., Agrell, E.: Performance prediction recipes for optical links. IEEE Photon. Technol. Lett. **33**(18), 1034–1037 (2021)

19. Jayaraj, N., Nagaraj, R.: Performance analysis of optical network for efficient transmission of multimedia data. In: 9th International Conference on Computing for Sustainable Global Development (INDIACom), pp. 87–92. New Delhi, India (2022)

Investigation of Highly Sensitive and Linearly Responsive SAW Based Gas Sensor for Better N₂ Detection

Nimmala Harathi[1], Binduswetha Pasuluri[2], Argha Sarkar[3](✉),
and Naveen Kumar Maurya[4]

[1] Department of Electronics and Instrumentation Engineering, Sree Vidyanikethan Engineering College, Tirupati 517102, Andhra Pradesh, India
[2] Department of Electronics and Communication Engineering, Vardhaman College of Engineering, Shamshabad 501218, Telengana, India
[3] School of Computer Science and Engineering, REVA University, Bangalore 560064, Karnataka, India
argha15@gmail.com
[4] Department of Electronics and Communication Engineering, Vishnu Institute of Technology, Bhimavaram 534202, Andhra Pradesh, India

Abstract. Surface acoustic wave sensors are becoming more and more essential in research. The research has advanced in such a way that SAW sensors are now used in many different fields. The application areas of SAW sensor include gas sensing, biosensor, measurement of many physical parameters like humidity, pressure, temperature and torque. In this study, the SAW sensor is presented as a nitrogen gas sensor. Nanomaterials are introduced to a conventional SAW sensor to strengthen the sensor's performance. The most frequently used nano - materials for applications in gas detection is ZnO. Comsol Multiphysics is used to design a 2D SAW-based gas sensor using ZnO as the sensing material. Finite element analysis (FEA) is used in sensor characterization. Testing is done to determine whether nitrogen gas is present or not in the sensor. The sensor's sensitivity varies depending on whether there is gas present or not while operating at the same frequency. The sensor's frequency range is 3 MHz. Nitrogen gas is present in concentrations varying from 10 ppm to hundreds of ppm. With a rise in concentration, the sensor showed excellent linearity.

Keywords: Surface Acoustic Wave Sensor · Gas Detection · Zinc Oxide · Safety levels

1 Introduction

Lord Rayleigh initially announced the existence of surface acoustic waves during the year 1885. [1]. The surface acoustic waves are therefore referred to as Rayleigh waves. The first surface acoustic wave device was developed in 1965. The SAW sensors are high frequency devices that operate at frequencies between 1 MHz and 300 GHz. Since 1885

© ICST Institute for Computer Sciences, Social Informatics and Telecommunications Engineering 2024
Published by Springer Nature Switzerland AG 2024. All Rights Reserved
P. Pareek et al. (Eds.): IC4S 2023, LNICST 537, pp. 205–215, 2024.
https://doi.org/10.1007/978-3-031-48891-7_17

there is wide variety of improvement s in the surface acoustic wave sensor which made them to use in all application areas like mining applications in the form of gas sensor [2], biomedical area as sensor which can detect HIV and antigens, [3] as chemical sensor, as pressure sensor etc. Harathi. N et al. [2, 4] built a SAW-based gas sensor for hydrogen gas detection, and tested it with gas concentrations varying from 0.1 ppm to 100 ppm. The sensor's response showed proportionality among hydrogen gas concentration as well as displacement. Argha Sarkar et. al. [5] used ZnO as the sensing layer in the implementation of a highly sensitive SAW-based ethylene gas sensor with various electrode orientations. Hasan M et al. [6] developed and simulated SAW-based hydrogen gas sensor with two different active layers and compared the simulation and fabrication results. Neeruganti Vikram et al. [7] constructed a high sensitive 2D model of SAW sensor with two IDT (Inter Digitated Electrodes) and tested different chemical solutions concentration ranging from 1 ppm to 100 ppm. Sarkar A et al. [8] created a nanorod-based SAW methane gas sensor to get rid of crystal intrinsic defects. Beyond this SAW sensors can be used as biosensor, chemical sensor, toque sensor and humidity sensor. The SAW sensors are used for frequency range of megahertz to giga hertz. Above all, the author employed COMSOL for simulation, and finite element modelling was used for analysis.

2   Functionality of SAW Sensor

The three classes of acoustic wave devices are as follows: i) BAW (Bulk Acoustic Wave devices) ii) APM (Acoustic Plate mode devices) iii) SAW (Surface Acoustic Wave devices). These devices can also be further segregated, as indicated in the Fig. 1 and abbreviations are provided in Table 1. Acoustic devices can be used for both liquid and gaseous environments [8, 9].

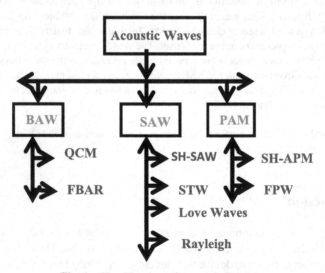

Fig. 1. Classification of Acoustic Devices.

Among all these acoustic devices the surface acoustic wave sensors are mostly preferred because of their operational flexibility, ease of usage and high sensitivity.

Table 1. Abbreviations of SAW Devices

S.No	Abbreviation
1	Quartz Crystal Microbalance (QCM)
2	Shear Horizontal Surface Acoustic Wave (SH-SAW)
3	Surface Transverse Wave (STW)
4	Shear Horizontal-Acoustic Plate Modulator (SH-APM)
5	Flexural Plate Wave (FPW)
6	Film Bulk Acoustic Resonators (FBAR)

The basic construction of SAW sensor is in Fig. 2. A piezoelectric substrate, which serves as the structure's foundation, two interdigitated electrodes the input IDT and output IDT—as well as a sensing layer are required for the development of SAW sensor.

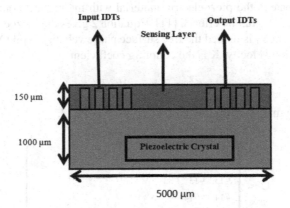

Fig. 2. Construction of SAW sensor

The input and output IDTs are enveloped inside the sensor layer, which is positioned slightly just above piezoelectric substrate. Opposing potentials are given to the IDTs. Because to the switching potentials at the input IDT, whenever an electrical signal is applied, stress forms over the piezoelectric substrate. By their very nature, piezoelectric materials have reversing piezoelectric properties. These crystals modify the physical dimensions of the crystal [10], converting input pressure into an electric signal, and converting input electric signal into input pressure. The IDTs converts electrical signals into acoustic signals at the input and acoustic signals into electrical signals at the output. On the surface of the sensor layer, an area known as the port length is where the generated acoustic signal travels. The three characteristics of the acoustic signal are frequency, phase and amplitude. Because of the unique properties of the sensing layer and piezoelectric substrate, any of the acoustic wave's characteristics can change as it travels through the sensing layer. The acoustic waves alter because of the physical characteristics that are spread across the sensor layer. The acoustic wave's changes have a significant effect on the measure and (physical parameter).

3   Analytical Design of Proposed SAW Sensor

3.1   Piezoelectric Substrate

The piezoelectric substrate serves as the SAW sensor's base. Various piezoelectric materials, such as lithium niobate, quartz, and lithium titanate, are easily accessible in the material catalogue. The SAW sensor's output displacement is greatly influenced by the piezoelectric substrate. To accomplish this, the material must have a high coupling coefficient. The Displacement of the SAW sensor is given in Eq. 1. Where "e" denotes the material's coupling matrix, "S" denotes its strain matrix, "E" is the electric potential somewhere at IDTs, and "ε" denotes the permittivity matrix.

$$d = eS + \varepsilon E \tag{1}$$

Lithium niobate is the piezoelectric material with the highest coupling coefficient and the least crystallographic effects. [11]. Equation 2 gives the piezoelectric material's coupling coefficient V_f is indeed the free surface phase velocity, while V_m would be the metal surface phase velocity. K is the coupling coefficient.

$$K^2 = 2(V_f - V_m)/V_f \tag{2}$$

The coupling matrix of lithium niobate is given in Eq. 3.

$$C = \begin{pmatrix} c_{11} & c_{12} & c_{13} & c_{14} & 0 & 0 \\ c_{12} & c_{11} & 13 & -c_{14} & 0 & 0 \\ c_{13} & c_{13} & c_{33} & 0 & 0 & 0 \\ c_{14} & -c_{14} & 0 & c_{44} & 0 & 0 \\ 0 & 0 & 0 & 0 & c_{44} & c_{14} \\ 0 & 0 & 0 & 0 & c_{14} & (c_{11} - c_{12})/2 \end{pmatrix} \tag{3}$$

The piezoelectric constants are given in Eq. 4.

$$e = \begin{pmatrix} 0 & 0 & 0 & 0 & e_{15} & -e_{22} \\ -e_{22} & e_{22} & 0 & e_{15} & 0 & 0 \\ e_{31} & e_{31} & e_{33} & 0 & 0 & 0 \end{pmatrix} \tag{4}$$

The permittivity matrix is given in Eq. 5

$$\varepsilon = \begin{pmatrix} \varepsilon_{11} & 0 & 0 \\ 0 & \varepsilon_{11} & 0 \\ 0 & 0 & \varepsilon_{33} \end{pmatrix} \tag{5}$$

Lithium niobate's chemical and physical properties such as elastic constants, piezo-electric constants, permeability constants, density are considered. The acoustic wave's velocity (V_o) and wavelength (λ) determine the SAW sensor's operating frequency. The operating frequency is given in Eq. 6. Equation 7 states that the wavelength is a function of the IDT dimensions, where W_e denotes the electrode width and w_{sp} denotes the electrode spacing.

$$f_0 = \frac{V_o}{\lambda} \tag{6}$$

$$\lambda = 2(W_e + W_{sp}) \tag{7}$$

3.2 Sensing Layer

The sensing layer is the heart of SAW sensor. For SAW sensors, there are a vast array of sensing layers available, including metal oxides like zinc oxide, indium oxide, tungsten trioxide, and tin oxide [12]. The sensor layer may be made of N- or P-type semiconductors. Depending on the type of semiconductor material used, an oxidation or reduction reaction occurs when the acoustic wave passes over the sensor layer, changing the conductivity of the layer. Zinc oxide, or ZnO, is the most utilised sensing layer due to its characteristics. ZnO is a semiconductor material of the N-type that can withstand high temperatures ZnO has a strong thermal properties and electron affinity. Inert gases have no effect on ZnO despite these benefits. ZnO is chosen for the sensing layer due to these benefits. ZnO is highly sensitive compared to other sensing layers like GaO, WO_3, InO_3 because of its high conductivity nature and its conductivity varies with the doping levels.

4 Design and Simulation of Proposed Sensor

With the aid of a lithium niobate piezoelectric substrate, a ZnO sensing layer, and aluminium IDTs, a two-dimensional SAW-based gas sensor is constructed in COMSOL Multiphysiscs software. Table 2 includes information on the constructional details. In Fig. 3, the suggested SAW sensor is presented.

Table 2. Geometrical details of Proposed sensor

S.No	Component's identification	Dimensions
1.	Piezoelectric Base	5000 μm × 1000 μm
2.	Interdigitated Transducers	5 μm × 140 nm
3.	Sensing layer	5000 μm × 150 μm
4.	Spacing of electrodes	5 μm
5.	Spacing between IDTs	4000 μm

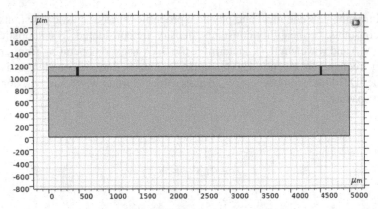

Fig. 3. Front View of developed Sensor

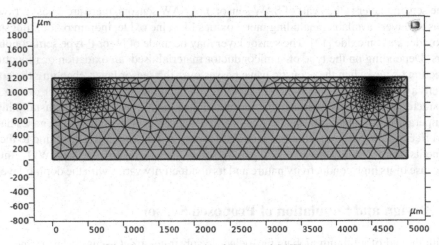

Fig. 4. Mesh analysis of Proposed SAW sensor

The fine meshing is completed in the shortest amount of time, 8 s. The Mesh analysis of the suggested sensor is shown in Fig. 4.

The proposed sensor has been modelled with and without gas applied across the sensing layer. With fluctuations in the displacement, the sensor obtained its maximum displacement at 3 MHz for both with and without gas. The Fig. 5 shows the maximum displacement of SAW sensor without gas. The response of the sensor with and without nitrogen gas is shown in Fig. 6. Gases with less of an odour and colour are nitrogen-based. The combustion process results in the emission of nitrogen into the environment. Nitrogen has a lower than 10 ppm safety limit. The mass density of the sensing layer changes as a result of nitrogen being deposited over it. The reactions of nitrogen gas with organic semiconductors alter the conductivity of the sensor layer, which reduces displacement when nitrogen gas is present. The reduction in the displacement is due to redox reaction between gas and sensing layer.

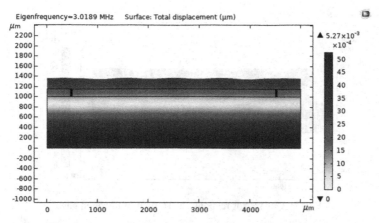

Fig. 5. Sensor displacement in the absence of nitrogen gas

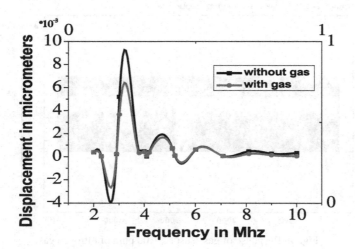

Fig. 6. Displacement of gas sensor

The density on the sensing layer rises as the nitrogen gas concentration rises from 10 parts per million to 100 parts per million, boosting displacement as the gas concentration rises. The sensor exhibited the linear. Figure 7 depicts the gas sensor's surface displacement at a nitrogen gas concentration of 10 ppm. Figure 8 shows the features of the surface displacement of the gas sensor at the maximum nitrogen gas concentration of 100 ppm. The sensor response for changing nitrogen gas concentration is shown in Fig. 9.

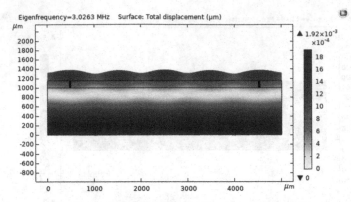

Fig. 7. Gas sensor response at 10 ppm of nitrogen gas

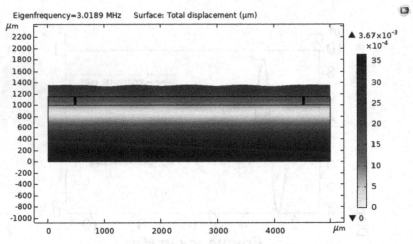

Fig. 8. Response of gas sensor at 100 ppm of nitrogen gas

The gas sensor's linear response to an increase in concentration from 10 ppm to 100 ppm is shown in Fig. 10.

Figure 11 shows the step wise linear characteristics of SAW based nitrogen gas sensor. Due to the low mass density on the sensor layer, the step change in displacement is modest at low concentrations. The stepwise linearity rises with mass density as concentration increases. The mass density of the sensing layer increases with the concentration of gas.

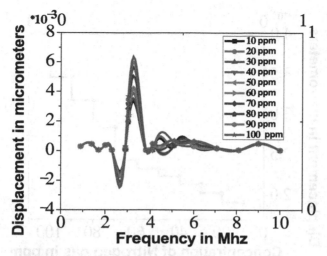

Fig. 9. Response of gas sensor with varying concentration

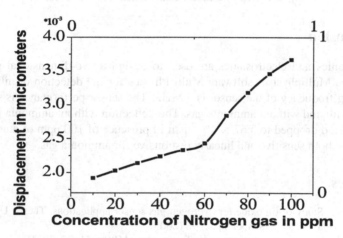

Fig. 10. Linear characteristics of the sensor

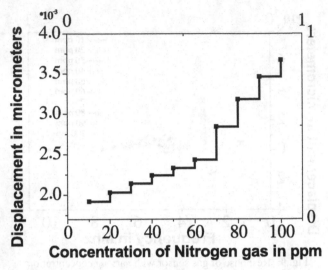

Fig. 11. Step wise Linear characteristics of the sensor

5 Conclusion

Solid mechanics and electrostatics are used to design a two-dimensional gas sensor in COMSOL Multiphysiscs software Multi Physics for the detection of nitrogen gas. The working frequency of the sensor is 3 MHz. The sensor performance is studied for deflection with and without ammonia gas. The deflection without ammonia gas is 5.27 $\times\ 10^{-3}$ μm and dropped to 3.67 $\times\ 10^{-3}$ μm in presence of 100 ppm of ammonia gas. The sensor is both sensitive and linearly responsive for ammonia gas.

References

1. Anand, M.: Study of tin oxide for hydrogen gas sensor applications. Thesis, University of South Florida, Graduate School, pp. 1–98 (2005)
2. Awang, Z.: Gas sensors: a review. Sens. Transducers **168**(4), 61–75 (2014)
3. Arabshahi, S., et al.: Simulation of surface acoustic wave NO2 gas sensor based on Zno/XY Linbo3 structure. IJERA **2**, 2120–2123 (2012)
4. Valluru, S.P.R.: Design of FIDT for 3D analysis of MEMS based gas sensor using SAW technology. In: Proceedings of the Comsol Conference, pp1–4. COMSOL, Pune (2015)
5. Sarkar, A., et al.: Design and optimization of ZnO nanostructured SAW-based ethylene gas sensor with modified electrode orientation. Adv. Sci. Technol. Eng. Syst **5**, 263–266 (2020)
6. Du, P., et al.: Simulation of ZnO enhanced SAW gas sensor. In: Proceedings of the Comsol Conference, pp 1–4. COMSOL, Rotterdam (2013)
7. Sarkar, A., et al.: Responsivity optimization of methane gas sensor through the modification of hexagonal nanorod and reduction of defect states. Superlattices Microstruct. **102**, 459–469 (2017)
8. Vellekoop, M. J.: Acoustic Wave Sensors. Theory, Design and Physico-Chemical Applications, by DS Ballantine, RM White, SJ Martin, AJ Ricco, ET Zellers, GC Frye and H. Wohltjen, published by Academic Press, San Diego, ISBN 0-12-077460-7, 436 p. Sens. Actuat. A Phys. **63**(1), 79–79 (1997)

9. Harathi, N., Kavitha, S., Sarkar, A.: ZnO nanostructured 2D layered SAW based hydrogen gas sensor with enhanced sensitivity. Mater. Today Proc. **33**, 2621–2625 (2020)
10. Sarkar, A., et al.: Methane-sensing performance enhancement in graphene oxide/Mg: ZnO heterostructure devices. J. Electron. Mater. **46**, 5485–5491 (2017)
11. Hasan, M.N., et al.: Simulation and fabrication of SAW-based gas sensor with modified surface state of active layer and electrode orientation for enhanced H2 gas sensing. J. of Electron. Mater. **46**, 679–686 (2017)
12. Harathi, N., et al.: PrGO decorated TiO2 nanoplates hybrid nanocomposite for augmented NO2 gas detection with faster gas kinetics under UV light irradiation. Sens. Actuat. B Chem. **358**, 131503 (2022)

Turnstile Diamond Dipole Nanoantenna Based Smart City Compatible Thin Film Solar Cell

Abhishek Pahuja[1]([⊠]) [iD], Sandeep Kumar[2] [iD], Vipul Agarwal[1] [iD],
Manoj Singh Parihar[3] [iD], and V. Dinesh Kumar[3] [iD]

[1] Department of Electronics and Communication Engineering, Koneru Lakshmaiah Education Foundation, Vijayawada 522501, Andhra Pradesh, India
pahuja.abhishek@kluniversity.in
[2] School of Engineering and Technology, BML Munjal University, Gurugram 122413, Haryana, India
[3] Department of Electronics and Communication Engineering, PDPM Indian Institute of Information Technology, Design and Manufacturing Jabalpur, Jabalpur 482001, Madhya Pradesh, India

Abstract. A unit cell design of a thin film solar cell incorporating turnstile diamond dipole nanoantenna as a means of light trapping structure is proposed and investigated. Diamond dipole nanoantenna (DDNA) is a transformed version of the conventional dipole nanoantenna whereby the arms of the dipole nanoantenna are replaced by diamond shaped nanoparticles. In contrast to the dipole nanoantenna, DDNA offers larger area for field confinement and it resonates in the maximum solar spectrum range. The reduction of reflection losses along with generation of localized surface plasmons leads to improved photovoltaic characteristics of the thin film solar cell. The suggested TFSC model offers 99% absorption with 1.52 times photocurrent calculated based on finite element approach.

Keywords: Thin film solar cells (TFSC) · Smart city · Plasmonics · Nanoantenna · Surface Plasmon Polaritons · Localized Surface Plasmons · Polarization

1 Introduction

In the growing age of smart cities and Internet of Things (IoT) there has been significant demand of portable, flexible and thinner energy storage devices to cope up the energy requirement of the various appliances [1, 2]. There are many applications where thinner batteries are desired such as smart phones, smart watches and many more. For such applications thin film solar cells (TFSCs) could be a suitable choice if it provides substantial efficiency that can compete with the existing energy storage solutions available in the market [3]. TFSC encounter commercial hurdles despite having a compact design as a consequence of being comparatively less efficient [4].

It has been observed that addition of a suitable resonant light trapping structure to the TFSC design could improve the conversion efficiencies of existing solar cell models

P. Pareek et al. (Eds.): IC4S 2023, LNICST 537, pp. 216–222, 2024.
https://doi.org/10.1007/978-3-031-48891-7_18

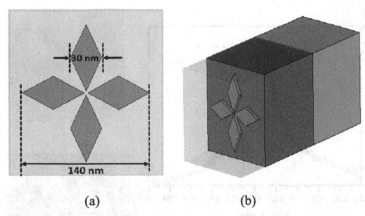

(a)   (b)

Fig. 1. (a) Dimensions of TDDNA (b) Proposed TFSC unit cell design showing TDDNA on the top of active layer surface

[5]. The common thickness of TFSCs ranges within hundreds of nanometers and some microns [3], under this regime amalgamation of nanotechnology becomes relevant for the quest of higher efficiency. Nanophotonic researchers have proposed various models of TFSCs with increased efficiencies [6, 7]. Researches have reported TFSCs models based on nanophotonics and Plasmonics phenomena such as localized surface plasmon (LSP) and surface plasmon polaritons (SPP) [8, 9]. Surface plasmon polaritons or SPP is the transverse magnetic mode electron cloud oscillation that is excited on the dielectric and metal interfaces. Whereas, localized surface plasmons being the localized form of SPPs that are confined in the proximity in a surrounded submicron region. One of the advantageous features of SPPs and LSPs is high energy confinement in the region where it is excited [10]. Based on these phenomena several TFSC models incorporating nanostructure have been reported [11]. The downscaled version of the conventional antenna which resonates at nanometric wavelength is known as nanoantenna [12] and plasmonic researchers have reported many works where efficient light trapping design for a thin film solar cell is attained by incorporating plasmonic nanoantennas such as Travelling wave antenna [13], needle shape [14], core-shell [15], Euler Spiral [16], Yagi-Uda [17]. Nanoantennas for TFSCs are designed in such a way that its resonance lies within the maximum solar irradiance range. The most desirable attributes of nanoantenna based TFSC are the multi-fold local field confinement and efficient trapping of the impinging sun light. In other words, the nanoantennas increases the effective aperture of a solar cell which results as higher absorption of solar energy. Subsequently, it produces more photocurrent and enhances the conversion efficiency. But there is still significant scope for further performance enhancement as the bandwidth of the nanoantenna are limited.

In this paper, a novel approach of nanoantenna based TFSC performance enhancement is presented in which a diamond dipole is used in turnstile manner. Turnstile positioning aids the performance by making the design polarization insensitive whereas the diamond shape increases the light confinement area. The efficacy of the solar cell

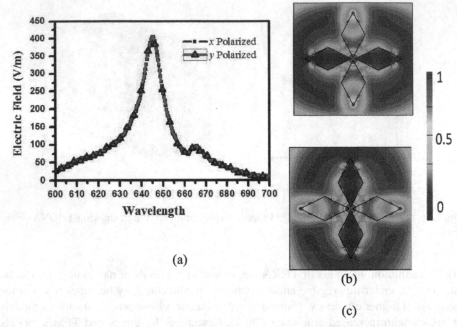

Fig. 2. (a) Electric field enhancement at the centre of the TDDNA and (b), (c) shows the distribution of electric field with respect to both the orthogonally polarized light component

as a whole is improved by higher localization of energy and light being more tightly focused by the adding the nanoantenna to the design.

2   Design Parameters and Computational Details

Similar to the conventional solar cell design, the unit cell (Fig. 1) has three layers. Amorphous Silicon as the absorber layer is sandwiched between the top Indium tin oxide (ITO) made anti-reflection coating (ARC) and bottom cathode which is made of silver. The ARC layer also serves as transparent anode whose thickness is calculated as 80 nm by considering λ wavelength light impinging normally to cancel the reflected light by maintaining 180° phase shift. Due to the fact that it is almost in the middle of the sunlight spectrum, as irradiance is at its highest, 600 nm wavelength light is taken into calculation.

$$h = \frac{\lambda}{4n_{ITO}} \tag{1}$$

The cathode layer and absorber layer both have thickness of 400 nm. As proposed modification, the turnstile dipole nanoantenna is positioned on the top surface of the

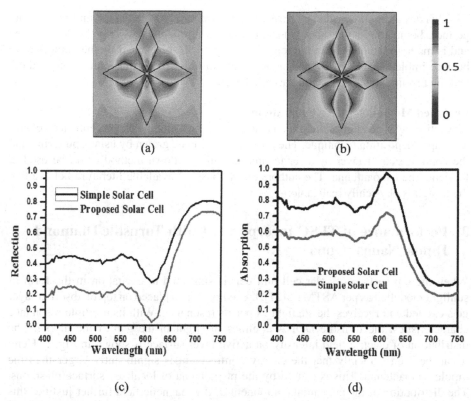

Fig. 3. (a), (b) shows the tangential component of the magnetic field and (c), (d) shows the comparison between the proposed solar cell and simple solar cell in terms Reflection and absorbance as a function of wavelength respectively.

amorphous silicon layer. Figure 1 depicts the design parameters of the proposed unit cell of TFSC.

Computational Domain

The time domain solver of CST Studio Suite [18] is used to conduct the presented study. The computational domain is discretised into mesh cells with $12 \times 12 \times 12$ nm^3 being the largest and smallest mesh cell of $2 \times 2 \times 2$ nm^3. The dispersive behaviour of silver at optical frequency is taken into consideration by including the Drude model is used [19], the permittivity of silver is modelled as:

$$\varepsilon_{Ag} = \varepsilon_0 \left\{ \varepsilon_\infty - \frac{f_p^2}{f(f + i\gamma)} \right\} \tag{2}$$

In the above equations, $\varepsilon_\infty = 5$, ε_0 is the permittivity of free space, f_p plasma frequency 2.175 PHz and γ is 4.35 THz as the collision frequency. All the above parameters are according to the Johnson and Christy model [20].

This design is surrounded by perfectly matched layers from the top and bottom and periodic boundaries at the sides. For excitation A uniform plane wave with 1 V/m is used and is made to incident from the top of the design. Results stated by the authors have been re-implemented using the same approach in order to test the simulation method, and the acquired results are under acceptable [8, 9].

Proposed Methodology for Fabrication
The active layer of amorphous silicon could be grown using the plasma enhanced chemical vapor deposition technique. The layer of ITO can be grown by using sputtering and the combination of layer-by-layer technique with top-down method could be used to fabricate the nanoantenna. The authors on the basis of available literature believe that the design is physically realizable.

3 Performance of TFSC Incorporated with Turnstile Diamond Dipole Nanoantenna

When the top surface of solar cell unit cell is subjected to excitation in the form of sunlight modelled as per ASTM1.5[21], the nanoantenna placed on top of absorber layer gets excited and receives the sunlight as per its resonance. With its multiple resonance characteristics, the turnstile DDNA confines the light in the gap of the dipole. The distribution of electric field intensity on active layer surface is shown in Fig. 2. Here, it can be seen evidently that the energy confinement is significantly larger than the dipole nanoantenna. This is caused by the phenomena of localized surface plasmons. The distribution of the tangential component of the magnetic field further justifies this phenomenon. To verify this the distribution of the tangential component of the magnetic field is shown in the Fig. 3. By observing Fig. 3, it can be concluded that there is higher concentration of surface current in the gap of the dipole because the current density is directly dependent on the magnetic field component. The main reason of this surface current is the excitation of localized surface plasmons. By comparing, this with the dipole nanoantenna characteristics, it can be said that there is higher LSP excitation in case of turnstile dipole nanoantenna.

The subsequent effect on the field enhancement is monitored by means of a probe which is mounted in the exact center of the dipole gap. The electric field measured by the probe shows resonant characteristics of the turnstile DDNA. In the absence of the nanoantenna, there are no bright spots of energy confinement on the surface or anywhere in the solar cell. But when, the nanoantenna are added to the design, then the composite design functions as a light-trapping structure as a result to the localized surface plasmons excitation. Therefore, minimizing the reflection losses of the thin film solar cell (Fig. 3c). Figure 3(d) depicts the comparison in terms of absorbance of the simple thin film solar cell with the TDDNA embedded thin film solar cell. It can be noted that there is significant increase in the absorption spectra around 640 nm wavelength. It results from the suggested turnstile dipole nanoantenna's ability to trap light.

Optical to Electrical Conversion
In the calculation of the conversion efficiency, it is believed that every absorbed photon is going to result in the generation of one pair of electrons and holes. The effectiveness

is best described as [5]:

$$\eta = \frac{\int_{300nm}^{\lambda_g} F_S(\lambda) A_{Si-a}(\lambda) \frac{\lambda}{\lambda_g} d\lambda}{\int_{300nm}^{4000nm} F_S(\lambda) d\lambda} \tag{3}$$

Here, λ_g is taken as 800 nm which is close to the band gap wavelength of amorphous silicon, $F_S(\lambda)$ is the photon flux density referred per the ASTM 1.5 solar spectrum. $A_{Si-a}(\lambda)$ is the absorbance as a function of wavelength. As the losses associated with silver can be assumed negligible then absorbance can be computed as:

$$A_{Si-a}(\lambda) = 1 - R(\lambda) \tag{4}$$

Further, the photocurrent density can be computed as [5]:

$$J_{SC} = e \frac{\lambda_g}{hc} \eta \int_{300nm}^{4000nm} F_s(\lambda) d\lambda = 87.66 \ \eta \frac{mA}{cm^2} \tag{5}$$

In order to show the merit of the proposed model, the photocurrent enhancement factor (PEF) is calculated as:

$$PEF = \frac{J_{sc}^{na}}{J_{sc}^{SSC}} \tag{6}$$

Here, J_{sc}^{na} and J_{sc}^{SSC} are the photocurrent densities of the unit cell, calculated by using Eqs. (3) and (4) for the solar cell comprising of turnstile DDNA and simple solar cell respectively. The PEF for the proposed model is compute as 1.52 which shows the merit of the turnstile DDNA embedded thin film solar cell in terms of photocurrent generation.

4 Conclusion

Compact energy storage systems are one of the major requirements of any smart city. Since smart cities are generally densely occupied and numerous applications such as monitoring vehicles, drones further need flexibility as well. These all parameters could be fulfilled by thin film solar cells. As reliance upon fossil fuels is not considered in internet of things, integration of renewable energy is not only beneficial by environmental point of view but also it is essential for effectively utilizing the space limitation as there would be numerous applications running in the smart city. A solution to overcome the lower efficiency of the existing TFSC has been proposed. Based on the theoretical study and simulation presented, it can be concluded that the proposed design overpowers the conventional thin film solar cell in terms of photocurrent enhancement factor and absorption of sunlight inside the active layer of thin film solar cell. The design exhibits 99% highest absorption with 52% increased photocurrent.

References

1. Kraemer, F.A., Palma, D., Braten, A.E., Ammar, D.: Operationalizing solar energy predictions for sustainable autonomous IoT device management. IEEE Internet Things J. **7**(12), 11803–11814 (2020)
2. Pareek, P., Maurya, N.K., Singh, L., Gupta, N., Reis, M.J.C.S.: Study of smart city compatible monolithic quantum well photodetector. In: Gupta, N., Pareek, P., Reis, M. (eds.) Cognitive Computing and Cyber Physical Systems. IC4S 2022. LNICST, , vol. 472. Springer, Cham (2023). https://doi.org/10.1007/978-3-031-28975-0_18
3. Lee, T.D., Ebong, A.U.: A review of thin film solar cell technologies and challenges. Renew. Sustain. Energy Rev. **70**, 1286–1297 (2017)
4. Best Research-Cell Efficiency Chart. https://www.nrel.gov/pv/cell-efficiency.html. Accessed 14 Feb 2023
5. Muhammad, M.H., Hameed, M.F.O., Obayya, S.S.A.: Broadband absorption enhancement in modified grating thin-film solar cell. IEEE Photon. J. **9**(3), 2700314 (2017)
6. Catchpole, K.R., Polman, A.: Design principles for particle plasmon enhanced solar cells. Appl. Phys. Lett. **93**, 191113 (2008)
7. Zhu, J., Hsu, C.-M., Yu, Z., Fan, S., Cui, Y.: Nanodome solar cells with efficient lightmanagement and self-cleaning. Nano Lett. **10**, 1979–1984 (2010)
8. Atwater, H.A., Polman, A.: Plasmonics for improved photovoltaic devices. Nat. Mater. **9**, 205–213 (2010)
9. Wu, J.L., Chen, F.C., Hasio, Y.S., Chien, F.C., Chen, P., Kuo, C.H., et al.: Surface plasmonic effects of metallic nanoparticles on the performance of polymer bulk heterojunction solar cells. ACS Nano **5**(2), 959–967 (2011)
10. Maier, SA.: Plasmonics: Fundamentals and Applications, 2007th edn. Springer-Verlag, India (2007)
11. Green, M.A., Pillai, S.: Harnessing plasmonics for solar cells. Nat. Photon. **6**, 130–132 (2012)
12. Muhlschlegel, P., et al.: Resonant optical antennas. Science **308**, 1607–1608 (2005)
13. Taghian, F., Ahmadi, V., Yousefi, L.: Enhanced thin solar cells using optical nano-antenna induced hybrid plasmonic travelling-wave. IEEE J. Lightwave Technol. **34**, 1267–1273 (2016)
14. Di, V.M., et al.: Plasmonic nano-antenna a-Si: H solar cell. Opt. Express **20**(25), 27327–27336 (2012)
15. Yu, Y., et al.: Dielectric core–shell optical antennas for strong solar absorption enhancement. Nano Lett. **12**(7), 3674–3681 (2012)
16. Pahuja, A., Parihar, M.S., Dinesh Kumar, V.: Investigation of Euler spiral nanoantenna and its application in absorption enhancement of thin film solar cell. Opt. Quant. Electron. **50**(11), 1–11 (2018). https://doi.org/10.1007/s11082-018-1665-z
17. Pahuja, A., Parihar, M.S., Dinesh Kumar, V.: Performance enhancement of thin-film solar cell using Yagi–Uda nanoantenna array embedded inside the anti-reflection coating. Appl. Phys. A **126**(1), 1–7 (2020). https://doi.org/10.1007/s00339-019-3250-0
18. CST Studio Suite. https://www.3ds.com/products-services/simulia/products/cst-studio-suite/. Accessed 5 Jan 2023
19. Raether, H.: Surface Plasmons on Smooth and Rough Surfaces and on Gratings, 1st edn. Springer, Berlin (1986)
20. Johnson, P.B., Christy, R.W.: Optical constant of the noble metals. Phys. Rev. Lett. **15**, 4370–4379 (1972)
21. ASTM, Reference Solar Spectral Irradiance: Air Mass 1.5 Spectra. http://rredc.nrel.gov/solar/spectra/am1.5. Accessed 5 Dec 2023

Efficient Quality Factor Prediction of Artificial Neural Network Based IsOWC System

Subhash Suman[✉] and Jitendra K. Mishra

Department of Electronics and Communication Engineering, Indian Institute of Information Technology Ranchi, Jharkhand 834010, India
subhash03.rs20@iiitranchi.ac.in

Abstract. This paper presents a novel approach utilizing an artificial network (ANN) for optical wireless communication (OWC) between satellites in geosynchronous earth orbit and lower earth orbit, covering a distance of 45000 km. The objective of this ANN based intersatellite optical wireless communication (IsOWC) system is to intelligently predict the quality factor considering different wavelengths. To enhance the transmission performance between these satellite systems, the mean squared error (MSE) is minimized using the Levenberg–Marquardt optimizer. Remarkably, after 25 epochs, the MSE value reaches an impressive 0.000373. The results demonstrate that the ANN-based learning outperforms other machine learning algorithms, exhibiting a significantly lower MSE. Furthermore, this system has a high convergence rate as well as resistant to outliers and overfitting. Even if the number of features is small, it can be predicted accurately. Such systems hold great promise for future wireless designs and integrations, spanning from satellite to terrestrial and underwater OWC systems.

Keywords: Optical Wireless Communication · Quality Factor · NRZ · Satellite Communication · Deep learning

1 Introduction

Over the last two decades, optical wireless communication (OWC) technology has increased dramatically due to its wide bandwidth, high data rate, requires less power, and easy implementation [1]. As a result, it has triggered research interest in 5G/6G and internet of things (IoT) applications that use light amplification by stimulated emission of radiation (LASER) as signal carriers [2]. Due to advancement in photonics based devices, semiconductor laser diode produces a monochromatic and coherent beam of optical light in the form of infrared, visible, and ultraviolet rays, all of which are part of the electromagnetic spectrum [3]. On the contrary to radio frequency (RF) bands, these optical frequency bands are very fancy for satellite communication and many other applications. In addition, these bands are easily implemented in the space link due to the absence of atmospheric turbulence (AT) [4].

Radio frequency (RF) based terrestrial systems garnered global attention in the past, showcasing notable achievements across various atmospheric turbulent channel models

P. Pareek et al. (Eds.): IC4S 2023, LNICST 537, pp. 223–230, 2024.
https://doi.org/10.1007/978-3-031-48891-7_19

[5]. Recently, the performance of these systems has been investigated by using advanced modulation and coding techniques, as well as diversity techniques in space, time, wavelength, and other approaches [6]. Nevertheless, the design of RF based communication system posed numerous challenges for these systems. As a result, there has been a shift towards adopting OWC-based space link systems to meet the demanding requirements of future wireless communication systems.

On the other hand, inter-satellite optical wireless communication (IsOWC) system is advocating as a promising application of OWC technology. In this system, two satellites revolve around the earth's axis within three different earth orbit paths, namely geostationary earth orbit (GEO), medium earth orbit (MEO), and lower earth orbit (LEO). Recently, the performance of IsOWC systems has been investigated using advanced techniques like wavelength division multiplexing (WDM) and multiple input multiple outputs (MIMO) [7]. But due to external chaos, the quality factor of the system has been degraded up to certain values. However, with continuous technological advancements, the performance of the IsOWC system has been further enhanced through the utilization of machine learning (ML) algorithms. Presently, more advanced ML algorithms are being integrated into IsOWC systems [8].

In this paper ANN based IsOWC system is investigated to intelligently predict the quality factor between two satellites. The input variables consist of the laser diode wavelengths, while the output variable is the quality (Q) factor. The optimization process aims to minimize the mean squared error (MSE) performance metric by gradually reaching a global minimum point, utilizing a very small learning rate.

2   System Model

The transmission setup of the proposed IsOWC system is shown in Fig. 1. The pseudo-random bit sequence generator generates a bit sequence of length 127 bits with a data rate of 2 Gbps, which is further encoded into pulse form through a non-return to zero pulses generator. The electrical pulse signal is modulated with a laser diode using a Mach-Zehnder modulator with 30 dB extinction ratio. The filtered electrical signal is analyzed through a bit error rate analyzer. The received power P_r for the IsOWC system can be expressed as [9]:

$$P_r = P_t \eta_t \eta_r \left(\frac{\lambda}{4} \pi R \right)^2 G_t G_r L_t L_r \tag{1}$$

where P_t, η_t, η_r, $(\lambda/4\pi R)^2$, λ, R, G_t, G_r, L_t, and L_r represent transmitted power, optical transmitter efficiency, optical receiver efficiency, free space path loss, operating wavelength, propagation distance, transmitter aperture gain, receiver aperture gain, transmitter pointing loss factor and receiver pointing loss factor, respectively. Moreover, scintillation is not considered in the present calculation due to the absence of atmospheric turbulence. The relationship between the diameter and aperture gain of an antenna is given as [10]

$$G = \left(\frac{\pi D}{\lambda} \right)^2 \tag{2}$$

putting Eq. (2) in (1), it can be expressed as:

$$P_r = P_t \eta_t \eta_r L_t L_r \left(\frac{D_t D_r}{2\pi \lambda R} \right)^2 \tag{3}$$

Equation (3) reveals that as the operating wavelength is decreased, the overall received optical power increases gradually, and therefore, the optical signal-to-noise (OSNR) ratio increases, which enhances the quality factor of the IsOWC system.

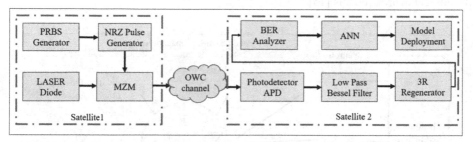

Fig. 1. Transmission set up of ANN-based IsOWC system

3 Machine Learning and Its Applications

3.1 Machine Learning

Machine learning (ML) is a straightforward algorithm that enables computers to learn autonomously, eliminating the need for explicit programming [11]. ML encompasses three main types of learning processes: i) supervised learning, ii) unsupervised learning, and iii) reinforcement learning. Supervised learning relies on labeled data, whereas unsupervised learning utilizes unlabeled data. On the other hand, reinforcement learning involves agents performing specific tasks in an environment and making decisions based on the outcomes. Furthermore, ML encompasses various statistical algorithms, including decision tree (DT), naive Bayes algorithm, gradient boosting algorithm, adaboosting algorithm, k-nearest neighbor (KNN), random forest (RF), and support vector machine (SVM), which are employed for prediction or classification purposes [12]. The entire ML process entails crucial steps such as feature extraction, data pre-processing, model training, validation, and model deployment, all of which are essential for developing a robust ML model [13].

Moreover, ML models has primarily relied on the feature selection method and data pre-processing techniques. Earlier, a variety of ML approaches were used to improve the IsOWC system's performance. In order to improve the performance of an IsOWC system for better optical wireless connectivity, this system has been shifted to advanced ML techniques. This is because the model is robust against outliers, and requires less time and memory space to complete its training process.

3.2 Deep Learning

DL is a subset of machine learning that build an intelligent machine capable of performing tasks that typically require human intelligence. It mainly focuses on regression and classification problems, image segmentation, and object detection [14].

The basic architecture of ANN contains single or multilayer perceptron (MLP) along with feed-forward and backpropagation paths. The single-layer perceptron of the ANN-based IsOWC system is shown in Fig. 2. MLPs are most commonly used for linear and non-linear data, which is motivated by the self-learning procedure, which is the same as biological neuron models [15].

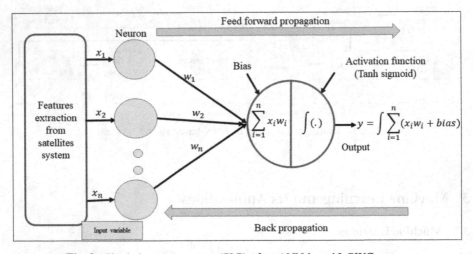

Fig. 2. Single layer perceptron (SLP) of an ANN-based IsOWC system

ANN model is widely used in regression and classification problems. While optimizing the error, the most important thing is to update weight and bias parameters. It converges to a global minimum point keeping the learning rate very low. Besides the use of ANN in the IsOWC system, the transmission performance of the IsOWC system will be further improved by using advanced algorithms like convolutional neural network (CNN), recurrent neural network (RNN), long-term short memory (LSTM), generative adversarial networks (GAN), transfer learning, vision transformers, and auto encoders.

4 Simulation Setup Description

4.1 Dataset Building

The supervised ANN based IsOWC systems are considered for the present investigation. The wavelength of laser diode, along with the Q-factor, is used for training the ANN model. Each variable contains 130 samples, where the wavelength ranges from 800 to 1600 nm and the Q- factor ranges from 5.01761 to zero. Out of 130 samples, 70% of samples are randomly selected for the training dataset, the remaining 15% for the testing dataset and the remaining 15% for validation dataset.

4.2 Performance Metric

Mean squared error (MSE) is commonly used metric for evaluating of regression of model. It quantifies the average squared difference between the predicted values and the actual values in a dataset. A lower MSE value indicates better model performance it means the predicted values are closer to the actual values on average. Therefore, the value of MSE directly reflects the performance of the system. For the proposed model, various parameter is taken for the simulation of an ANN based IsOWC system, as shown in Table 1.

Table 1. Parameters used for simulation of an ANN based IsOWC system

Parameter	Value
Number of ANN layer	03
Number of hidden neurons	10
Number of epochs	31
Optimizer	Levenberg-Marquardt (LM)
Performance metrics	Mean squared error (MSE)

5 Results and Discussions

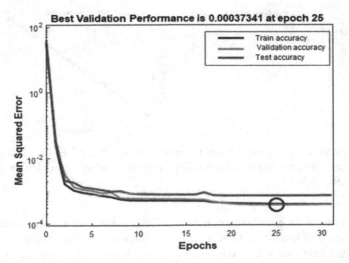

Fig. 3. Mean squared error vs. epochs of an ANN-based IsOWC system

The transmission performance of an IsOWC system is investigated using the ANN technique. Figure 3 shows the variations of MSE with epochs of training, validation, and testing accuracy of the given ANN based IsOWC system. It can be seen from Fig. 3 that the MSE value of different curves, i.e., the blue line (training accuracy), the green line (validation accuracy), and the red line (test accuracy), exponentially decreases as the number of epochs increases. After several iterations, the MSE value is obtained as 0.000373 with respect to the validation dataset, which clarifies that the model has predicted very well on the test or unseen datasets. After the four iterations of a proposed ANN model, the MSE value of the validation accuracy is slowly increased, and after that, it remains constant.

Afterwards, in order to investigate the coefficient of determination (R) analysis also known as R-squared among three training, validation, and testing parameter, the output predicted versus target graph is plotted as shown in Fig. 4. It can be observed from Fig. 4 that the overall (R) is obtained as 99.98%, which signifies that most of the datasets follow the best-fit line. Therefore, the model has shown that it strongly correlates with the target variable and predicted output.

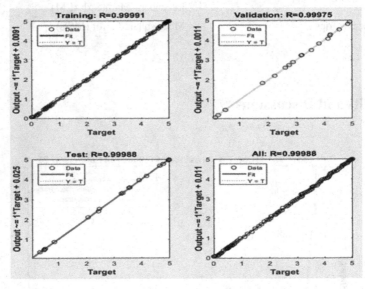

Fig. 4. Output vs. target for the different parameters of an ANN model (a) training parameter (upper left), (b) validation parameter (upper right), (c) test parameter (lower left), and (d) overall value of R (lower right)

Figure 5 shows the error histogram with 20 bins of different training, validation, and testing accuracy of the given ANN model. The error can be represented in histogram form with different rectangular sizes having a blue box (training accuracy), green box (validation accuracy), and red box (test accuracy). A zero-error line is set in the middle of the graph and signifies no error among these three parameters. In Fig. 5, one bin has a high number of instances showing that training and validation accuracy are very close to the zero-error line, representing that the mean squared error value is optimized correctly to a significant level. Furthermore, the number of instances decreases on either side of the zero-error line.

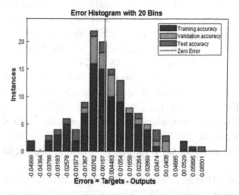

Fig. 5. Instances vs. errors of an ANN based IsOWC system

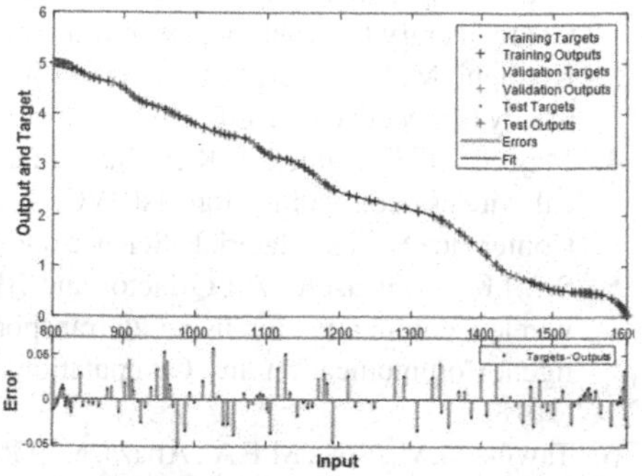

Fig. 6. Output and target vs. wavelength of training, validation, and test parameter (upper part) and error vs. input (lower part)

Figure 6 shows the fit function of the proposed ANN based IsOWC system. The training target and outputs (represented as a blue dot and plus sign), validation target and outputs (represented as a green dot and plus sign), and test target and outputs (represented as a light red dot and plus sign) follow the wavelength vs. quality factor graph. However, the error vs. input graph is plotted at the lower part of the function fit graph. It shows that at which wavelength the quality factor is maximum or minimum. Also, this graph shows three ranges of wavelength where the fit function is invalid, i) 900–1000 nm, ii) 1300–1350 nm, and iii) above 1600 nm. At these wavelengths, error is more which signifies that system has faced more disturbance due to the vibration of satellites or pointing error. It can be seen that the proposed ANN based IsOWC system performs better than the previous reported results [16].

6 Conclusion

The ANN based IsOWC system is used to intelligently estimate the quality factor without the requirement of channel state information. Depending on the wavelength selected, ANN based regression model assists in determining signal quality in the presence of atmospheric attenuation, transmission range, beam divergence, interference and crosstalk, and receiver sensitivity. Based on these findings, the suggested ANN approaches have a high potential for automating satellite communication systems. The LM optimizer occupies less memory and converges quicker than other optimizer.

Finally, these ANN based IsOWC will be integrated with underwater networks as well as terrestrial networks to satisfy the high expectations of 5G beyond communication and to develop a truly intelligent network.

References

1. Patnaik, B., Sahu, P.K.: Inter-satellite optical wireless communication system design and simulation. IET Commun. **6**(16), 2561–2567 (2012)
2. Kaur, N., Soni, G.: Performance analysis of inter-satellite optical wireless communication (IsOWC) system at 980 nm and 1550 nm wavelengths. In: 2014 International Conference on Contemporary Computing and Informatics (IC3I), pp. 1245−1250. IEEE, Mysore (2014)
3. Khalighi, M.A., Uysal, M.: Survey on free space optical communication: a communication theory perspective. IEEE Commun. Surv. Tutorials **16**(4), 2231–2258 (2014)
4. Majeed, M.H., Ahmed, R.K., Alhumaima, R.S.: Performance analysis of inter-satellite optical wireless communication (IsOWC) system with multiple transmitters/receivers. In: IOP Conference Series: Materials Science and Engineering, vol. 1076, p. 012052 (2021)
5. Sri, I.K., Srinivasulu, A.: Q-factor and BER performance analysis of inter-satellite optical wireless communication using 26 transponders. In: 2017 International Conference on Intelligent Communication and Computational Techniques (ICCT), pp. 165–168. IEEE, Jaipur (2018)
6. Tawfik, M.M., Sree, M.F.A., Abaza, M., Ghouz, H.H.M.: Performance analysis and evaluation of inter-satellite optical wireless communication system (IsOWC) from GEO to LEO at range 45000 km. IEEE Photonics J. **13**(4), 1–6 (2021)
7. Alom, M.Z., et al.: A state-of-the-art survey on deep learning theory and architectures. Electronics **8**(3), 292 (2019)
8. Tian, Q., et al.: Turbo-coded 16-ary OAM shift keying FSO communication system combining the CNN-based adaptive demodulator. Opt. Express **26**(21), 27849 (2018)
9. Sahu, P.P., Singh, M.: Multi channel frequency hopping spread spectrum signaling using code M-ary frequency shift keying. Comput. Electr. Eng. **34**(4), 338–345 (2008)
10. Kaur, S.: Analysis of inter-satellite free-space optical link performance considering different system parameters. Opto-Electron. Rev. **27**(1), 10–13 (2019)
11. Zhu, G., Liu, D., Du, Y., You, C., Zhang, J., Huang, K.: Toward an intelligent edge: wireless communication meets machine learning. IEEE Commun. Mag. **58**(1), 19–25 (2020)
12. Ghanem, Z., Alsaraira, A., Al-Tarawneh, L., Sarareh, O.A.: Comparative analysis of ml-schemes in OWC systems. J. Electr. Eng. Technol. (IJEET) **12**(8), 115–132 (2021)
13. Dahrouj, H., et al.: An overview of machine learning-based techniques for solving optimization problems in communications and signal processing. IEEE Access **9**, 74908–74938 (2021)
14. Minaee, S., Boykov, Y., Porikli, F., Plaza, A., Kehtarnavaz, N., Terzopoulos, D.: Image segmentation using deep learning: a survey. IEEE Trans. Pattern Anal. Mach. Intell. **44**(7), 3523–3542 (2022)
15. Tan, M.C., Khan, F.N., Al-Arashi, W.H., Zhou, Y., Lau, A.P.T.: Simultaneous optical performance monitoring and modulation format/bit-rate identification using principal component analysis. J. Opt. Commun. Netw. **6**(5), 441–448 (2014)
16. Algedir, A.A., Elganimi, T.Y.: Machine learning models for predicting the quality factor of FSO systems with multiple transceivers. In: 2020 2nd Global Power, Energy and Communication Conference (GPECOM), pp. 308–311. Izmir, Turkey (2020)

Tunable UWB Metasurface Absorber for Smart City Compatible IoT Applications

Naveen Kumar Maurya[1], Sadhana Kumari[2], Prakash Pareek[1(✉)],
Jayanta Ghosh[3], and Manuel J. Cabral S. Reis[4]

[1] Department of Electronics and Communication Engineering,
Vishnu Institute of Technology, Bhimavaram 534202, Andhra Pradesh, India
prakash.p@vishnu.edu.in
[2] Department of Electronics and Communication Engineering,
BMS College of Engineering, Bengaluru 560019, Karnataka, India
[3] Department of Electronics and Communication Engineering,
National Institute of Technology, Patna 800005, Bihar, India
[4] Engineering Department, UTAD/IEETA, 5001-801 Vila Real, Portugal

Abstract. This work presents an ultra-wideband tunable graphene-based metasurface absorber for the terahertz (THz) gap region of the electromagnetic (EM) spectrum. The proposed absorber provides an absorption bandwidth (BW) of 7.8 THz (fractional BW = 195%) with absorptivity $A(f) \geq 90\%$, i.e., from 0.1 to 7.9 THz. The impedance matching between free space and the absorber's surface has been achieved by engraving different shapes of slots on the top graphene layer. The working principle behind the UWB absorption mechanism has also been studied with the help of parametric studies and field plots. The thickness of the metasurface is only 2 μm, i.e., $\lambda_g/958.3$, where λ_g has been computed at 0.1 THz, thus, maintaining the ultra-thin nature required for the metasurface design in the THz regime. The absorber's periodicity is also quite less, i.e., 6 μm ($\lambda_g/319.43$), which is sufficient to achieve an effective homogeneity condition. The four-fold symmetry in the design makes the structure polarization insensitive to the incoming plane wave. The metasurface also works well for a wide incidence angle (θ) under both transverse electric (TE) and transverse magnetic (TM) polarizations. The $A(f) \geq 80\%$ has been achieved for θ up to 45°. In addition, the absorber provides full-width at half-maxima (FWHM) BW in the complete frequency range, i.e., from 0.1 to 7.9 THz. Hence, the proposed metasurface absorber is found suitable for suppressing/absorbing unwanted electromagnetic radiation in a close indoor environment for smart city-enabled Internet of Things (IoT) applications.

Keywords: Metasurface · Graphene · Ultra-wideband · Terahertz

Published by Springer Nature Switzerland AG 2024. All Rights Reserved
P. Pareek et al. (Eds.): IC4S 2023, LNICST 537, pp. 231–240, 2024.
https://doi.org/10.1007/978-3-031-48891-7_20

1 Introduction

A recent worldwide survey by United Nations revealed that a major chunk of the world's population will migrate to cities in the coming years [1]. Due to the exponential increase in the population of cities triggered by massive urbanization, city administrators are looking for viable solutions that effectively monitor the citizen to enhance their living standards. In this context, the smart city emerged as the potential key to quenching this quest. The smart city ecosystem allows the city administration to effectively and precisely implement various aspects of the city, like security, transportation, cleaning and sanitation, etc., through proper monitoring and sensing [2]. Indeed, the functioning of smart city applications relies on gathering data from various devices/sensors wirelessly, analyzing these data, implementing intelligent control measures, and securely disseminating information [3].

Hence, to tackle smart cities' complex and diverse challenges, the internet is crucial in facilitating communication, sharing and processing information, transferring and analyzing data, and enabling distributed computing. The emergence of the Internet of Things (IoT) and the widespread integration of web technologies for cloud computing in urban settings have demonstrated the effectiveness of Internet-based solutions in effectively addressing societal issues [4]. In order to realize this vision, the next generation of IoT devices must be created with the capability to operate autonomously through wireless medium [5] and support numerous wireless standards [6,7]. Additionally, precise sensing plays a critical role in enabling autonomy in smart cities and researchers worldwide have attempted various devices and technologies to improve the sensitivities of IoT sensors [8].

Metamaterials/metasurfaces are unique materials that possess properties that are not naturally occurring and are composed of resonators arranged in a uniform pattern on a dielectric substrate [9]. Interestingly, metamaterials/metasurfaces exhibit diverse properties, including the ability to selectively absorb specific frequencies [10–13] or detect even the slightest variations in specific parameters [12,14–16]. Thus, metasurfaces hold significant potential for utilization in various smart city compatible IoT applications due to their distinctive characteristics such as absorbers, modulators, sensors, switches, etc. [9,12,14,15,17]. At the same time, these artificially engineered materials have the capacity to absorb undesired frequencies, which is crucial to avoid cross-talk between wireless sensor nodes and provide end users with the best performance. These intriguing characteristics make them valuable for supporting IoT technology in challenging conditions, primarily through their precise sensing and absorption applications [18]. Since the population is increasing at an astonishing rate [1], the demand for higher data rates will also be going to increase exponentially. This motivates the researchers to shift towards the terahertz (THz) range of electromagnetic (EM) spectrum due to the availability of ultra-wideband (UWB) frequency spectrum and non-ionizing nature of THz radiation [19]. Hence, the future smart city compatible IoT applications will incorporate THz technology for UWB communication [20,21].

This work focuses on the design and analysis of frequency tunable metasurface for absorption over an UWB band in the THz regime. The tunable frequency response enables the proposed absorber to be used as a sensor, modulator, and switch apart from its usage as an absorber. The rest of the paper has been organized as follows: The design procedure has been presented in Sect. 2. Section 3 focus on the results and discussions. Finally, Sect. 4 concludes this work.

2 Metasurface Design

In the THz regime, graphene-based designs have gotten tremendous attention due to their ability to generate plasmonic resonances by exciting surface plasmon polaritons (SPPs) [9,22–24]. Interestingly, these resonances could be made frequency reconfigurable by varying complex surface conductivity of the graphene (σ_{graphene}) defined by Eq. 1, which is the summation of intra-band (σ_{intra}) and inter-band (σ_{inter}) conductivities given by Eq. 2 and 3, respectively [25]. In the THz gap region of the EM spectrum (0.1–10 THz), the contribution of σ_{inter} could be ignored [9,22]. Therefore, σ_{graphene} could be independently represented by σ_{intra}. As a result, by varying chemical potential (μ_c) in Eq. 2, σ_{graphene} could be varied. Thereby imparting plasmonic resonances reconfigurable frequency characteristics in the graphene-based devices.

$$\sigma_{\text{graphene}} = \sigma_{\text{intra}} + \sigma_{\text{inter}} \tag{1}$$

$$\sigma_{\text{intra}} = -j\frac{e^2 k_B T}{\pi \hbar^2 (\omega - j\Gamma)}\left[\frac{\mu_c}{k_B T} + 2\ln(e^{-\mu_c/k_B T} + 1)\right] \tag{2}$$

$$\sigma_{\text{inter}} = \frac{-je^2}{4\pi\hbar}\ln\left(\frac{2|\mu_c| - (\omega - j\Gamma)\hbar}{2|\mu_c| + (\omega - j\Gamma)\hbar}\right) \tag{3}$$

The schematic diagram of the proposed graphene-based tunable UWB metasurface absorber is shown in Fig. 1a and the corresponding reflection and absorption curves have been depicted in Fig. 1b. The structure of the unit cell comprises of SiO_2 ($\epsilon_r = 3.9$ and tan $\delta = 0.0006$) substrate backed by continuous monolayer graphene (MLG) sheet having a thickness of 0.335 nm. The top surface of the unit cell has also been designed using a 0.335 nm thick MLG sheet. Since the SiO_2 substrate is backed by the MLG sheet having thickness greater than the skin depth, the transmission coefficient ($|S_{21}|$) becomes zero and the absorptivity ($A(f)$) defined by Eq. 4 depends only on the reflection coefficient ($|S_{11}|$) represented by Eq. 5. Hence, in this work, the UWB absorption response has been obtained by achieving impedance matching between the absorber's top surface and free space.

Interestingly, while designing the metasurface unit cell in this work, the reflections at the interface between the absorber's surface and free space have been reduced by incorporating slots of different shapes (viz. square loop and circular) and sizes on the top of the MLG sheet, as shown in Fig. 1a. It is worthy to mention here that the square-shaped slots with two different sizes have been used here

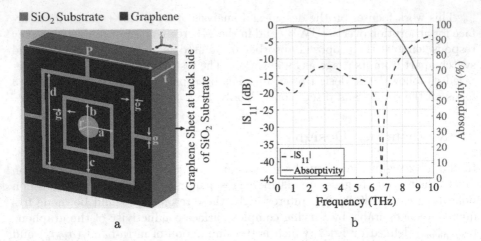

Fig. 1. (a) Perspective view of the proposed UWB absorber, $P = 6$, $t = 2$, $a = 1.2$, $b = 1$, $c = 2$, $d = 4$, $g = 0.2$, $r = 0.5$ (unit:μm) and (b) $|S_{11}|$ and absorptivity.

to excite two distinct plasmonic resonances. The larger square slot resonates at 1.06 THz, whereas the smaller square loop slot resonates at 3.14 THz. In addition, the central circular slot resonates at 6.63 THz. Subsequently, the three distinct plasmonic resonances generated by square and circular slots have been combined together to provide a UWB absorptivity response. The proposed UWB absorber covers the THz frequency spectrum from 0.1 to 7.9 THz with $A(f) \geq 90\%$, which corresponds to 195% fractional bandwidth (FBW) as shown in Fig. 1b. The periodicity and thickness of the metasurface are $\lambda_g/319.43$ and $\lambda_g/958.3$, respectively, where λ_g is the guided wavelength computed at 0.1 THz, thus, satisfying both the effective homogeneity condition and the ultra-thin nature required for designing a metasurface absorber in the THz regime.

$$A(f) = 1 - |S_{11}|^2 - |S_{21}|^2 \tag{4}$$

$$A(f) = 1 - |S_{11}|^2 \tag{5}$$

3   Results and Discussions

To have a better understanding regarding the excitation of localized SPP (LSPP) wave, generation of resonant modes and loss mechanism taking place inside the absorber structure, magnitude E-field and magnitude H-field have been numerically computed using CST Microwave Studio in the xy-plane as shown in Figs. 2 and 3, respectively. The |E|-field distribution depicted in Fig. 2a clearly shows that the outer square slot ring is mostly excited at 1.06 THz while the inner square slot ring and central circular slot remain unenergized. This confirms the electric excitation of the metasurface absorber by the incoming plane wave, which

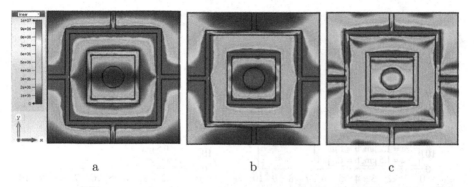

Fig. 2. |E|-field density at: (a) 1.06 THz, (b) 3.14 THz, and (c) 6.63 THz in xy-plane.

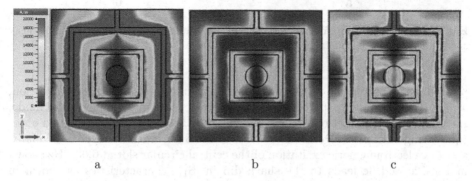

Fig. 3. |H|-field density at: (a) 1.06 THz, (b) 3.14 THz, and (c) 6.63 THz in xy-plane.

leads to the generation of the LSPP wave at the interface of the graphene and dielectric substrate near the vicinity of the outer square ring. The LSPP wave is non-propagative in nature and mostly oscillates at its position. This causes energy localization and thereby leads to losses in the form of conductor loss induced by graphene and dielectric loss induced by the SiO_2 substrate. It is worthy to mention here that the perimeter of the outer square ring is more compared to the inner square ring and circular slot. As a result, the outer square ring generates plasmonic resonance at the lowest operating frequency, i.e., at 1.06 THz. The |H|-field density has also been computed at 1.06 THz to analyze the response of the proposed metasurface with respect to the H-field vector of the incident plane wave as shown in Fig. 3a. From Fig. 3a, it is evident that the induced H-field is mostly concentrated inside the outer square loop and it is orthogonal to the |E|-field distribution shown in Fig. 2a. Hence, perfect absorption takes place at 1.06 THz with a sharp dip in $|S_{11}|$ characteristics (see Fig. 1b) through proper electromagnetic excitation of the metasurface absorber.

Similarly, the |E|-field and |H|-field densities have been computed at 3.14 THz as shown in Figs. 2b and 3b, respectively. The field plots reveal that the inner square ring is excited in its fundamental mode, i.e., one wavelength loop mode

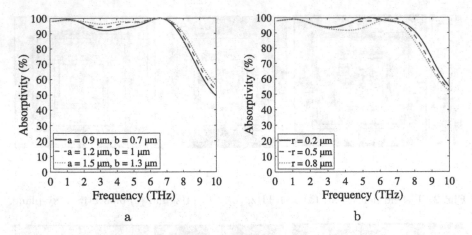

Fig. 4. Parametric study for the variation in: (a) parameters a and b, simultaneously such that $(a\text{-}b) = g$ is constant at 0.2 µm and (b) r.

[22], which is responsible for the absorption mechanism at 3.14 THz. In addition, at 3.14 THz, the outer square ring has also been excited in its higher-order mode. As a result, two distinct resonances generated by the inner square ring and outer square ring combined together to provide a UWB absorptivity response. Finally, a strong electromagnetic excitation of the central circular slot at 6.63 THz shown in Figs. 2c and 3c leads to the sharp dip in $|S_{11}|$ characteristics as shown in Fig. 1b. The strong confinement of the electric and magnetic fields near the central circular slot clearly indicates the generation of the LSPP wave, which is responsible for the absorption of the EM wave towards the upper side of the UWB absorption band. Here also, the outer and inner square rings are excited in their higher-order loop modes, which justifies the recombination of different modes to provide a UWB absorption response.

The observations made above by analyzing the $|E|$-field and $|H|$-field densities have been further verified with the help of the parametric study of the design variables as shown in Fig. 4. Due to brevity, the effect of variation in the perimeter of the outer square ring has been ignored. This is because, at 1.06 THz, only the outer square loop is energized and the other two slots remain unexcited, as shown in Figs. 2a and 3a, which clearly show that at 1.06 THz, absorption is taking place due to the excitation of the outer square loop only. In the first parametric study, parameters a and b have been varied simultaneously such that a-b, i.e., g (see Fig. 1a) is always constant at 0.2 µm while keeping all other parameters unchanged as shown in Fig. 4a. With an increase in parameters a and b, the perimeter of the inner square ring increases. As a result, the level of absorption near the center of the absorption band (i.e., around 3.14 THz) gets modulated without any change in the absorption peaks at 1.06 THz and 6.63 THz as shown in Fig. 4a. The same conclusion has been drawn using the $|E|$-field and $|H|$-field density shown in Figs. 2b and 3b, respectively, where we have claimed that the inner square ring is mainly responsible for absorption near 3.14 THz. In the next study, the radius of the central circular slot (r) has been varied from 0.2 to 0.8 μm, as shown in Fig. 4b. With an increase in r, it is evident that the absorption curve is getting changed mostly near the 6.63 THz, which again justifies our claim that r is responsible for absorption near the higher side of the absorption band.

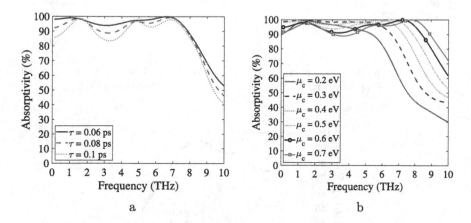

Fig. 5. Parametric study for the variaton in: (a) τ and (b) μ_c of the graphene.

The effect of graphene's electrical parameters viz. τ and μ_c on the $A(f)$ has also been analyzed using a parametric study as shown in Fig. 5. For the higher value of τ, the absorption coefficient changes from UWB to the multiband nature. Hence, in this work a lower value of τ, i.e., 0.06 ps, has been considered to obtain an ultra-wideband response over a multiband response. To test the tunable frequency characteristics of the proposed graphene-based metasurface absorber, the complex surface conductivity of graphene has been varied by changing μ_c as shown in Fig. 5b. From Fig. 5b, it is evident that with an increase in the value of μ_c, $A(f)$ response is clearly showing frequency tunability with a blue shift in absorption curves. Moreover, absorption BW also increases for higher values of μ_c as shown in Fig. 5b. For $\mu_c = 0.5$ eV, the absorber provides BW of 7.8 THz with $A(f) \geq 94\%$. With a further increment in μ_c beyond 0.5 eV, the absorption BW is increasing but at the cost of a decrease in the level of absorptivity. Hence, $\mu_c = 0.5$ eV has been selected in this work.

The orientation of the E-field vector and the angle of incidence of incoming plane waves can severely affect the absorption coefficient of a metasurface absorber. To examine this, the effect of the polarization angle (ϕ) and incidence angle (θ) has been studied, as in Fig. 6. Figure 6a shows that with the change in ϕ from 0° to 45°, the absorption coefficient remains unchanged, which confirms the polarization insensitivity of the proposed absorber. For the perfect absorption of EM waves in indoor wireless environments for massive IoT applications, polarization insensitivity is one of the most crucial factors. In an indoor wireless environment, multiple reflections and scattering of EM waves

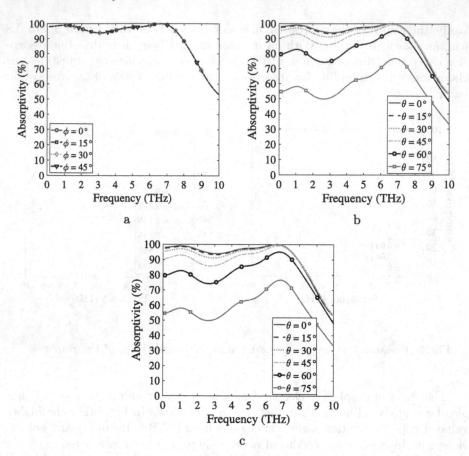

Fig. 6. (a) $A(f)$ for different polarization angles (ϕ). $A(f)$ for different incidence angles (θ) under: (b) TM polarization ($\phi = 0°$) and (c) TE polarization ($\phi = 90°$).

occur, often changing the polarization angle of the incident plane wave. Therefore, making absorber polarization insensitive can helps to absorb EM waves having different polarization angles as well. Similarly, the effect of θ under both transverse magnetic (TM) and transverse electric (TE) has also been analyzed, as shown in Fig. 6b and c. It is found that the proposed absorber provides $A(f) \geq 80\%$ up to $\theta = 45°$. At the same time, the proposed absorber also covers the BW from 0.1 to 7.9 THz with $A(f) \geq 50\%$, thus, providing very wide full-width half-maxima (FWHM) bandwidth, which is very much required for massive IoT applications in multipath-rich wireless environment taking place in closed indoor environment.

The efficacy of the proposed work has been brought to the attention by comparing it with current state-of-the-art designs present in the literature as shown in Table 1. The proposed work provides the highest absorption BW of 195%, which is far better than the others, as shown in Table 1. At the same time, the thickness and periodicity of the proposed design are also quite less, as shown in Table 1. Hence, the proposed metasurface could be one of the good solutions for absorber application over ultra-wideband in the THz regime.

Table 1. Comparison Study

Reference	Freq Range (THz) For A(f) ≥ 90%	Fractional BW (%)	Thickness	Periodicity	Design Configuration
[10]	2.06–11.8	140.86	$\lambda_0/55.9$	$\lambda_0/21.7$	Graphene-Dielectric-Graphene
[11]	1.07–2.88	91.64	$\lambda_0/11$	$\lambda_0/14$	Graphene-Dielectric-Gold
[12]	4–8	67	$\lambda_0/8.39$	$\lambda_0/9.38$	Hybrid-Dielectric-Gold
[13]	2.2–4.6	70.59	$\lambda_0 11.36$	$\lambda_0/13.64$	Graphene-Dielectric-Gold
This Work	0.1–7.9	195	$\lambda_0/1500$	$\lambda_0/500$	Graphene-Dielectric-Graphene

4 Conclusions

This work presents a frequency-reconfigurable UWB metasurface absorber for the THz regime. The tunability in the design has been achieved by using graphene as a design material. Interestingly, the tunable absorption characteristic also allows the proposed solution to be used for sensing and switching applications. The structural symmetry in the design makes the metasurface independent from the polarization of the incident plane wave. In addition, the proposed absorber also provides a good absorption response under a wide incident angle over the UWB band. Hence, the designed tunable metasurface absorber could be used in smart city scenarios to suppress unwanted interference over the UWB band that takes place in indoor wireless environments, particularly for IoT applications.

References

1. Da Silva, I.N., Flauzino, R.A.: Smart Cities Technologies, 1st edn. Intechopen, London (2016)
2. Pareek, P., Maurya, N.K., Singh, L., Gupta, N., Reis, M.J.C.S.: Study of smart city compatible monolithic quantum well photodetector. In: Gupta, N., Pareek, P., Reis, M. (eds.) IC4S 2022. LNCS, SITE, vol. 472, pp. 215–224. Springer, Cham (2023). https://doi.org/10.1007/978-3-031-28975-0_18
3. Iqbal, A., Olariu, S.: A survey of enabling technologies for smart communities. Smart Cities 4(1), 54–77 (2020)
4. Bellini, P., Pantaleo, G.: Special issue on the internet of things (IoT) in smart cities (2023)
5. Rafiq, I., Mahmood, A., Razzaq, S., Jafri, S.H.M., Aziz, I.: IoT applications and challenges in smart cities and services. J. Eng. 2023(4), e12262 (2023)
6. Maurya, N.K., Bhattacharya, R.: Design of compact dual-polarized multiband MIMO antenna using near-field for IoT. AEU-Int. J. Electron. Commun. 117, 153091 (2020)

7. Maurya, N.K., Bhattacharya, R.: CPW-fed dual-band compact Yagi-type pattern diversity antenna for LTE and WiFi. Prog. Electromagn. Res. C **107**, 183–201 (2021)

8. Sisinni, E., Saifullah, A., Han, S., Jennehag, U., Gidlund, M.: Industrial internet of things: challenges, opportunities, and directions. IEEE Trans. Industr. Inf. **14**(11), 4724–4734 (2018)

9. Maurya, N.K., Ghosh, J., Sumithra, P.: Design of graphene-based tunable ultrathin UWB metasurface for terahertz regime. Optik, 170753 (2023)

10. Ghosh, S.K., Das, S., Bhattacharyya, S.: Graphene based metasurface with near unity broadband absorption in the terahertz gap. Int. J. RF Microwave Comput. Aided Eng. **30**(12), e22436 (2020)

11. Yadav, V.S., Kaushik, B.K., Patnaik, A.: Broadband THz absorber for large inclination angle TE and TM waves. IEEE Photonics J. **13**(5), 1–7 (2021)

12. Lv, Y., Liu, W., Tian, J., Yang, R.: Broadband terahertz metamaterial absorber and modulator based on hybrid graphene-gold pattern. Physica E **140**, 115142 (2022)

13. Zakir, S., et al.: Polarization-insensitive, broadband, and tunable terahertz absorber using slotted-square graphene meta-rings. IEEE Photonics J. **15**(1), 1–8 (2022)

14. Shen, H., et al.: Multi-band plasmonic absorber based on hybrid metal-graphene metasurface for refractive index sensing application. Results Phys. **23**, 104020 (2021)

15. Khan, M.S., Giri, P., Varshney, G.: Generating multiple resonances in ultrathin silicon for highly selective THz biosensing. Physica Scripta **97**(8), 085009 (2022)

16. Shalini, V.B.: A polarization insensitive miniaturized pentaband metamaterial THz absorber for material sensing applications. Opt. Quant. Electron. **53**, 1–14 (2021)

17. Shabani, M., Karimi, G.: Compact single-band and multiband terahertz plasmonic absorbers using hybrid graphene-metal resonators with switching and modulation capability. Optik, 171010 (2023)

18. Amiri, M., Tofigh, F., Shariati, N., Lipman, J., Abolhasan, M.: Review on metamaterial perfect absorbers and their applications to IoT. IEEE Internet Things J. **8**(6), 4105–4131 (2020)

19. Tonouchi, M.: Cutting-edge terahertz technology. Nat. Photonics **1**(2), 97–105 (2007)

20. Farooq, M.S., Nadir, R.M., Rustam, F., Hur, S., Park, Y., Ashraf, I.: Nested bee hive: a conceptual multilayer architecture for 6G in futuristic sustainable smart cities. Sensors **22**(16), 5950 (2022)

21. Niu, M.: Intelligent Electronics and Circuits: Terahertz, Its, and Beyond, 1st edn. Intechopen, London (2022)

22. Maurya, N.K., Kumari, S., Pareek, P., Singh, L.: Graphene-based frequency agile isolation enhancement mechanism for MIMO antenna in terahertz regime. Nano Commun. Netw. 100436 (2023)

23. Ram, G.C., Sambaiah, P., Yuvaraj, S., Kartikeyan, M.: Graphene based tunable bandpass filter for terahertz spectroscopy of polymers. Optik **268**, 169792 (2022)

24. Ram, G.C., Sambaiah, P., Yuvaraj, S., Kartikeyan, M.: Tunable bandstop filter using graphene in terahertz frequency band. AEU-Int. J. Electron. Commun. **144**, 154047 (2022)

25. Neto, A.C., Guinea, F., Peres, N.M., Novoselov, K.S., Geim, A.K.: The electronic properties of graphene. Rev. Mod. Phys. **81**(1), 109 (2009)

Optimal Time Splitting in Wireless Energy Harvesting-Enabled Sensor Networks

Dipen Bepari[1], Soumen Mondal[2], Prakash Pareek[3], and Nishu Gupta[4(✉)]

[1] Department of Electronics and Communication Engineering,
National Institute of Technology, Raipur 492010, Chhattisgarh, India
dbepari.etc@nitrr.ac.in
[2] Department of Electronics and Communication Engineering,
National Institute of Technology, Durgapur, WB, India
[3] Department of Electronics and Communication Engineering,
Vishnu Institute of Technology, Bhimavaram 534202, Andhra Pradesh, India
prakash.p@vishnu.edu.in
[4] Department of Electronic Systems, Faculty of Information Technology and
Electrical Engineering,
Norwegian University of Science and Technology, 2815 Gjøvik, Norway
nishu.gupta@ntnu.no

Abstract. This research paper presents a study on wireless energy harvesting (WEH) protocols and their impact on the performance of sensor networks. A time switching (TS)-based WEH protocol is proposed, which allows sensor nodes to switch between energy harvesting and data transmission modes. The primary objective of this research is to maximize the uplink (UL) sum throughput while considering the constraint of a minimum downlink (DL) throughput. To achieve this, an optimization problem is formulated, and the Karush-Kuhn-Tucker (KKT) conditions and Lagrangian multiplier are employed to solve the optimization problem. Additionally, a UL-DL channel gain-based unequal sensor node operating time scheme is introduced. The results of the study demonstrate that increasing the DL threshold data rate enhances UL performance in terms of sum throughput and outage. Moreover, the proposed channel gain-based unequal operating time scheme outperforms the equal sensor node operating time approach.

Keywords: Wireless energy harvesting · Sensor networks · Sum throughput · Outage

1 Introduction

Wireless energy harvesting (WEH) has been recognized as a critical element of modern communication systems, particularly in the context of 5G and beyond. With the expansion of wireless devices and services, energy consumption has become a major challenge for sustainability, and WEH furnishes a potential solution [1]. WEH can indeed be implemented in various applications, including the

© ICST Institute for Computer Sciences, Social Informatics and Telecommunications Engineering 2024
Published by Springer Nature Switzerland AG 2024. All Rights Reserved
P. Pareek et al. (Eds.): IC4S 2023, LNICST 537, pp. 241–252, 2024.
https://doi.org/10.1007/978-3-031-48891-7_21

Internet of Things (IoT), Wireless Sensor Networks (WSNs), and cognitive radio (CR) systems. In the IoT, WEH can power small devices and sensors, empowering them to operate without the need for batteries or wired power sources. This allows for increased flexibility and mobility in IoT deployments [2]. In WSN, WEH can be used to power the sensors, eliminating the need for frequent battery replacements. This enhances the sustainability and maintenance-free operation of the sensor network [3]. Wireless energy harvesting also plays a significant role in the field of CR, offering significant advantages and addressing critical challenges [4]. The CR, as a technology that enables intelligent and adaptive wireless communication, requires a continuous and reliable power source to operate efficiently. However, WEH is crucial as it overcomes the limitations of traditional battery power, enhancing the scalability, flexibility, and lifespan of CR devices. This technology allows devices to extract energy from the environment, such as from ambient sources like RF signals, light, or vibrations. Radio Frequency (RF) WEH involves harvesting energy from ambient RF signals that already exist in the environment, such as from Wi-Fi, cellular, or other wireless communication networks. The energy can then be used to power the device or recharge its battery, reducing reliance on external power sources and improving the device's lifespan.

Power splitting (PS) and time switching (TS) are the two most popular energy harvesting protocols used to manage the harvested energy [5,6]. The PS protocol divides energy into two parts, where one part is directly used to power the device, and another part is stored for information transmission. This makes a balance between immediate energy requirements and energy storage. On the other hand, TS involves switching between two different time periods for energy utilization. During the first time period, the harvested energy fulfills immediate power requirements. In the second time period, the harvested energy is stored in a battery for transmission. A comprehensive analysis of Simultaneous Wireless Information and Power Transfer (SWIPT) is Demonstrated in [7]. The authors cover a wide range of features related to SWIPT, including TS, PS, and antenna beamforming. It examines the advantages, limitations, and trade-offs associated with each technique, providing a comprehensive understanding of their capabilities. The recent developments in materials, wireless power transfer standards, and integration with other technologies are expected to drive the growth of WEH techniques.

1.1   Related Work

Hosein et al. proposed a Time Division Multiple Access (TDMA) based protocol in WEH networks, where each time slot is divided into two intervals: one for energy absorption and the other for data transmission by the sensors [8]. In their proposed model, a sensor can transmit its information if the amount of energy it has harvested surpasses its power consumption requirements. They focus on achieving energy-efficient resource allocation while considering constraints on time scheduling parameters and transmission power consumption. Another protocol, named TSAPS, is introduced for EH relay networks. It combines elements from the Traditional TS and adaptive power splitting (APS) techniques [6]. The study aims to derive a closed-form outage probability expression of the TSAPS protocol and analyze its effective transmission rate in scenarios involving random relay selection and opportunistic relay selection. Saman et al. propose a new hybrid protocol that combines PS and TS EH protocols [9]. An optimization problem is formulated to determine the optimal PS and TS ratios, aiming to maximize throughput in information transfer from the source to the destination for both decode-and-forward (DF) and amplify-and-forward (AF) relaying schemes [9]. Chao et al. implement EH in cooperative spectrum sharing within CR systems. Primary users (PUs) harvest energy from their access points (APs), while APs and secondary users (SU) are powered by a stable power supply [10]. Cooperation from SUs in wireless energy transfer enhances EH efficiency for PUs, and SU assistance in primary data transmission improves link robustness. Nguyen et al. analyze the impact of relay transceivers in terms of outage probability and throughput of cognitive network with an energy-harvesting relay. Two wireless power transfer policies and two bidirectional relaying protocols are considered in the network configuration [11]. In a proposed protocol for an underlay cognitive relay network, secondary nodes harvest energy from the primary network while sharing its licensed spectrum. Outage probability expression is derived, considering constraints on maximum transmit power, peak interference power, and interference power from primary users to the secondary network [12]. An integrated model is proposed for cooperative dual-hop DF relay transmission, combining information relay and wireless power supply through a TS protocol based on RF energy harvesting [13]. The relay node assists communication between an energy-constrained source and destination while supplying power to them. In a WEH relay sensor network, Nirati et al. aim to maximize system throughput using DF relaying and TS for energy harvesting and transmission. The harvested energy charges the battery of a common control unit, which is then distributed among the relay nodes for transmission [14].

1.2   Motivations and Contributions

In the above-mentioned article, TS-based protocol is implemented to perform two operations; UL and DL operations. However, in this study, a TS protocol is considered to operate three operations. Along with UL and DL, an additional dedicated EH mode exists. The importance of optimal EH time for UL sum throughput maximization under the constraint of guaranteed DL quality of service (QoS) also was not addressed previously. Furthermore, the literature commonly assumes equal time for the operation of sensor nodes, which may not be a suitable approach considering the frequent variation of channel gain. The key contributions of this research paper are as follows:

- Proposal of a TS-based WEH protocol that allows sensor nodes to efficiently switch between energy harvesting and data transmission modes, optimizing the utilization of available energy resources.

– An optimization problem is formulated to maximize the UL sum throughput of the sensor network. The problem considers the constraint of a minimum DL throughput, ensuring a balance between UL and DL performance.
– A channel gain-based approach is proposed to allocate operating time among sensor nodes unequally. This approach utilizes the variations in channel gains to enhance the overall throughput performance of the sensor network compared to an equal operating time allocation.

The rest of the paper is organized as follows: Sect. 2 presents the system model for wireless energy harvesting (WEH) and introduces the equal time-based operating time as well as the proposed channel gain-based unequal operating time under the problem formulation in Sect. 3. Section 4 discusses the simulation results. Finally, Sect. 5 concludes the research work.

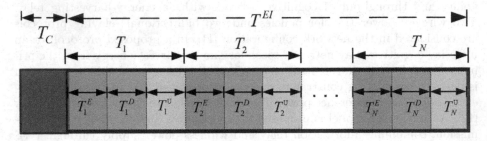

Fig. 1. Operating time frame for energy harvesting, UL, and DL transmission for each sensor node.

2   System Model

A typical multi-user wireless-powered communication system has been considered to analyze the proposed work. The system consists of one base station (BS) and N sensor nodes (SNs) uniformly distributed within the transmission range of BS. The BS is connected to a reliable power supply and maintains continuous communication with the SNs. However, the SNs have limited energy resources, meaning their ability to transmit information and perform regular operations is constrained by the available battery power. To overcome this limitation, the SNs first harvest energy from the BS and then utilize a portion of that harvested energy for information transmission. The surplus energy is stored in a supercapacitor, which is used to sustain regular operations. For energy harvesting and information transmission, a time switching (TS) based protocol is employed. The BS is equipped with multiple antennas to efficiently transfer energy and information, while the SNs are restricted to a single antenna due to size limitations. Additionally, all the SNs operate in half-duplex mode, meaning they cannot transmit and receive simultaneously.

The SNs exhibit the ability to switch flexibly between energy harvesting and information decoding during downlink (DL) transmissions. To prevent interference among the SNs, a Time Division Multiple Access (TDMA) protocol is adopted. The operation cycle of the BS, denoted as T_T, is divided into two phases: the control phase (T_C) and the transmission phase (T_{EI}) as depicted in Fig. 1. During the T_C phase, the BS estimates the channel gain and synchronizes the SNs. The T_{EI} phase is dedicated to energy harvesting and information transmission and is further divided into N time slots $(T_1, T_2, ..., T_N)$. Representing the time assignment vector as $\mathbf{t} = [T_1, T_2, ..., T_N]^T$, it is evident that $\mathbf{t^T 1} = T_{EI}$. Each time slot T_i, within the operating cycle of the ith SN, consists of three durations: $\mathbf{T^E} = \{T_i^E | i = 1, 2, ..., N\}$ for energy harvesting during DL, $\mathbf{T^D} = \{T_i^D | i = 1, 2, ..., N\}$ for information receiving during DL, and $\mathbf{T^U} = \{T_i^U | i = 1, 2, ..., N\}$ for uplink (UL) transmission. During the energy harvesting and DL information receiving times of the jth SN, the ith SN (where $i \neq j$) also detects RF signals due to the broadcast nature of the BS. Consequently, the ith SN is capable of harvesting energy during $T_j^{DL} = T_j^E + T_j^D$ and also during its dedicated T_i^E time slot. It is important to note that SNs cannot perform energy harvesting during the UL time slot.

The UL and DL channels are considered as quasi-static independent block fading channels i.e., channel power gain remains constant during each frame time and may change independently from frame to frame. The DL channel from $BS - to - SN_i, g_i \sim \mathcal{CN}(0, \sigma_{BS_i}^2)$ and UL channel $SN_i - to - BS, h_i \sim \mathcal{CN}(0, \sigma_{S_i B}^2), (i = 1, 2, ..., N)$ are Rayleigh faded channels and normalized channel power gains are symbolized as $|g_i|^2$ and $|h_i|^2$ respectively. The instantaneous channel gains of g_i and h_i are exponentially distributed with mean λ_x and λ_y respectively. It is also assumed that perfect channel state information is available. The UL and DL channel noise is symbolized as n_b and n_s, and respective noise powers σ_U^2 and σ_D^2 are assumed to be identical. For simplicity, let $\sigma_U^2 = \sigma_D^2 = \sigma^2$.

3   Problem Formulation

$y_D(k)$, the DL received signals at the ith SN and $y_D(k)$, the UL transmitted signal from ith SN at kth time instant are given by (1) and (2), respectively.

$$y_D(k) = \sqrt{P_i^B d_i^{-m}} g_i S_b(k) + n_s \tag{1}$$

$$y_U(k) = \sqrt{P_i d_i^{-m}} h_i S_s(k) + n_b \tag{2}$$

where, P_i^B and P_i are the transmitted power from the BS and SN respectively, d_i represent the distance between the BS and the ith SN, m denote the path loss exponent, and $S_b(k)$ and $S_s(k)$ represent the normalized information signal from the BS. and SN.i.e $E\{|S_b(k)|^2\} = 1$ and $E\{|S_s(k)|^2\} = 1$. When the jth SN uses $y_D(k)$ signal for energy harvesting during T_j^E and information receiving during T_j^D, ith SN $(i \neq j)$ uses for energy harvesting during T_j^{DL}. It is noticeable that

ith SN is also in energy harvesting mode during T_i^E. Hence, T_i^{EH}, the total time for energy harvesting by ith SN can be denoted as

$$T_i^{EH} = \sum_{j=1, j\neq i}^{N} T_j^{DL} + T_i^E = \sum_{i=1}^{N} T_i^{DL} - T_i^D \tag{3}$$

Using (1), Eh_i, the energy harvested by ith node can be expressed as

$$Eh_i = \frac{\eta T_i^{EH} P_i^B |g_i|^2 + \sigma^2}{d_i^m} \tag{4}$$

The symbol η denotes the efficiency of the rectifier circuit which converts the received radio signal to direct current. For the sack of simplicity, η is considered same for all SNs.

3.1 Equal Time Distribution

In this section, we will analyze outage when the data rate on any of UL and DL falls below R_{th}, a threshold data rate. It is assumed that an equal time frame is assigned to all the SNs, i.e., $T_1 = T_2 = \dots = T_N = T$. We also assumed $T^E = \alpha_i T, T^D = T^U = (1 - \alpha_i)T/2$. The ith SN utilizes harvested energy, expressed in (4), to DL transmission during $(1 - \alpha_i)T/2$. Hence, transmit power of ith SN is expressed as

$$P_i = \frac{2Eh_i}{(1 - \alpha_i)T} = \frac{2\left(\eta T_i^{EH} P_i^B |g_i|^2 + \sigma^2\right)}{d_i^m (1 - \alpha_i)T} \tag{5}$$

It is noticeable that only a fraction of the time of the received signal is exploited by SN for DL and the harvested energy during $\mathbf{T^{EH}}$ is fully exploited for UL transmission. R_i^D and R_i^U are the achievable DL and UL throughput of SN_i expressed in (6) and (7), respectively.

$$R_i^D = \frac{(1 - \alpha_i)T}{2} \log_2 \left(1 + \frac{P_i^B |g_i|^2}{\sigma^2}\right) \tag{6}$$

$$R_i^U = \frac{(1 - \alpha_i)T}{2} \log_2 \left(1 + \frac{P_i |h_i|^2}{\sigma^2}\right) \tag{7}$$

It is noticeable that the operation of SN is uncorrelated to each other. Hence, sum of the maximum achievable throughput of each SN is the maximum achievable throughput of the system. Therefore,

$$\text{Objective:} \quad \text{maximize} \quad R_i^U = \sum_{i=1}^{N} \frac{(1 - \alpha_i)T}{2} \log_2 \left(1 + \frac{P_i |h_i|^2}{\sigma^2}\right) \tag{8a}$$

$$\text{Constraints:} \quad R_i^D = \frac{(1 - \alpha_i)T}{2} \log_2 \left(1 + \frac{P_i^B |g_i|^2}{\sigma^2}\right) \geq R_{th}, \forall i \tag{8b}$$

$$0 < \alpha_i < 1 \tag{8c}$$

The constraint (8b) ensures minimum QoS requirement for downlink data rate. The optimization problem (8a) is a convex problem, which is proved in Appendix I.

Theorem 1. *The fraction of energy harvesting time α_i, $\forall i \in \mathcal{N}$ for the optimal solution of throughput maximization problem (8a) with constraints (8b) and (8c), can be expressed as*

$$\alpha_i = 1 - \frac{2R_{th}}{T \log_2 \left(1 + \frac{P_i^B |g_i|^2}{\sigma^2}\right)}, \qquad \forall i \in \mathcal{N}, \tag{9}$$

Proof. The proof is reproduced from standard literature in Appendix II.

It should be noted that for $\alpha_i > 0$, it is necessary to have $\frac{2R_{th}}{T \log_2 \left(1 + \frac{P_i^B |g_i|^2}{\sigma^2}\right)} < 1$, which determines the minimum transmit power of the BS for a particular R_{th} as $P_i^B > \frac{\sigma^2}{|g_i|^2} \left(2^{\frac{2R_{th}}{T}} - 1\right)$. The BS needs to transmit signals with varying power levels for each subscriber node (SN), and this can be expressed as:

$$P_i^B = \frac{\sigma^2}{|g_i|^2} \left(2^{\frac{2R_{th}}{T}} - 1\right) + P_b \tag{10}$$

where P_b represents the additional power beyond the minimum transmit power.

3.2   Unequal Time Distribution

In the previous subsection, we assumed an equal operating time frame for each of all SNs i.e., $T_1 = T_2 = \ldots = T_N = T$. Here we propose an operating time distribution scheme that decides T_i, $\forall i \in N$, the single operating time frame of each SN based on their UL and DL channel gain quality as

$$T_i = T^{EI} \frac{g_i h_i}{\sum_{i=1}^{N} g_i h_i}. \tag{11}$$

Now the fraction of energy harvesting time α_i, $\forall i \in \mathcal{N}$ can be expressed as

$$\alpha_i = 1 - \frac{2R_{th}}{T_i \log_2 \left(1 + \frac{P_i^B |g_i|^2}{\sigma^2}\right)}, \qquad \forall i \in \mathcal{N}. \tag{12}$$

It is acknowledged that the problem of maximizing throughput, as expressed in Eq. (8a), can be formulated to determine the optimal T_i. However, this specific task is left for future research and exploration.

4   Simulation Results

In this section, we have examined the optimal uplink (UL) sum throughput and outage probability of the proposed WEH system model through simulation. The

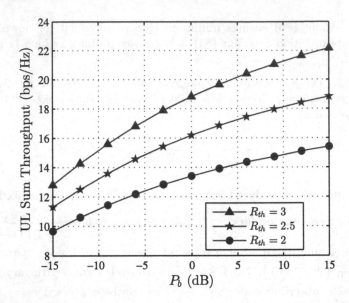

Fig. 2. Achievable UL sum throughput as a function of P_b, the additional transmit power

simulation process was conducted using Matlab software, taking into account the following crucial parameters for the analysis; $N = 10$, $m = 2.7$, $\eta = 0.9$, $d = \{4, 6, 9, 5, 10, 7, 8, 10, 5, 9\}$ meter. The simulation results were obtained by averaging over 10^5 independent Monte Carlo trials. It is important to highlight that the minimum data rate for the DL transmission is guaranteed. Therefore, the simulation results focus on the performance analysis of the UL communication.

The achievable UL sum throughput as a function of P_b, the additional transmit power for varying DL throughput is shown in Fig. 2. As expected, the UL sum throughput improves with an increase in P_b. It is worth noting that the UL sum throughput also strongly depends on the DL threshold of the system. Interestingly, the UL sum throughput improves as the DL threshold increases. This can be explained by the fact that an increase in R_{th} leads to an extended UL transmission time T^U, resulting in enhanced system performance.

Figure 3 illustrates the relationship between the outage probability and P_b the additional transmit power of the BS for the nearest and farthest SN. As expected, the outage probability decreases with an increase in transmit power, and the nearest user exhibits better outage performance compared to the farthest user. When the operating time frame for each SN is increased (from $T = 1\,\mathrm{s}$ to $1.5\,\mathrm{s}$ and $2\,\mathrm{s}$), the SNs have more time to harvest energy, resulting in an improved outage performance. As the minimum UL rate requirement, Ru_{th}, increases, the outage performance degrades. However, it is worth noting that the outage performance improves when the minimum DL rate, R_{th}, increases. This is because as R_{th} increases, α_i decreases, which leads to an increase in both

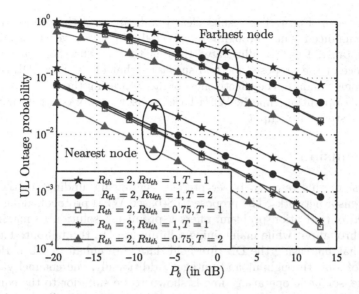

Fig. 3. UL outage performance of the nearest and farthest user as a function of additional transmit power of BS for the following parameters, DL threshold $R_{th} = \{2, 3\}$ bps/Hz, UL threshold $Ru_{th} = \{0.75, 1\}$ bps/Hz, equal operating time frame $T = \{1, 2\}$ s.

DL and UL transmission time (i.e., T^U and T^D). As a result, the UL transmission rate also improves.

The superiority of the proposed channel gain-based unequal SN operating time over equal SN operating time in terms of UL sum throughput is

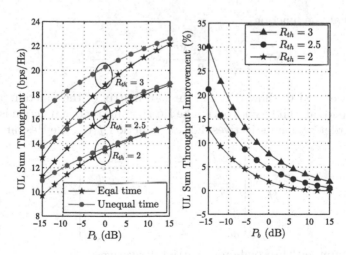

Fig. 4. (a) UL sum throughput comparison between equal operating time frame and proposed unequal operating time frame for $R_{th} = \{2, 2.5, 3\}$ bps/Hz, (b) UL sum throughput improvement achieved by proposed unequal operating time frame over the equal operating time frame for $R_{th} = \{2, 2.5, 3\}$ bps/Hz.

demonstrated in Fig. 4. The results clearly indicate that the proposed method achieves enhanced sum throughput, especially in the lower transmit power region. Notably, Fig. 4(b) illustrates that as the DL threshold increases, the performance of the proposed method surpasses that of the lower DL threshold. This outcome validates the effectiveness of incorporating the channel gain-based unequal SN operating time in WEH-based sensor networks, as it significantly improves system throughput.

5   Conclusion

In conclusion, this research paper presents significant findings in the field of WEH in sensor networks. The proposed TS-based WEH protocol, along with the formulated optimization problem, offers a promising solution to maximize the UL sum throughput while maintaining a minimum DL throughput. The results indicate that increasing the DL threshold data rate enhances UL performance in terms of sum throughput and outage. Additionally, the channel gain-based unequal sensor node operating time is shown to be superior to the equal operating time, further improving system performance. These findings highlight the potential of WEH protocols and unequal operating time allocation to enhance the performance of sensor networks, contributing to the advancement of wireless communication systems. In the future, we will extend this work to the cognitive radio systems and analyze the system's performance. Furthermore, there is a scope to introduce non-orthogonal multiple access (NOMA)-based [15] transmission and cell-free massive MIMO [16] in the WEH systems.

Appendix-I

To prove that the function

$$f(\alpha) = \sum_{i=1}^{N} \frac{(1 - \alpha_i)T}{2} \log_2 \left(1 + \frac{P_i |h_i|^2}{\sigma^2}\right) \tag{13}$$

is convex, we need to show that the Hessian matrix of the function is positive semidefinite for all valid values of α. The Hessian matrix of a function is a matrix of second-order partial derivatives. For our function $f(\alpha)$, the Hessian matrix is defined as:

$$H = \begin{bmatrix} \frac{\partial^2 f}{\partial \alpha_1^2} & \frac{\partial^2 f}{\partial \alpha_1 \partial \alpha_2} & \cdots & \frac{\partial^2 f}{\partial \alpha_1 \partial \alpha_N} \\ \frac{\partial^2 f}{\partial \alpha_2 \partial \alpha_1} & \frac{\partial^2 f}{\partial \alpha_2^2} & \cdots & \frac{\partial^2 f}{\partial \alpha_2 \partial \alpha_N} \\ \vdots & \vdots & \ddots & \vdots \\ \frac{\partial^2 f}{\partial \alpha_N \partial \alpha_1} & \frac{\partial^2 f}{\partial \alpha_N \partial \alpha_2} & \cdots & \frac{\partial^2 f}{\partial \alpha_N^2} \end{bmatrix} \tag{14}$$

Let's compute the second-order partial derivatives:

$$\frac{\partial^2 f}{\partial \alpha_i \partial \alpha_j} = -\frac{T}{2} \log_2 \left(1 + \frac{P_i |h_i|^2}{\sigma^2}\right) \delta_{ij}, \tag{15}$$

where δ_{ij} is the Kronecker delta function. Notice that the second-order partial derivatives are constant with respect to α, which means the Hessian matrix is a constant matrix. Specifically, all diagonal elements are the same, and all off-diagonal elements are zero. The Hessian matrix of our function is then:

$$H = -\frac{T}{2} \log_2 \left(1 + \frac{P|h|^2}{\sigma^2} \right) I, \tag{16}$$

where I is the identity matrix. Since the Hessian matrix is a constant matrix with negative values on the diagonal (due to the negative logarithm term), it is negative definite, which implies it is also positive semidefinite. Therefore, the function $f(\alpha)$ is convex, as the Hessian matrix is positive semidefinite for all valid values of α.

Appendix-II

To solve the optimization problem using the Lagrangian and KKT method, we first define the Lagrangian function as follows:

$$L(\alpha, \lambda, \mu) = \sum_{i=1}^{N} \frac{(1-\alpha_i)T}{2} \log_2 \left(1 + \frac{P_i|h_i|^2}{\sigma^2} \right) + \sum_{i=1}^{N} \lambda_i \left(\frac{(1-\alpha_i)T}{2} \log_2 \left(1 + \frac{P_i^B|g_i|^2}{\sigma^2} \right) - R_{th} \right)$$

$$+ \sum_{i=1}^{N} \mu_i(\alpha_i)(1-\alpha_i) \tag{17}$$

where $\alpha = [\alpha_1, \alpha_2, \ldots, \alpha_N]$ is the vector of variables, $\lambda = [\lambda_1, \lambda_2, \ldots, \lambda_N]$ and $\mu = [\mu_1, \mu_2, \ldots, \mu_N]$ are the Lagrange multipliers for the inequality constraints and bound constraints, respectively. Next, to find the stationary points we need partial derivatives of the Lagrangian with respect to α_i, λ_i, and μ_i and setting them to zero:

$$\frac{\partial L}{\partial \alpha_i} = \frac{T}{2} \log_2 \left(1 + \frac{P_i|h_i|^2}{\sigma^2} \right) - \lambda_i \frac{T}{2} \log_2 \left(1 + \frac{P_i^B|g_i|^2}{\sigma^2} \right) - 2\mu_i\alpha_i = 0 \tag{18}$$

$$\frac{\partial L}{\partial \lambda_i} = \frac{(1-\alpha_i)T}{2} \log_2 \left(1 + \frac{P_i^B|g_i|^2}{\sigma^2} \right) - R_{th} = 0 \tag{19}$$

$$\frac{\partial L}{\partial \mu_i} = \alpha_i(1-\alpha_i) = 0 \tag{20}$$

Note that to find the α_i from $\frac{\partial L}{\partial \alpha_i}$, we need to find the values of the Lagrange multipliers λ_i and μ_i. This can be done iteratively using numerical methods, such as gradient descent or Newton's method. However, we can solve for α_i from the derivative of the Lagrangian with respect to λ_i. Finally, solving for α_i:

$$\alpha_i = 1 - \frac{2R_{th}}{T \log_2 \left(1 + \frac{P_i^B|g_i|^2}{\sigma^2} \right)} \tag{21}$$

References

1. Kamalinejad, P., Mahapatra, C., Sheng, Z., Mirabbasi, S., Leung, V.C., Guan, Y.L.: Wireless energy harvesting for the internet of things. IEEE Commun. Mag. **53**(6), 102–108 (2015)
2. Ma, D., Lan, G., Hassan, M., Hu, W., Das, S.K.: Sensing, computing, and communications for energy harvesting IoTs: a survey. IEEE Commun. Surv. Tutor. **22**(2), 1222–1250 (2019)
3. Singh, J., Kaur, R., Singh, D.: Energy harvesting in wireless sensor networks: a taxonomic survey. Int. J. Energy Res. **45**(1), 118–140 (2021)
4. Singla, J., Mahajan, R., Bagai, D.: A survey on energy harvesting cognitive radio networks. In: 2018 6th Edition of International Conference on Wireless Networks and Embedded Systems (WECON), pp. 6–10. IEEE, November 2018
5. Zhu, Z., Huang, S., Chu, Z., Zhou, F., Zhang, D., Lee, I.: Robust designs of beamforming and power splitting for distributed antenna systems with wireless energy harvesting. IEEE Syst. J. **13**(1), 30–41 (2018)
6. Singh, V., Ochiai, H.: An efficient time switching protocol with adaptive power splitting for wireless energy harvesting relay networks. In: 2017 IEEE 85th Vehicular Technology Conference (VTC Spring), pp. 1–5. IEEE, June 2017
7. Ponnimbaduge Perera, T.D., Jayakody, D.N.K., Sharma, S.K., Chatzinotas, S., Li, J.: Simultaneous wireless information and power transfer (SWIPT): recent advances and future challenges. IEEE Commun. Surv. Tutor. **20**(1), 264–302 (2018)
8. Azarhava, H., Niya, J.M.: Energy efficient resource allocation in wireless energy harvesting sensor networks. IEEE Wirel. Commun. Lett. **9**(7), 1000–1003 (2020)
9. Atapattu, S., Evans, J.: Optimal energy harvesting protocols for wireless relay networks. IEEE Trans. Wirel. Commun. **15**(8), 5789–5803 (2016)
10. Zhai, C., Liu, J., Zheng, L.: Cooperative spectrum sharing with wireless energy harvesting in cognitive radio networks. IEEE Trans. Veh. Technol. **65**(7), 5303–5316 (2015)
11. Nguyen, D.K., Jayakody, D.N.K., Chatzinotas, S., Thompson, J.S., Li, J.: Wireless energy harvesting assisted two-way cognitive relay networks: Protocol design and performance analysis. IEEE Access **5**, 21447–21460 (2017)
12. Liu, Y., Mousavifar, S.A., Deng, Y., Leung, C., Elkashlan, M.: Wireless energy harvesting in a cognitive relay network. IEEE Trans. Wirel. Commun. **15**(4), 2498–2508 (2016)
13. Biswas, S., Bepari, D., Mondal, S.: Relay selection and performance analysis of wireless energy harvesting networks. Wirel. Pers. Commun. **114**, 3157–3171 (2020)
14. Nirati, M., Oruganti, A., Bepari, D.: Power allocation in wireless energy harvesting based relaying sensor networks. In: 2019 4th International Conference on Recent Trends on Electronics, Information, Communication and Technology (RTEICT), pp. 491–495, March 2020
15. Bepari, D., et al.: A survey on applications of cache-aided NOMA. IEEE Commun. Surv. Tutor (2023)
16. Gangadhar, C., Chanthirasekaran, K., Chandra, K.R., Sharma, A., Thangamani, M., Kumar, P.S.: An energy efficient NOMA-based spectrum sharing techniques for cell-free massive MIMO. Int. J. Eng. Syst. Model. Simul. **13**(4), 284–288 (2022)

Design and Parametric Study of Monopole Blade Antenna for UHF-Band Aerospace Applications

Sadhana Kumari[1], Naveen Kumar Maurya[2($\boxtimes$)], Tripta[3], and Argha Sarkar[4]

[1] Department of ECE, BMS College of Engineering, Bangalore, Karnataka 560019, India
[2] Department of ECE, Vishnu Institute of Technology, Bhimavaram, Andhra Pradesh 534202, India
naveen.maurya222@gmail.com
[3] Department of ECE, Government Engineering College, Vaishali, Bihar 844115, India
[4] School of Computer Science and Engineering, REVA University, Bangalore, Karnataka 560064, India

Abstract. This paper provides an idea to design a wideband antenna suitable for UHF-band aircraft application. The blade antenna (BA) has been simulated for different elementary parameters using CST simulator. Antenna performance characteristic with the parametric variation of the antenna model have been studied and demonstrated. The simulation results of the designed antenna show 3.18 dBi gain, and 53-degree half power beamwidth (HPBW) over the entire frequency range 0.9 GHz to 2.0 GHz. Further, the designed antenna a very good return loss performance with VSWR <2.2 in the entire frequency band. This antenna can find its application in aircraft GPS navigation system and datalink operation as well.

Keywords: Blade antenna · Monopole antenna · Omnidirectional

1 Introduction

In the receiver chain for any wireless system, antenna plays a significant role in the complete performance. Its function is to ensure a proper transition between a transmission line and free space. It necessitates enough impedance matching and gain at the same time. The rapid development in wireless communication requires broadband antennas to assist high data rate performance of wireless systems such as aircraft system, radar systems, and satellite communication system. At the same time, omnidirectional radiation from an antenna is preferred for 360-degree radiation coverage. Dipole and monopole antennas (whether electric or magnetic) are well known for their omnidirectional radiation coverage [1–6]. In recent times, monopole antennas have got tremendous attention over its dipole counterpart due to compactness in size [4–6]. As we know bandwidth achieved by a conventional λ/4 monopole antenna is narrow, a wide impedance bandwidth is possible only when the structure of antenna is manipulated using different techniques. One such solution is; by diminishing the length to diameter ratio of the antenna [7], modifying antenna geometry with respect to angle; triangular sheet for planar and conical structure for non-planar are the widely used structure for this technique.

© ICST Institute for Computer Sciences, Social Informatics and Telecommunications Engineering 2024
Published by Springer Nature Switzerland AG 2024. All Rights Reserved
P. Pareek et al. (Eds.): IC4S 2023, LNICST 537, pp. 253–261, 2024.
https://doi.org/10.1007/978-3-031-48891-7_22

Different types of broadband antennas suitable for airborne application are available out of those many; one of the most accepted antennas is the BA, and the reasons are; low air drag, lightweight and low-cost [8, 9]. A conventional BA is often a monopole antenna, that employs the first two listed techniques to achieve a broader bandwidth. A broadband monopole antenna with the inverted hat is reported in [10]. This antenna covers a frequency range 0.5 GHz to 2.0 GHz with a very low gain and complex structure. In [11], a broadband reconfigurable antenna that covers VHF/UHF/L-bands is reported for aircraft application. This antenna is modified in biconical shape and frequency reconfigurability is achieved by PIN diode. A broadband (0.5–2 GHz) BA, printed on RO5880 with a dielectric constant 2.2 and substrate thickness 1.6 mm, was reported in [12]. In this antenna, an inductor coil filled with air core, and shaping of the triangular shape help to attain extra bandwidth of 0.5 GHz from a BA reported in [12]. Another BA fabricated with a 2 mm thick Aluminum sheet, designed with an oblique edge to achieve a broadband (0.03–0.60 GHz), is reported in [13]. The oblique edge used in this paper helps to reduce the height of the antenna. However, these antennas require a large ground plane. As a matter of fact, the broadband in these two papers has been achieved by employing a large ground plane, while in many applications such as unmanned aerial vehicles, to have a low radar cross-section airplane surface is made up of dielectric materials. Recently wideband slotted BA dipole form is reported in [14], and it does not employ any ground plane. However, in contrast to the conventional monopole blade antenna, this antenna requires more implementation area and also does not support an aerodynamic profile.

In this paper, a detailed analysis of a BA with an oblique edge and compact ground size, covering 0.90–2.0 GHz, has been introduced. This paper is arranged as: Sect. 1 mentions a brief introduction to the monopole BA for airborne applications, and Sect. 2 describes the design method of the BA. Sect. 3 consists of the various simulation results. Finally, the paper is concluded in Sect. 4. The antenna is simulated for a 2 mm thick aluminum sheet. The final optimized dimensions of this BA is 100×56 mm^2 with an elliptic ground plane with major and minor diameter dimensions of 85 mm and 30 mm, respectively.

2 Antenna Design with Parametric Study

The physical outline of the antenna with current distribution are shown in Fig. 1. It is well identified fact that in the BA the current distribution is intense at the border of the monopolar patch (Fig. 1 (b)).

Therefore, the resonant frequency of the antenna can be tuned by tuning the height of the BA. So, H1 decides the lower resonating frequency (f_{01}). The effect on the S-parameter for different values of H1 is plotted in Fig. 2. This figure clearly indicates a lowering of f_{01} with the increment in the value of this parameter. Figure 3 shows the S-parameter magnitude (dB) versus frequency for the swept θ. The result indicates that for $\theta = 80^0$, a good impedance match is found for f_{02}. Parametric analysis has been done to attain broader bandwidth with a good impedance match over the entire band.

However, $\theta = 90^0$ has been chosen for covering the entire band with a minimum reflection of -10 dB at least. Figure 4 shows the S-parameter magnitude (dB) versus frequency for the swept W. The result indicates that as we increase the value of the W

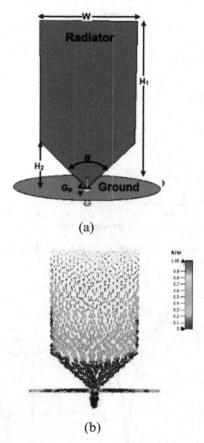

(a)

(b)

Fig. 1. Monopole BA: (a) physical geometry outline (b) surface current distribution.

there will be hardly any change in the resonating frequency rather than getting a good match @ f_{02}. However, due to fabrication constraints, this parameter is selected to be 2 mm.

The effect of different design parameters along with the final optimized value (for H1 = 42 mm and elliptic ground plane with major and minor diameters 85 mm and 30 mm, respectively) of the antenna parameter is listed in Table 1.

Final optimized S-parameter is shown by blue line curve in Fig. 5 of the BA with different value of Gp. Further, the VSWR and the smith chart plot of the input impedance of the final design are shown in the Fig. 6.

Figure 6 (b) illustrates that in the frequency range where the current path length becomes less than 0.25 λ, input impedance becomes capacitive. The radiation pattern at $\Phi = 0^0$ and 90^0 plane are plotted Fig. 7. The gain pattern characteristics for different frequency are listed in Table 2.

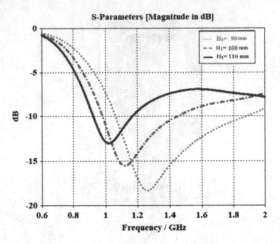

Fig. 2. Scattering parameter of the BA with the variation of H_1.

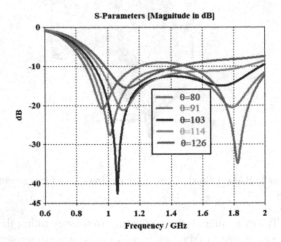

Fig. 3. *S*- parameter of the BA with the variation in oblique angle.

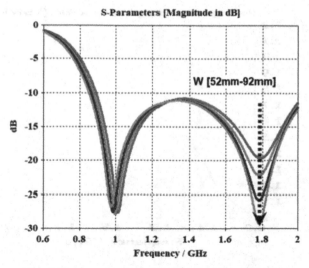

Fig. 4. *S*-parameter of the BA with the variation of *W*.

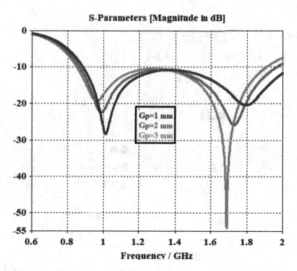

Fig. 5. *S*-parameter of the BA with the variation of G_p.

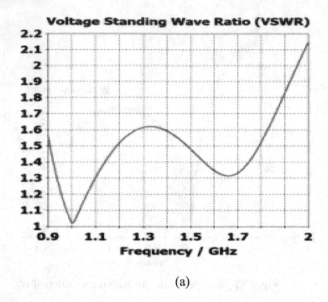

(a)

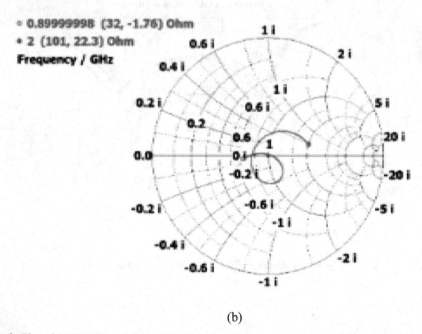

(b)

Fig. 6. Plot of (a) VSWR and (b) smith chart plot of the antenna input impedance of optimized antenna.

Table 1. Parametric Study Summary.

	H1 $\uparrow$	W $\uparrow$	θ $\downarrow$	G_p $\uparrow$
	f_{01}	Improve the matching @ f_{02}, not any change in f_{01} and f_{02}	Improve the matching @ f_{02} with a slight shift in f_{01}	Improve the matching @ f_{02} with a slight shift in f_{01} and f_{02}
Optimized Value	100 mm	56 mm	90^0	2 mm

Table 2. Gain Characteristics for $\Phi = 0^0 / 90^0$.

Frequency (GHz)	Main Lobe Direction (degree)	Main Lobe Magnitude (dBi)	3 dB Angular width	Frequency (GHz)
1	83/83	1.71/2.37	86.5/81	83/83
1.3	70/74	1.64/2.82	76.6/72	1.3
1.6	58/64	2.61/4.01	62.5/61.3	1.6
1.9	50/58	3.18/4.55	53.4/53	1.9

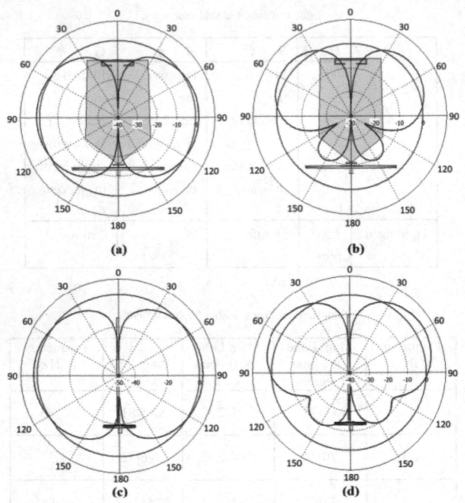

Fig. 7. Polar plot of radiation characteristic, (a) 0.9 GHz, $\Phi = 0^0$, (b) 1.9 GHz, $\Phi = 0^\circ$, (c) 900 MHz, $\Phi = 90^\circ$, (d) 1.9 GHz, $\Phi = 90^\circ$.

3 Conclusion

A wideband monopole BA for UHF band application (0.9–2.0 GHz) has been designed. The effect of various design parameters have been studied and illustrated to design an optimized BA. The simulation results show VSWR <2.2 over the entire bandwidth. An omnidirectional radiation pattern with a maximum gain of 3.18 dBi at 1.9 GHz is achieved. This antenna can be used for aerospace application.

References

1. Balanis, C.A.: Antenna theory: a review. In: Proceedings of the IEEE, vol. 80, no. 1, pp. 7–23. IEEE (1992)
2. Maurya, N.K., Kumari, S., Pareek, P., Singh, L.: Graphene-based frequency agile isolation enhancement mechanism for MIMO antenna in terahertz regime. Nano Commun. Netw. **35**, 100436 (2023)
3. Reddy, R.S., Prasad, N., Deevi, N.: Evaluation and testing of dual band reconfigurable monopole antenna for cognitive radio applications. In: Vasudevan, H., Gajic, Z., Deshmukh, A. (eds.) Proceedings of International Conference on Wireless Communication . Lecture Notes on Data Engineering and Communications Technologies, vol. 36, pp. 65–73. Springer (2020). https://doi.org/10.1007/978-981-15-1002-1_8
4. Maurya, N.K., Bhattacharya, R.: Design of compact dual-polarized multiband MIMO antenna using near-field for IoT. AEU-Int. J. Electron. Commun. **117**, 153091 (2020)
5. Varma, D.R., Murali, M., Krishna, M.V.: Design of wearable microstrip patch antenna for biomedical application with a metamaterial. In: Chowdary, P.S.R., Anguera, J., Satapathy, S.C., Bhateja, V. (eds.) Evolution in Signal Processing and Telecommunication Networks. LNEE, vol. 839, pp. 421–434. Springer, Singapore (2022). https://doi.org/10.1007/978-981-16-8554-5_40
6. Maurya, N.K., Bhattacharya, R.: CPW-fed dual-band compact Yagi-type pattern diversity antenna for LTE and WiFi. Progress Electromagn. Res. C **107**, 183–201 (2021)
7. Balanis, C.A: Antenna Theory: Analysis and Design. John Wiley & Sons (2016)
8. Ahirwar, S., Sairam, C., Kumar, A.: Broadband blade monopole antenna covering 100–2000 MHz frequency band. In: 2009 Applied Electromagnetics Conference (AEMC), pp. 1–4. IEEE (2009)
9. Sairam, C., Khumanthem, T., Ahirwar, S., Singh, S.: Broadband blade antenna for airborne applications. In: 2011 Annual IEEE India Conference, pp. 1–4. IEEE (2011)
10. Zhao, J., Peng, T., Chen, C.C., Volakis, J.: Low-profile ultra-wideband inverted-hat monopole antenna for 50 MHz–2 GHz operation. Electron. Lett. **45**(3), 142–144 (2009)
11. Rhee, C.Y., Kim, J.H., Jung, W.J., Park, T., Lee, B., Jung, C.W.: Frequency re-configurable antenna for broadband airborne applications. IEEE Antennas Wirel. Propag. Lett. **13**, 189–192 (2014)
12. Sairam, C., Khumanthem, T., Ahirwar, S., Kumar, A.: Design and development of broadband blade monopole antenna. In: 2008 International Conference on Recent Advances in Microwave Theory and Applications, pp. 150–151. IEEE (2008)
13. Arand, B.A., Shamsaee, R., Yektakhah, B.: Design and fabrication of a broadband blade monopole antenna operating in 30 MHz–600 MHz frequency band. In: 2013 21st Iranian Conference on Electrical Engineering (ICEE), pp. 1–3. IEEE (2013)
14. Nosrati, M., Jafargholi, A., Pazoki, R., Tavassolian, N.: Broadband slotted blade dipole antenna for airborne UAV applications. IEEE Trans. Antennas Propag. **66**(8), 3857–3864 (2018)

Transmission Losses Due to Surface Reflections in Deep Water for Multipath Model

Veera Venkata Ramana Kandi[1] ⓘ, Pulugujju Rajesh[2] ⓘ, S V Kiranmayi Sridhara[3] ⓘ,
P U V S N Pavan Kumar Nalam[4] ⓘ, B Srinivasa Seshagiri Rao[5] ⓘ,
M Ravi Sankar[6] ⓘ, and Ch. Venkateswara Rao[7(✉)] ⓘ

[1] Department of ECE, GIET Engineering College, Rajahmundry, India
[2] Department of ECE, B V Raju Institute of Technology, Narsapur, Medak, India
[3] Department of ECE, Aditya College of Engineering and Technology, Surampalem, India
[4] Department of ECE, B V C College of Engineering, Rajamundry, India
[5] Physics and Electronics Department, B V Raju College, Bhimavaram, India
[6] Department of ECE, Sasi Institute of Technology and Engineering, Tadepalligudem, India
[7] Department of ECE, Vishnu Institute of Technology, Bhimavaram, India
venkateswararao.c@vishnu.edu.in

Abstract. There are several factors which introduces transmission losses in deep water such as: surface reflections; surface ducts; bottom bounce; convergence zones; deep sound channel; reliable acoustic path; and ambient noise. Hence, it is crucial to model the acoustic channel characteristics and evaluate the effect of transmission losses by considering aforementioned factors inorder to employ the network for specific application. This study primarily aims to estimate the transmission losses caused by surface reflections in deep water environments using a multipath acoustic channel model. The simulation is conducted, considering the impact of absorption, sound speed, temperature, and salinity. The depth of the network scenario is varied to analyze the effects of these factors on the transmission losses. It is evident from simulation results, the acoustic velocity increased by 250 m/s when the depth varies from 100 m to 7000 m and temperature decreased from 30 °C to 4 °C. Similarly, when the salinity increased from 30 ppt to 35 ppt, the acoustic velocity has been increased by 7.14% in deep water. An increase in transmission loss of 5 dB has been attained when the wind speed (W) increased from 4 m/s to 12.5 m/s. Similarly, the transmission losses are increased by 8 dB when the angle of incidence (Theta) increased from 20° to 30°.

Keywords: Absorption · Attenuation · Acoustic channel · Deepwater · Sound speed · Temperature · Transmission loss · Salinity · Surface reflection

1 Introduction

A network known as an underwater acoustic sensor network (UASN) consists of underwater devices equipped with sensors, which establish communication through acoustic signals [1]. Its primary objective is to monitor and gather data in underwater environments, including oceans, lakes, and rivers [2]. In contrast to conventional wireless sensor

© ICST Institute for Computer Sciences, Social Informatics and Telecommunications Engineering 2024
Published by Springer Nature Switzerland AG 2024. All Rights Reserved
P. Pareek et al. (Eds.): IC4S 2023, LNICST 537, pp. 262–272, 2024.
https://doi.org/10.1007/978-3-031-48891-7_23

networks [3], which employ radio waves for communication, UASNs utilize acoustic signals as their preferred communication medium [4, 5]. This preference arises from the fact that sound waves can travel much farther and faster in water compared to radio waves, making acoustics highly suitable for underwater communication.

Acoustic transmission characteristics are substantially influenced by the parameters of the underwater medium, including temperature, salinity, pressure, turbidity, dissolved gases, and seabed properties [6]. These factors play a crucial role in shaping how sound waves propagate and interact in underwater environments. Variations in temperature can cause sound to refract or bend, leading to changes in the transmission path and potentially affecting the accuracy of localization and communication systems [7]. Salinity, which refers to the salt content in water, also affects sound velocity. Higher salinity generally leads to increased sound speed. Variations in salinity can alter the density and com-pressibility of the water medium, thereby affecting the transmission path and the time it takes for sound to propagate from the source to the receiver [8]. Underwater pressure increases with depth, and this pressure variation affects the speed of sound propagation [9]. Higher pressures can lead to increased sound velocity due to the compressibility of water. Understanding pressure changes is crucial for accurately estimating the range, localization, and transmission characteristics of acoustic signals in different underwater depths.

The presence of suspended particles affects the transmission loss, signal-to-noise ratio, and overall clarity of the received acoustic signal. Dissolved gases, such as nitrogen and oxygen, can affect acoustic transmission. Gases can introduce additional scattering and absorption, impacting the propagation of acoustic waves. The concentration and distribution of dissolved gases in water can vary, influencing the transmission charac-teristics of acoustic signals [10]. In deep water, the water depth is much greater than the wavelength of the sound wave, and the surface acts as a nearly perfect reflective boundary for sound waves. This means that a significant portion of the sound energy incident on the surface gets reflected back into the water. These reflected sound waves can interfere with the original transmitted sound waves, leading to complex acoustic phenomena [11]. This will lead surface reflections, which inherently leads to transmis-sion losses. Understanding these underwater medium parameters is crucial for designing and optimizing underwater acoustic systems. Accurate knowledge of these parameters helps in modeling sound propagation, estimating transmission losses, and developing algorithms and techniques to improve communication, localization, and sensing capa-bilities in underwater environments. Hence, this work focuses on effect on sound speed, salinity, and temperature with respect depth on acoustic transmission, thereby evaluating the transmission losses due to surface reflections for a multipath acoustic channel model.

2 Related Work

The advent of Underwater Acoustic Sensor Networks (UASNs) has generated significant interest across various fields, owing to their versatile applications in underwater moni-toring, environmental sensing, marine exploration, and military surveillance. Within the realm of UASNs, addressing transmission losses becomes a crucial aspect for achieving reliable and efficient communication in underwater environments. To this end, extensive research papers and studies have been devoted to investigating the factors influencing transmission losses in UASNs. These factors encompass path loss models, absorption and scattering effects, multipath propagation phenomena, channel estimation, equalization techniques, network topology, and routing considerations. By delving into these aspects, researchers aim to enhance the performance and efficacy of UASNs, striving to mitigate transmission losses and optimize communication strategies. The present research significantly contributes to the advancement of underwater communication systems, enabling them to function successfully amidst challenging underwater conditions and unlocking their full potential across diverse applications.

In [12], the authors have investigated the fundamental physics of wave propagation, specifically focusing on acoustic, electromagnetic (EM), and optical communication carriers. In [13], the authors have extensively studied the impact of propagation characteristics on underwater communication. In [14], the authors have focused on the relationship between propagation loss, ambient noise, and channel capacity in underwater communication. To address inaccuracies in sound speed estimation in oceans and seas, a mathematical model [15] has been proposed. This model provides a conversion framework between atmospheric pressure and depth, as well as depth and atmospheric pressure, aiding in sound speed determination. In [16], the authors have presented an experimental setup that investigates the impact of underwater medium parameters.

A method [17] has been proposed to enhance localization accuracy in underwater environments. To simulate underwater networks effectively, a specifically developed acoustic channel model [18] is employed. In [19], researchers have conducted real-time measurements of route loss in underwater acoustic channels. In [20], a deep learning-based framework has been introduced to enhance accuracy and throughput in channel modeling. The authors have provided a detailed account of the statistical properties of the channel model in [21]. Moreover, a novel technique for frame boundary estimation in UASN has been proposed in [22]. In [23], the authors especially address the clustering in UASN by focusing on the integration of three essential approaches in the context of IoT applications. In [24], the authors investigated how water absorption affected the hybrid phenol formaldehyde (PF) composites' mechanical characteristics.

3 Methodology

This comprehensive approach provides valuable insights into the complex acoustic environment of shallow and deep water, and enabling better understanding and modeling of underwater acoustic propagation.

3.1 Sound Speed

The transmission of sound through water differs significantly from electromagnetic (EM) waves, primarily due to its slow speed. Mackenzie's empirical formula, denoted by (1), provides a means to calculate sound velocity.

$$c(T, S, z) = a_1 + a_2T + a_3T^2 + a_4T^3 + a_5(S - 35) + a_6z + a_7z^2 + a_8T(S - 35)a_9Tz^3 \quad (1)$$

3.2 Acoustic Propagation Loss

Propagation loss of sound refers to the reduction in the strength or intensity of sound waves as they travel through a medium or propagate in a given environment [27]. The loss associated with cylindrical spreading is expressed using Eq. (2), while spherical spreading loss is represented by Eq. (3).

$$L_{CS} = 10 \times \log(R_t) \tag{2}$$

$$L_{SS} = 20 \times \log(R_t) \tag{3}$$

3.3 Absorption Loss

Sound waves in a medium, such as air or water, experience absorption, where the energy of the sound wave is converted into heat [28]. The absorption is frequency-dependent, with higher frequencies generally being absorbed more rapidly which is represented using (4). Where, α is absorption coefficient in underwater, it represents the rate at which sound energy is converted into other forms, such as heat, due to the inherent properties of the water medium. The absorption coefficient is frequency-dependent, meaning that different frequencies of sound waves are absorbed to varying degrees. Higher frequencies generally experience greater absorption than lower frequencies. The absorption coefficient is represented using (5). Where, A_1 and A_2 represent the contributions of boric acid and magnesium sulphate components, respectively, in sea water. Similarly, P_1, P_2, and P_3 denote the depth pressure components for boric acid, magnesium sulphate, and pure water, respectively. The relaxation frequency for boric acid, denoted as f_1 (in kHz), is given by Eq. (6). In Eq. (6), S represents salinity (in parts per 1000), and T represents temperature in degrees Celsius. The relaxation frequency for magnesium sulfate, denoted as f_2 (in kHz), is given by Eq. (7).

$$L_{ab} = (\alpha \times R_t) \times 10^{-3} \tag{4}$$

$$\alpha = \frac{A_1 P_1 f_1 f^2}{f^2 + f_1^2} + \frac{A_2 P_2 f_2 f^2}{f^2 + f_2^2} + A_3 P_3 f^2 \tag{5}$$

$$f_1 = 2.8 \left(\frac{S}{35}\right)^{0.5} \times 10^{\left[4 - 1245/(273 + T)\right]} \tag{6}$$

$$f_2 = \frac{8.17 \times 10^{\left[8 - 1990/(273 + T)\right]}}{1 + 0.0018(S - 35)} \tag{7}$$

3.4 Transmission Loss Due to Surface Reflections

Transmission loss due to surface reflections in underwater environments refers to the reduction in sound intensity caused by the reflection of sound waves at the interface between water and the surface, such as the water-air interface or the water-seabed interface. These reflections result in energy being redirected away from the desired propagation path, leading to a decrease in received sound level. The transmission loss (assuming direct path) resulting from cylindrical spreading and absorption can be mathematically expressed using (8). The transmission loss due to surface reflections can be expressed using (9) by considering the wind speed (w) and angle of incidence (θ). Finally, the transmission losses when an acoustic wave transmits in a multipath environment can be expressed as sum of transmission loss due to direct path and surface reflections which can be represented using (10).

$$TL_{dp} = 20\,log_{10}R_t + \alpha R_t \times 10^{-3} \tag{8}$$

$$TL_{SR} = 10 \times log\left[\frac{1 + \left(f/f_1^2\right)}{1 + \left(f/f_2^2\right)}\right] - (1 + (90 - w)/60)\left(\frac{\theta}{30}\right)^2 \tag{9}$$

$$TL_{Multi-path} = TL_{dp} + TL_{SR} \tag{10}$$

4 Simulation Parameters

To simulate the transmission losses of an UASN, several parameters need to be considered for the simulation model which helps in accurately predicting the transmission losses. Table 1 provides the detailed list of parameters used for simulation along with their ranges.

Table 1. Execution Parameters

Parameter	Range
Depth (meters)	100–7000
Temperature (°C)	27–4
Salinity (ppt)	30–37
Frequency (kHz)	0.1–100
pH	7.8
R_t (meters)	100
Wind speed (w m/s)	4–12.5
Angle of incidence (Theta)	20–36

5 Simulation Results

The speed of sound in deep water significantly impacts the transmission of underwater acoustics. As depicted in Fig. 1, a noticeable trend is observed regarding sound speed in relation to changes in the depth of the scenario and temperature. At a specific temperature and depth (T = 30 °C, D = 100 m), the initial sound speed is measured to be 1450 m/s. However, as the depth increases to 7000 m and the temperature decreases to 4 °C, the sound speed significantly rises to 1650 m/s, as evident in Fig. 1. This variation in sound speed, influenced by both temperature and depth, emphasizes the significant relationship between these parameters and their impact on sound propagation characteristics in the given underwater environment. Understanding these variations is essential for accurate acoustic communication and exploration in deep water environments.

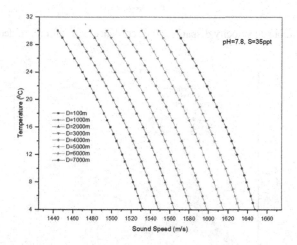

Fig. 1. Acoustic velocity disparities in accordance with temperature in deep water.

Variations in salinity significantly influence the acoustic velocity in deep water. Changes in salt concentration alter the water's density and compressibility, leading to modifications in the sound speed. Higher salinity levels typically result in increased sound speed, while lower salinity levels correspond to decreased sound speed.

These salinity-induced changes in sound speed have implications for underwater acoustic communication, sonar systems, and environmental monitoring in deep-water environments. As the depth increases and salinity levels rise, the sound speed also increases. At a specific salinity (S = 33 ppt), different sound speed profiles are observed along the depth, ranging from 1540 m/s to 1650 m/s, as depicted in Fig. 2. This demonstrates the direct correlation between depth, salinity, and the resulting variations in sound speed. The frequency of sound waves plays a crucial role in determining the transmission loss in deep-water environments. Figure 3, demonstrate the transmission loss by varying salinity, and temperature with respect to depth in deep water.

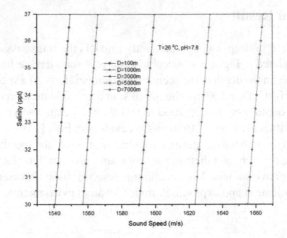

Fig. 2. Acoustic velocity disparities in accordance with salinity in deep water.

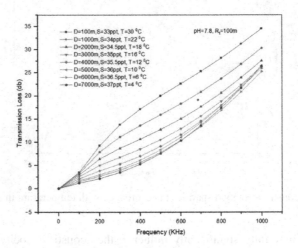

Fig. 3. Frequency *vs* Transmission losses in deep water.

The presence of different chemicals in underwater environments can lead to sound absorption. Chemical components such as dissolved gases, salts, and organic matter have distinctive absorption characteristics, influencing the propagation of sound waves in the water medium. These chemicals interact with sound waves, absorbing energy at specific frequencies and resulting in a reduction in sound intensity over distance (see Fig. 4).

The transmission losses due to surface reflections are purely depends on frequency, windspeed and angle of incidence. Figure 5, shows the transmission losses due to surface reflections under various wind speed values at a particular depth in deep water. It is observed that, the transmission losses are directly proportional to wind speed and frequency. Similarly, the angle of incidence at which the acoustic wave reflected back from

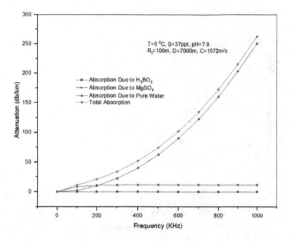

Fig. 4. Absorption in deep water due to various chemical compositions in deep water.

the surface also had an impact on transmission losses. Figure 6, shows the transmission losses due to surface reflections under various angle of incidence values at a particular depth in deep water. As the angel of incidence increases, the transmission losses are gradually decreased.

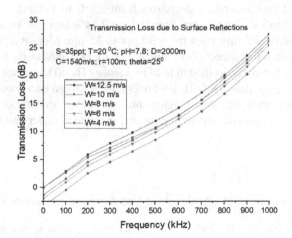

Fig. 5. Transmission Losses in deep water for multipath model with varying wind speed.

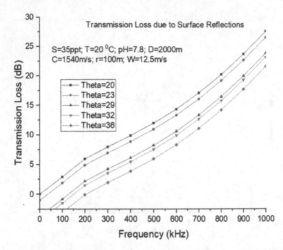

Fig. 6. Transmission Losses in deep water for multipath model with varying theta.

6 Conclusion

The primary objective of this study is to assess transmission losses resulting from surface reflections in deep water environments using a multipath acoustic channel model. The simulation explores the impact of absorption, sound speed, temperature, and salinity on these losses, while varying the depth of the network scenario to analyse their effects. The simulation results reveal noteworthy observations. When the depth varies from 100 m to 7000 m and the temperature decreases from 30 °C to 4 °C, the acoustic velocity increases by 250 m/s in deep water. Similarly, a 7.14% increase in acoustic velocity is observed when the salinity rises from 30 ppt to 35 ppt. Moreover, the study shows that an increase in wind speed (W) from 4 m/s to 12.5 m/s leads to a 5 dB rise in transmission loss. Similarly, as the angle of incidence (Theta) increases from 20° to 30°, transmission losses escalate by 8 dB. By understanding these transmission loss factors, this research contributes valuable insights into optimizing acoustic communication in deep water environments and improving the performance of underwater acoustic sensor networks.

References

1. Sozer, E.M., Stojanovic, M., Proakis, J.G.: Underwater acoustic networks. IEEE J. Ocean. Eng. **25**(1), 72–83 (2000)
2. Akyildiz, I.F., Pompili, D., Melodia, T.: Underwater acoustic sensor networks: research challenges. Ad Hoc Netw. **3**(3), 257–279 (2005)
3. Venkateswara Rao C., Padmavathy, N.: Effect of link reliability and interference on two-terminal reliability of mobile ad hoc network. In: Advances in Data Computing, Communication and Security. Lecture Notes on Data Engineering and Communications Technologies, vol. 106, pp. 555–565 (2022)
4. Rao, C.V., Padmavathy, N., Chaturvedi, S.K.: Reliability evaluation of mobile ad hoc networks: with and without interference. In: IEEE 7th International Advance Computing Conference, pp. 233–238 (2017)

5. Barbeau, M., Garcia-Alfaro, J., Kranakis, E., Porretta, S.: The sound of communication in underwater acoustic sensor networks: (position paper). In: Ad Hoc Networks: 9th International Conference, AdHocNets Niagara Falls, ON, Canada, pp. 13–23 (2017)
6. Akyildiz, I.F., Pompili, D., Melodia, T.: Challenges for efficient communication in underwater acoustic sensor networks. ACM SIGBED Rev. Spec. Issue Embed. Sens. Netw. Wirel. Comput. **1**(2), 3–8 (2004)
7. Stojanovic, M., Preisig, J.: Underwater acoustic communication channels: propagation models and statistical characterization. IEEE Commun. Mag. **47**(1), 84–89 (2009)
8. Jindal, H., Saxena, S., Singh, S.: Challenges and issues in underwater acoustics sensor networks: a review. In: International Conference on Parallel, Distributed and Grid Computing Solan, pp. 251–255 (2014)
9. Ismail, N.S.N., Hussein, L., Syed, A., Hafizah, S.: Analyzing the performance of acoustic channel in underwater wireless sensor network. In: Asia International Conference on Modelling & Simulation, pp. 550–555 (2010)
10. Wanga, X., Khazaiec, S., Chena, X.: Linear approximation of underwater sound speed profile: precision analysis in direct and inverse problems. Appl. Acoust. **140**, 63–73 (2018)
11. Ali, M.M., Sarika, J., Ramachandran, R.: Effect of temperature and salinity on sound speed in the central Arabian sea. Open Ocean Eng. J. **4**, 71–76 (2011)
12. Kumar, S., Prince, S., Aravind, J.V., Kumar, G.S.: Analysis on the effect of salinity in underwater wireless optical communication. Mar. Georesour. Geotechnol. **38**(3), 291–301 (2020)
13. Preisig, J.: Acoustic propagation considerations for underwater acoustic communications network development. Mob. Comput. Commun. Rev. **11**(4), 2–10 (2006)
14. Sehgal, A., Tumar, I., Schonwalder, J.: Variability of available capacity due to the effects of depth and temperature in the underwater acoustic communication channel. In: Oceans 2009-Europe, Bremen, pp. 1–6 (2009)
15. Leroy, C.C., Parthiot, F.: Depth-pressure relationships in the oceans and seas. J. Acoust. Soc. Am. **103**(3), 1346–1352 (1998)
16. Yuwono, N.P., Arifianto, D., Widjiati, E., Wirawan.: Underwater sound propagation characteristics at mini underwater test tank with varied salinity and temperature. In: 6th International Conference on Information Technology and Electrical Engineering (ICITEE), pp. 1–5 (2014)
17. Shi, H., Kruger, D., Nickerson, J.V.: Incorporating environmental information into underwater acoustic sensor coverage estimation in estuaries. In: MILCOM 2007 - IEEE Military Communications Conference, pp. 1–7 (2007)
18. Morozs, N., Gorma, W., Henson, B.T., Shen, L., Mitchell, P.D., Zakharov, Y.V.: Channel modeling for underwater acoustic network simulation. IEEE Access **8**, 136151–136175 (2020)
19. Lee, H.K., Lee, B.M.: An underwater acoustic channel modeling for Internet of Things networks. Wireless Pers. Commun. **116**(3), 2697–2722 (2020). https://doi.org/10.1007/s11277-020-07817-x
20. Onasami, O., Feng, M., Xu, H., Haile, M., Qian, L.: Underwater acoustic communication channel modeling using reservoir computing. IEEE Access **10**, 56550–56563 (2022)
21. Zhu, X., Wang, C.X., Ma, R.: A 2D non-stationary channel model for underwater acoustic communication systems. In: IEEE 93rd Vehicular Technology Conference (VTC2021-Spring), pp. 1–6 (2021)
22. Kotipalli, P., Vardhanapu, P.: Frame boundary detection and deep learning-based doppler shift estimation for FBMC/OQAM communication system in underwater acoustic channels. IEEE Access **10**, 17590–17608 (2022)
23. Venkata Lalitha, N., et al.: IoT based energy efficient multipath power control for underwater sensor network. Int. J. Syst. Assur. Eng. Manag. 1–10 (2022)
24. Sekhar, S., et al.: Effects of water absorption on the mechanical properties of hybrid natural fibre/phenol formaldehyde composites. Sci. Rep. **11**(1), 13385 (2021)

25. Etter, P.C.: Underwater Acoustic Modeling and Simulation. CRC press (2018)
26. Padmavathy, N., Venkateswara Rao, C.H.: Reliability evaluation of underwater sensor network in shallow water based on propagation model. J. Phys. Conf. Ser. **1921**(1), 012018 (2021)
27. Venkateswara Rao, C., Swathi, S., Charan, P.S.R., Santhosh Kumar, C.V., Pathi, A.M.V., Praveena, V.: Evaluation of sound propagation, absorption, and transmission loss of an acoustic channel model in shallow water. In: Congress on Intelligent Systems, pp. 455–465 (2023)
28. Padmavathy N., Venkateswara Rao, C.H.: Effect of undersea parameters on reliability of underwater acoustic sensor network in shallow water. In: IOP Conference Series: Materials Science and Engineering, vol. 1272, no. 1, p. 012011 (2022)
29. Venkateswara Rao, C., et al.: Analysis of acoustic channel model characteristics in deep-sea water. In: International Conference on Cognitive Computing and Cyber Physical Systems, pp. 234–243 (2023)
30. Venkateswara Rao, C., et al.: Comparison of acoustic channel characteristics in shallow and deep-sea water. In: International Conference on Cognitive Computing and Cyber Physical Systems, pp. 256–266 (2023)

Performance Evaluation of Multicast Routing Protocols in Mobile Ad Hoc Networks

N. Padmavathy(✉) ⓘ, K. Srinivas ⓘ, and P. Srinivas

Department of Electronics and Communication Engineering,
Vishnu Institute of Technology (A), Bhimavaram, AP, India
padmavathy.n@vishnu.edu.in

Abstract. Wireless Ad hoc networks are networks connecting mobile devices that are self-contained. It's these networks simplicity and ease of implementation make them the best choice for frontline communications, incident management, and other related applications when dependable infrastructure is not easily accessible. The applicability of such networks has been challenging due to partial bandwidth, energy restrictions, and unexpected system topologies. Recent years have seen resurgence in this area's research. Specifically in routing, security and multicast concerns. The paper focuses on multicast routing in peer-to-peer networks in this paper. In this work, the source packet dumping, a fresh multicast routing system has been presented. Based on restrictions on hop distance, the connectivity routes between providers and group members are established as a single hop/multihop. In order to ensure effective data dissemination, a probabilistic data forwarding mechanism has been suggested. The simulation results demonstrated the performance of suggested routing protocol comparison to factors that define an ad hoc network. It is evident from simulation results, that the suggested protocol delivers effective data distribution and is resistant to topology changes.

Keywords: Ad hoc network · Connectivity · Dynamic topology · Hop distance · Routing Protocols

1 Introduction

Ad hoc networks are nodes that self-organize and communicate with one another directly rather than through a fixed infrastructure [1]. As nodes move and come into range of one another, these networks can spontaneously form. Smart phones, laptops, and other portable devices, as well as sensors and actuators, can all function as nodes in an ad hoc network. The key characteristics of ad hoc networks include decentralization, changeable topology, infrastructure-less, resource constraints, and multi-hop. Ad hoc networks can be used for a variety of purposes, including automotive networks, wireless sensor networks, and disaster relief efforts [2–4]. Ad hoc networks, however, can present a number of technological difficulties, including routing, security, energy conservation, and quality of service (QoS).

P. Pareek et al. (Eds.): IC4S 2023, LNICST 537, pp. 273–287, 2024.
https://doi.org/10.1007/978-3-031-48891-7_24

Routing protocols in mobile ad hoc networks (MANETs) are designed to handle the dynamic topology of the network, where nodes can move around and change their position frequently. And also responsible for discovering and maintaining routes to destinations, even as nodes move around or leave the network. Routing protocols can help improve the performance of the network by reducing delays, minimizing packet loss, and optimizing the use of network resources. Routing protocols can select the most efficient path for data transmission and avoid congested areas, which can lead to better network performance. Routing protocols are responsible for ensuring reliable communication between nodes in the network. Routing protocols can detect link failures, re-establish routes, and prevent data loss, which can improve the overall reliability of the network. Routing protocols can adapt to the changing network conditions, such as changes in topology, traffic load, and link quality. Routing protocols can adjust their behavior to optimize network performance, which can lead to better throughput and reduced latency.

The MANET ideal is a good fit for the multicast standard since hosts typically work together as a group in an ad hoc environment to complete a task. Additionally, the multicast paradigm boosts network efficiency through widespread data dissemination, making it perfect for MANETs and other networks with limited capacity. Consequently, multicast communicating in ad hoc networks is really essential. Hence, a novel multicast routing protocol for wireless ad hoc networks in this paper is proposed. These protocols' effectiveness as multicast protocols was assessed and contrasted with the effectiveness of conventional flooding. A variety of operational values for different characteristics that define a MANET, such as mobility speed, traffic load, network density, etc., based on trade-off curves has been considered. And also multicasting in a MANET with multihop connectivity [5] is significantly more complex than in wired networks due to aforementioned characteristics, addressing and deployment considerations etc.

Routing protocols are essential in mobile ad hoc networks (MANETs) because it enable nodes to communicate with each other and exchange data, even in the absence of a centralized infrastructure. MANETs are characterized by their dynamic topology, where nodes can move around and join or leave the network at any time. As a result, routing becomes a complex task in MANETs, and specialized routing protocols are needed to ensure efficient and reliable communication.

2 Literature Review

Multicast routing is an important communication paradigm in mobile ad hoc networks (MANETs) that enables efficient group communication between multiple nodes. In this literature review, we will discuss some of the commonly used multicast routing protocols for MANETs are; On-Demand Multicast Routing Protocol (ODMRP): Ad Hoc Multicast Routing Protocol (AMRoute): Core-Assisted Mesh Protocol (CAMP): Zone-Based Hierarchical Link State (ZHLS). Comprehensive multitree routing proto-cols for connected/fixed networks include DVMRP [6], MOSPF [7], CBT [8], PIM-SM and PIM-DM [9]. Currently few researchers used Energy Constraint Secure Routing Protocol [10] used tree-based protocols [11–13] to reduce power constraint and dynamic connectivity.

Mesh protocols [14, 15, 18, 19] are used to establish and maintain mesh structures, or a group of network nodes that connect all the teammates. The Neighbour-Supporting Multicast mechanism (NSMP) [16, 17], a novel ad hoc multicast routing mechanism adopts a mesh topology to increases the resilience against mobility. In addition, NSMP also make use of node localization to lower the cost of route failure recovery and mesh maintenance. Due to redundant data transmission, this network's structure is more resistant to network dynamics thanks to the mesh of nodes that offers numerous paths to the group's members. According to a UCLA comparison mesh techniques are being studied for ad hoc multicast routing systems more resistant to topology changes since there are more channels to the destination [20], but performance is compromised because of terminated transmitting data. Grid protocols are said to be less efficient compared to tree-based protocols. In fact, interactive routing trees are the most efficient structure for moving data. Decision trees are subject to structural changes. During the tree update process, common topology modifications could lead to significant data loss and extreme control exchange. Mesh- based protocols have higher levels of dependability for a variety of mobility speeds. Extensions of unicast protocols include [11, 21–24] and other multicast routing protocols. The two functionalities are sufficiently different to merit independent attention, thus it is still debatable whether it is a smart idea to merge them [22].

An improved Dynamic Source Routing (DSR) Protocol with a new schema called as O-DSR was proposed in [25] to maximise the network lifetime [26–28] of mobile nodes and also overcoming the congestion simultaneously. LLECP-AOMDV lowers node energy consumption, and has a shorter average end-to-end delay [26]. The modified O-DSR algorithm called as MDSR [29] along with Ant Colony Optimization (ACO) finds the best path and optimises total weight (cost, delay, and hop count) of the network.

3   Multicast Routing Using Source Grouped Flooding

A new multicast routing system has been proposed for wireless mobile ad hoc networks that build a network of node which is built from a source called a flood set to deliver information from that source. Flood sets differ from routing sets in that the former are constructed using distance and hop enumeration metrics, while the latter are constructed using the inverse shortest path mechanism. Although the flooding cluster is a source crafted mesh, the routing group is also a grouping based lattice of nodes.

3.1   Flooding Group Creation

Each source in this protocol specifies routes to all other participants in the multicast group as needed. For the origin, the flood set is the result of the origin request phase and the set response phase. The resulting distance constraints dictate the creation of flood sets for sources, as given in (1) and (2)

$$D_{sn} \leq D_{sm} \tag{1}$$

$$D_{mn} \leq D_{sm} \tag{2}$$

The value of D_{sn} denotes the count of intermediary steps from input to central node. The number of hops that exist between the origin point as well as a multicast group participant is referred as D_{sm}. In between multicast attendee and the middle node, measured in hops is represented using D_{mn}. In the network, nodes decide whether to join the dive group (group of nodes operating as a cluster to standardize an IP routing protocol that is application specific) according to the distance using (1). During the request-reply process, the nodes acquire these distance metrics. A source periodically broadcasts a join request message whenever it has packets to send. The address of the multicast team with the count of a hop pitch is both included in the join request message. The flooding group is updated as follows by this recurring communication. Upon receiving a unique join request, a node will increase the hop count of the origin and save it in its own memory before relaying the packet.

Upon receipt of a request to join, a multicast group member takes note of the distance between the source D_{sm} and itself, adds a standard delay, and subsequently issues a response in the form of a "join" as a reply message. The reply to the accept request includes details regarding both the multicast group and the hop count between the source and destination nodes. Typically, the TTL in this message corresponds to the hop count of the origin (D_{sm}). Also, a node only forwards the response message if it's not being used to download things right now, it won't help with fixing the route, otherwise the information is lost. So, while doing something, given update command, the node reader only issues the first response message for each resource. As a result, the protocol generates for each source a flood set consisting of nodes subject to the hop distance (see Fig. 1).

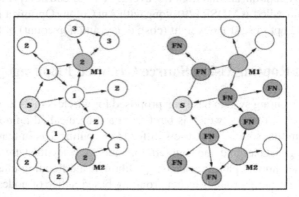

Fig. 1. Flooding Group Creation.

3.2 Flooding Group Update and Soft-State

The join request message will be periodically rebroadcast, reinforcing the flooding group for apiece source. This route restores takes into consideration topology variations brought on by member discovery and mobility. The protocol records what group members remember, but one's memory may change over time. To become part of or exit a group, members do not send public messages.

3.3 Detection of Duplicate Packets

Every packet from apiece source contains a broadcast order number. The broadcast structure number, which uniquely identifies each packet produced by the source and is made up of the source address and a counter assessment. Any node that gets a packet by a categorization 18 number higher than the significance that is currently stored processes the packet and updates the cache. If not, the package will be removed because it is a duplicate.

3.4 Data Forwarding

A source has new, active routes to every associate of the multicast group after the flooding group has been created. Only nodes in that source's forwarding group will assume that the packet was sent by that source. Using the distinct source broadcast ID, all duplicate data packets are detected and deleted. The data forwarding mechanism and potential issues brought on by redundant data transmissions are revealed in Fig. 2.

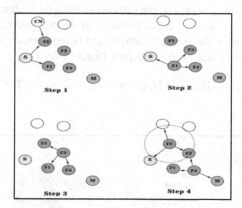

Fig. 2. Contention and Collision during Data Forwarding.

3.5 Hop Count Data Forwarding (HCDF)

Care has been taken to add a hop count pitch to the data packets to reduce the MAC layer clogging and collisions caused by redundant data transmission. The increased data passing mechanism is shown in Fig. 3. The hop count distances discovered through the request response switch phase are represented by the numerical numbers inside the nodes. The axes represent where a given broadcast will go, while the arrow values represent the number of packet hops. Steps 1 through 3 correspond to Fig. 3.

The hop count of the packet that node F2 receives from node F3 is set to 2 in step 4. Even though the packet is original, node F2 won't forward it in this scheme since the hop count it has stored is less than the hop count in the data packet. Thus, extra communication that could cause conflict and a collision is prevented

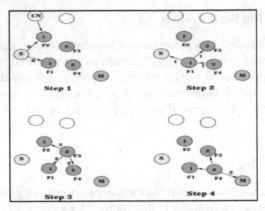

Fig. 3. Hop count based Data Forwarding.

3.6 Size Controlling of the Flooding Group

If the distance between that node and the start point, plus the distance between that node and the end point, is less than or equal to the total distance between the start and end point, then that node is on the shortest path, which is represented using (3) a controlled dive group (shortest path) has been represented using Fig. 4.

$$D_{sn} + (D_{sm} - TTL_{rep}) \leq D_{sm} \rightarrow D_{sn} \leq TTL_{rep} \tag{3}$$

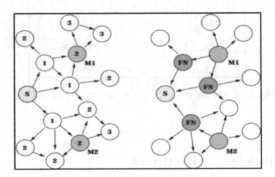

Fig. 4. Creation of Controlled Flooding Group.

3.7 Probabilistic Data Forwarding

The group flooding offers several routes after the source to the group's participants. To reduce data overhead a probabilistic data forwarding mechanism has been proposed and presented how to figure out the chance of a packet being sent again (P_{send}). The probability that a node chooses to retransmit a packet is P_{send}, and the probability that a node chooses to discard the packet is $(1 - P_{send})$ (Fig. 6).

$$P_{send} = \frac{1}{1+N} \tag{4}$$

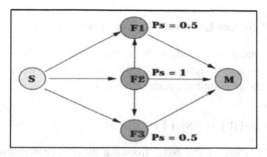

Fig. 5. Probabilistic data forwarding.

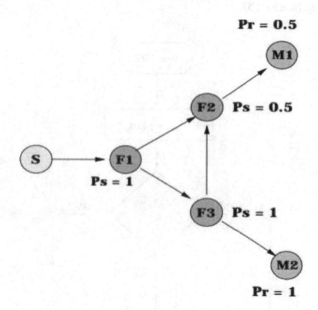

Fig. 6. Non-Guaranteed data delivery.

All duplicate data packets, regardless of hop count number, are dropped after a data packet has been retransmitted which is represented using (4). Where, N is the amount of packets that are the same and expected to leap the same number of times on that layer.

4 Algorithm

Based on grouped flooding protocol, an algorithm for evaluating the performance of MANET has been developed based on flooding protocol design.

4.1 Basic Source Grouped Flooding Protocol (BSGFF)

By using a rule called distance constraint (1) the algorithm creates groups that make you feel like you're inside them. The algorithm is described in the flowchart below, which also describes the methods used to generate flood sets and transfer data

4.2 Shortest Path Source Grouped Flooding Protocol

Using distance constraint (3), this algorithm creates the flooding groups with the shortest paths. This algorithm is employed to determine whether shortest path flooding groups are advantageous based on shortest path (see Fig. 4).

4.3 Probabilistic BSGFF (PBSGFF)

This algorithm makes use of the basic flooding group's probabilistic data advancing mechanism. Similar to Fig. 7, the routing has been established based on probabilistic data forwarding (see Fig. 5).

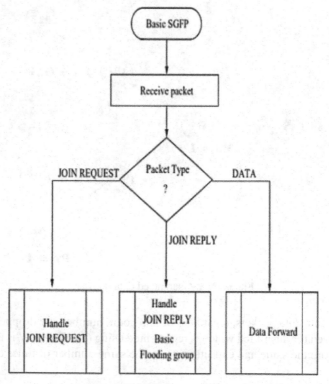

Fig. 7. BSGFF Flow Diagram.

5   Simulation Parameters

The performance investigations of the multicast protocols through simulation have been performed using Matlab 2018a in Windows 10 environment at speed of 3.91 GHz. The simulation parameters adaption for the study are listed in Table 1,

Table 1. Simulation Parameters

Parameters	Specification
Simulation area	$1000000 \ m^2$
Network Size	1 to 70 Nodes
No. of Source Nodes	1 – 20 Nodes
Mobility	Static Network
Node Speed	0–5 m/s
Transmission Range	250 m
Run time	100 secs
	Flooding
	Basic-sgfp
Protocol	Sp-sgfp
	p-sgfp
	psp-sgfp

6   Simulation Results

The proposed algorithms were simulated using MATLAB. The simulation represented a network of nodes that were dispersed at random over an area of 1000×1000 Sq. Metres. As a simulation parameter, the network thickness, or the amount of nodes in the network, was changed. The reach of the radio on each point was 250 m and the amount of information it could send was 1 million bits per second. If there is no special space for sending messages in our setup, when two or more messages are sent at the same time, they are not delivered. Nodes have two-way connectivity and broadcast are the communication medium. The extent of each replication was 100 s. For each situation, multiple runs were performed using various seed values, and the gathered data was be around throughout these runs. The OPNET routing layer protocols used for the multicast algorithms were developed separately. The goodput of proposed algorithm for various network attributes have been analyzed.

The source grouped protocols appear to work best in the mobility range of 0−5 m/s. The protocols are most efficient (best Goodput) and effective in this range (least total overhead). We can see that the most effective protocol for maintaining good throughput within 10% of flooding is the psp-sgfp scheme (see Fig. 8).

The trade-off among Goodput and overall slide for various sources sending packets to the multicast group is depicted in Fig. 9. It can see that the overall overhead for every

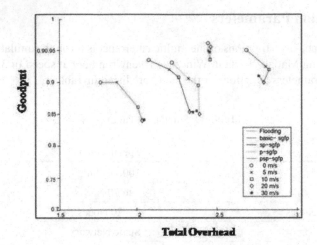

Fig. 8. Total Overhead Vs Goodput for different mobility speeds.

scheme is essentially unchanged. With an increase in sources, the output falls off linearly. For 1–5 network sources, all schemes exhibit minor variations in Goodput. The Goodput decreases slightly for 10 sources. Therefore, for 1–10 sources, the protocols' efficacy is comparatively stable. The most effective scheme is the psp-sgfp scheme. Additionally, it achieves a good output that is 6% less than flooding's.

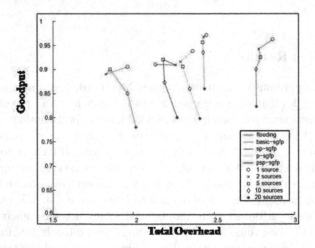

Fig. 9. Total Overhead Vs Goodput for different number of sources.

Figure 10 shows good performance and general tradeoffs for different multiplex group sizes. Flood mode records a stable output as the pool size increases. Full flooding occurs when there are 40 (or around 1.2) members in the cluster, indicating that flooding is effective when 70% or more of the network nodes are cluster members. Overall, flood values for flood events categorized by source are good, falling between 6 and 10%. Flood continues to be superior to psp-sgfp. Psp-sgfp seems to be the most effective of these techniques for MANETs when more than 40% of nodes are cluster members.

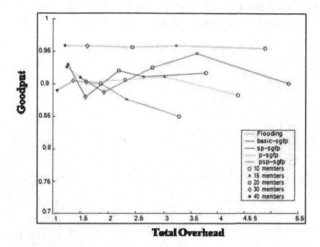

Fig. 10. Total Overhead Vs Goodput for different number of members.

In the Fig. 11, it is clearly depicted that, how producing more or less affects the cost depending on how often updates are made. The floods analyze has not been done because it doesn't happen during softening periods. However when it is updated over interval longer, fewer control packets are generated. This helps in decreasing the amount of overall overhead. The figure shows that even with different update intervals, the important information stays mostly the same. The best times for sending loits of packets to fix a problem are every 6, 8, or 10 s.

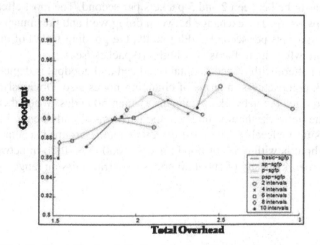

Fig. 11. Total Overhead Vs Goodput for different refresh intervals.

The system called psp-sgfp worked best when the time to refresh was either 8 or 10 s, even when the loads were very low at 1.5. Execution of unused pool individuals must be deferred until another way overhaul cycle until the asset enrollment asks has been replied. This implies that on the off chance that the bunch enrollment is exceptionally energetic; a lower esteem for the upgrade interim ought to be utilized.

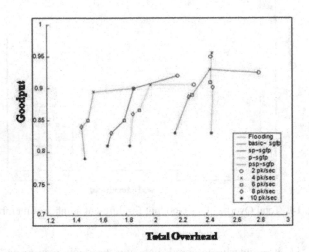

Fig. 12. Total Overhead Vs Goodput for different traffic load.

Figure 12, displays the exchange between doing well and how much work it takes, which is measured in packets per second. Diverse Traffic load in packets/s. Increase, the total amount of flooding mentioned above remains unchanged. When using source-grouped flooding schemes, the overall overhead originally drops as the load rises before slowly stabilizing at a constant overhead. The ideal traffic loads for source grouped schemes appear to be between 2 and 5 packets per second. The most effective scheme is once again psp-sgfp. The exchange between doing well and how much work it takes is measured in packets per second. Additionally, the goodput for all of the schemes is almost the same when the traffic is very high (10 packets/sec).

The graph pattern with respect to total overhead and goodput is depicted in Fig.13, as the network compactness in terms of changing nodes size. Obviously, all schemes perform better when the network contains more than 50 nodes. When there are 60 and 70 nodes in the network, the flooding scheme performs slightly better, but the overall overhead goes up noticeably. The most effective source-grouped scheme is psp-sgfp, and its throughput is within 8% of flooding's. A 1000 m × 1000 m network seems to have a good network density of about 50 nodes with transmission range as 250 m.

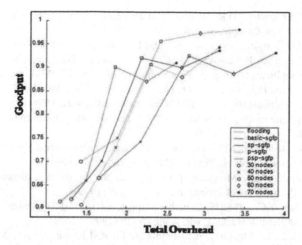

Fig. 13. Total Overhead Vs Goodput for different network density.

7 Conclusion

The design of multicast routing protocols faces significant challenges due to the significant limitations of mobile-targeted wireless networks, namely mobility, bandwidth, and power limitations. Hence, a multicast routing system for a MANET should accomplish effective data distribution and be resistant to changes in topology. In this work a source collection dump methods for multicast routing in MANET have been proposed. For each source, flood groups are generated using a finite distance diagram. The flooding group is an efficient and resistant to mobility each source, multiple path, and mesh construction. We showed that the protocol is more effective when a probabilistic data forwarding mechanism is used, which is based on probabilities obtained from the network. Moreover, the shortest path flooding strategy increases protocol effectiveness by reducing the need for rebroadcasts. It's understood that psp-sgfp (Source Shortest Probable Path Packet Flooding Protocol) is 8% faster than flooding.

References

1. Leiner, B.M., Neison, D.L., Tobagi, F.A.: Issues in packet radio network design. In: Proceedings of the IEEE, Special Issue on Packet Radio Networks, vol. 75, no. 1, pp. 6–20 (1987).
2. Jublin, J., Tornow, J.D.: The DARPA packet radio network protocol. In: Proceedings of the IEEE, vol. 75, no. 1, pp. 21–32 (1987)
3. Clausen, T., Dearlove, C., Dean, J.: Mobile adhoc networks IETF chapter, 88 (2011). http://www.ietf.org/html.charters/manet-chapter.html
4. Paul, S.: Multicasting on the Internet and its Applications. Springer Publishers, New York, p. 421 (1998)
5. Maragatharajan, M., Balakannan, S.P.: Analysis of multicast routing protocols for secure MANET. In: 2017 IEEE International Conference on Intelligent Techniques in Control, Optimization and Signal Processing, pp. 1–6 (2017)

6. Deering, S.E., Cheriton, D.R.: Multicast routing in datagram internetworks and extended LANS. ACM Trans. Comput. Syst. **8**(2), 85–110 (1990)

7. Moy, J.: Multicast routing extensions for OSPF. Commun. ACM **37**(8), 61–66 (1994)

8. Ballardie, T., Francis, P., Crowcroft, J.: Core based trees (CBT): an architecture for scalable inter-domain multicast routing. In: Proceedings of ACM SIGCOMM, pp. 85–95 (1993)

9. Deering, S., Estrin, D.L., Farinacci, D., Jacobson, V., Liu, C.G., Wei, L.: The PIM architecture for wide-area multicast routing. IEEE/ACM Trans. Netw. **4**(2), 153–162 (1996)

10. Moy, J.: Link state routing. In: Steenstrup, M.E. (ed.) Routing in Communications Network. Prentice Hall, vol. xiv, p. 399 (1995)

11. Malkin, G.S., Steenstrup, M.E.: Distance-vector routing. In: Steenstrup, M.E., (ed.) Routing in Communications Network. Prentice Hall, vol. xiv, p. 399 (1995)

12. Royer, E., Perkins, C.E.: Multicast operation of ad hoc on- demand distance vector routing protocol. In: Proceedings of MobiCom, pp. 207–218 (1999)

13. Wu, C.W., Tay, Y.C.: AMRIS: a multicast protocol for Ad Hoc wireless networks. In: Proceedings of IEEE MILCOM, vol. 1, pp. 25–29 (1999)

14. Liu, M., Talpade, R., McAuley, A., Bommaiah, E.: AM Route: Ad Hoc Multicast Routing Protocol. Technical Report 8 (1999)

15. Lee, S.J., Gerla, M., Chiang, C.C.: On-demand multicast routing protocol. In: Proceedings of IEEE WCNC, vol. 3, pp. 1298–1304 (1999)

16. Lee, S., Kim, C.: Neighbor supporting ad hoc multicast routing protocol. In: Proceedings of the ACM/IEEE Workshop on Mobile Ad hoc Networking and Computing (MOBIHOC), pp. 37–44 (2000)

17. LAN/MAN Standards Committee of IEEE Computer Society. Wireless LAN Medium Access Protocol (MAC) and Physical Layer (PHY) Specification. IEEE, STD 802.11 Edition (1999)

18. Opnet modeler version 7.0

19. Madruga, E.L., Garcia-Luna-Aceves, J.J.: Scalable multicasting: the core assisted mesh protocol. ACM/Baltzer Mob. Netw. Appl. J. Spec. Issue Manag. Mobility **6**, 151–165 (1999)

20. Sinha, P., Sivakumar, R., Bharghavan, V.: MCEDAR: multicast core extraction distributed ad hoc routing. In: Proceedings of the Wireless Communications and Networking Conference, vol. 3, pp. 1313–1317 (1999)

21. Lee, S.J., Su, W., Hsu, J., Gerla, M., Bagordia, R.: A performance comparison study of Ad hoc wireless multicast protocols. In: Proceedings of IEEE INFOCOM, Tel Aviv, vol. 2, pp. 565–574 (2000)

22. Lusheng, J., Corson, M.S.: A lightweight adaptive multicast algorithm. In: Proceedings of the IEEE Globecom, vol. 2, pp. 1036–1042 (1998)

23. Sabaresan, V., Godfrey Winster, S.: Multicast routing protocols for mobile ADHOC networks. Int. J. Eng. Adv. Technol. **9**(1S) (2019)

24. Baker, R., Ali, M.A.: A survey of multicast routing protocols in Ad-hoc networks. Gazi Univ. J. Sci. **24**(3), 451–462 (2011)

25. Rejab, H., Touil, S., Achour, W.: O-DSR: optimized DSR routing protocol for mobile ad hoc network. Int. J. Wirel. Mob. Netw. **7**(4), 37–47 (2015)

26. Zhang, D.G., Chen, L., Zhang, J.: A multi-path routing protocol based on link lifetime and energy consumption prediction for mobile Edge computing. IEEE Access **8**(1), 69058–69071 (2020)

27. Yirga, H.G., Taye, G.D., Melaku, H.M.: An optimized and energy efficient ad-hoc on-demand distance vector routing protocol based on dynamic forwarding probability (AODVI). J. Comput. Netw. Commun. **2022**, 13 (2022)

28. Prasad, R., Shivashankar, P.: Enhanced energy efficient secure routing protocol for mobile ad-hoc network. Glob. Trans. Proc. **3**(2), 412–423 (2022)

29. El-Sayed, H.H., Younes, A., Alghamdi, F.A.: Multi objective multicast DSR algorithm for routing in mobile networks with cost, delay, and hop count. Complex. Sp. Issue Dyn. Anal. Learn. Robust Control Complex Syst. 1–8 (2021)

Analysis of Acoustic Channel Characteristics in Shallow Water Based on Multipath Model

Y. Durgachandramouli[1] , A. Sailaja[2] , P. Joel Josephson[3] , T. Nalini Prasad[4] ,
K. Eswara Prasad[5] , M. Ravi Sankar[2] , and Ch. Venkateswara Rao[6]([⊠])

[1] Department of ECE, Aditya College of Engineering and Technology, Surampalem, India
[2] Department of ECE, Sasi Institute of Engineering and Technology, Tadepalligudem, India
[3] Department of ECE, Malla Reddy Engineering College, Secunderabad, India
[4] Department of ECE, SRKR Engineering College, Bhimavaram, India
[5] Physics and Electronics Department, B V Raju College, Bhimavaram, India
[6] Department of ECE, Vishnu Institute of Technology, Bhimavaram, India
venkateswararao.c@vishnu.edu.in

Abstract. In the shallow water environment, the water surface, and the seafloor act as reflective boundaries for the sound waves. When a sound wave encounters these boundaries, it undergoes reflection, bouncing back and forth between the surface and the bottom. As a result, the sound energy is distributed among various paths, leading to multipath arrivals at the receiver. The repeated reflections contribute to the complexity of the sound propagation in shallow water. This multipath propagation can cause interference and fading, making the received signals challenging to decode, interpret accurately, and transmission losses. Therefore, proper modelling of channel is essential inoreder to deploy a network with high accuracy. In this work, we have developed and analyzed an acoustic multipath channel model to investigate the impact of mixed layer depth and near field anomaly on transmission losses in underwater environments. The main focus is on understanding how various underwater medium parameters, such as temperature, salinity, depth, and pH, affect the transmission losses. It is evident from simulation results; acoustic velocity has increased by 30 m/s when the temperature reduced from 30 °C to 14 °C and 7 m/s when the salinity increased from 30 ppt to 35 ppt. Transmission losses are increased by 58.8% when the mixed layer depth (MLD) increased from 10 m to 95 m. Whereas, these losses are reduced by 43.7% when the near field anomaly (K_L) increased from 7 dB to 20 dB.

Keywords: Acoustic Channel · Multipath · Reflection · Sound Speed · Temperature · Transmission Loss · Salinity

1 Introduction

Underwater Acoustic Sensor Networks (UASNs) are specialized networks that utilize underwater acoustic communication for data gathering, monitoring, and collaboration in underwater environments [1]. UASNs consist of a collection of autonomous or semi-autonomous underwater sensor nodes that work together to perform various tasks such

© ICST Institute for Computer Sciences, Social Informatics and Telecommunications Engineering 2024
Published by Springer Nature Switzerland AG 2024. All Rights Reserved
P. Pareek et al. (Eds.): IC4S 2023, LNICST 537, pp. 288–297, 2024.
https://doi.org/10.1007/978-3-031-48891-7_25

as oceanographic data collection, environmental monitoring, underwater surveillance, and underwater exploration [2]. UASNs play a crucial role in expanding our understanding of underwater ecosystems, oceanography, and marine resources. They enable real-time monitoring of underwater phenomena, such as temperature, salinity, water quality, marine life, and seabed conditions. By collecting and transmitting data from diverse underwater locations, UASNs provide valuable insights into underwater processes and enable timely decision-making for various applications [3]. The unique characteristics of underwater environments pose significant challenges for UASNs [4, 5]. Underwater acoustic communication is the primary means of data transmission in UASNs since radio frequency signals do not propagate efficiently in water due to high absorption and attenuation. Acoustic signals can travel long distances, but they suffer from limited bandwidth, high propagation delays, multipath fading, and significant signal attenuation, which can affect the performance and reliability of underwater communication systems [6, 7].

To address these challenges, UASNs require specialized hardware and networking protocols [8]. Acoustic modems are used as communication interfaces, employing techniques such as frequency modulation, spread spectrum, and advanced error correction coding to enhance the reliability and data rate of underwater acoustic communication. UASNs also employ advanced signal processing algorithms and networking protocols specifically designed for underwater environments to mitigate the effects of multipath propagation, interference, and limited bandwidth. Multipath propagation is a significant phenomenon that affects UASNs and their communication performance [9]. It is primarily caused by the interaction of sound waves with the underwater environment, including factors such as reflections, refractions, and scattering. Through a sequence of repeated reflections from both the water's surface and the seabed, sound travels vast distances in shallow water. Due to these reflections, sound waves might propagate through many routes before they are detected by a receiver.

The seabed and the water surface serve as sound waves reflecting limits in an environment with shallow water. These barriers cause a sound wave to reflect, causing it to bounce back and forth between the surface and the bottom. The sound energy is thus dispersed along a number of routes, resulting in multipath arrivals to the receiver. Sound transmission in shallow water is difficult due to the recurrent reflections. The arrival timings, amplitudes, and phases of the signals that are received can vary depending on the trajectories that the sound waves follow, as well as their lengths and travel periods [10]. The interference and fading that might result from this multipath propagation make it difficult to decode and correctly understand the signals that are received. Designing and enhancing underwater communication systems and acoustic signal processing algorithms requires an understanding of and consideration for the multipath propagation phenomena in shallow water [11].

2   Related Work

The emergence of Underwater Acoustic Sensor Networks (UASNs) has garnered considerable interest in various fields, thanks to their wide range of applications in underwater monitoring, environmental sensing, marine exploration, and military surveillance. In the context of UASNs, understanding and addressing transmission losses are vital for achieving reliable and efficient communication in underwater environments. Extensive research papers and studies have been dedicated to exploring the factors that influence transmission losses in UASNs. These include path loss models, absorption and scattering effects, multipath propagation phenomena, channel estimation and equalization techniques, as well as network topology and routing considerations. By investigating and comprehending these factors, researchers aim to enhance the performance and effectiveness of UASNs by mitigating transmission losses and optimizing communication strategies.

This research contributes to the advancement of underwater communication systems, enabling them to operate successfully in challenging underwater conditions and fulfill their potential in diverse applications. In [12], the authors have investigated the fundamental physics of wave propagation, specifically focusing on acoustic, electromagnetic (EM), and optical communication carriers. In [13], the authors have extensively studied the impact of propagation characteristics on underwater communication. In [14], the authors have focused on the relationship between propagation loss, ambient noise, and channel capacity in underwater communication. To address inaccuracies in sound speed estimation in oceans and seas, a mathematical model [15] has been proposed. This model provides a conversion framework between atmospheric pressure and depth, as well as depth and atmospheric pressure, aiding in sound speed determination. In [16], the authors have presented an experimental setup that investigates the impact of underwater medium parameters.

A method [17] has been proposed to enhance localization accuracy in underwater environments. To simulate underwater networks effectively, a specifically developed acoustic channel model [18] is employed. In [19], researchers have conducted real-time measurements of route loss in underwater acoustic channels. In [20], a deep learning-based framework has been introduced to enhance accuracy and throughput in channel modeling. The authors have provided a detailed account of the statistical properties of the channel model in [21]. Moreover, a novel technique for frame boundary estimation in UASN has been proposed in [22]. In [23], the authors especially address the clustering in UASN by focusing on the integration of three essential approaches in the context of IoT applications. In [24], the authors investigated how water absorption affected the hybrid phenol formaldehyde (PF) composites' mechanical characteristics.

3 Methodology

This comprehensive approach provides valuable insights into the complex acoustic environment of shallow and deep water, and enabling better understanding and modeling of underwater acoustic propagation.

3.1 Sound Speed

The transmission of sound through water differs significantly from electromagnetic (EM) waves, primarily due to its slow speed [25, 26]. Mackenzie's empirical formula, denoted by (1), provides a means to calculate sound velocity.

$$c(T, S, z) = a_1 + a_2 T + a_3 T^2 + a_4 T^3 + a_5(S - 35) + a_6 z + a_7 z^2 + a_8 T(S - 35) a_9 T z^3 \qquad (1)$$

3.2 Acoustic Propagation loss

Propagation loss of sound refers to the reduction in the strength or intensity of sound waves as they travel through a medium or propagate in a given environment [27]. The loss associated with cylindrical spreading is expressed using Eq. (2), while spherical spreading loss is represented by Eq. (3).

$$L_{CS} = 10 \times \log(R_t) \tag{2}$$

$$L_{SS} = 20 \times \log(R_t) \tag{3}$$

3.3 Absorption Loss

Sound waves in a medium, such as air or water, experience absorption, where the energy of the sound wave is converted into heat [28]. The absorption is frequency-dependent, with higher frequencies generally being absorbed more rapidly which is represented using (4). Where, α is absorption coefficient in underwater, it represents the rate at which sound energy is converted into other forms, such as heat, due to the inherent properties of the water medium. The absorption coefficient is frequency-dependent, meaning that different frequencies of sound waves are absorbed to varying degrees. Higher frequencies generally experience greater absorption than lower frequencies. The absorption coefficient is represented using (5). Where, A_1 and A_2 represent the contributions of boric acid and magnesium sulphate components, respectively, in sea water. Similarly, P_1, P_2, and P_3 denote the depth pressure components for boric acid, magnesium sulphate, and pure water, respectively. The relaxation frequency for boric acid, denoted as f_1 (in kHz), is given by Eq. (6). In Eq. (6), S represents salinity (in parts per 1000), and T represents temperature in degrees Celsius. The relaxation frequency for magnesium sulfate, denoted as f_2 (in kHz), is given by Eq. (7).

$$L_{ab} = (\alpha \times R_t) \times 10^{-3} \tag{4}$$

$$\alpha = \frac{A_1 P_1 f_1 f^2}{f^2 + f_1^2} + \frac{A_2 P_2 f_2 f^2}{f^2 + f_2^2} + A_3 P_3 f^2 \tag{5}$$

$$f_1 = 2.8 \left(\frac{S}{35} \right)^{0.5} \times 10^{[4 - 1245/(273+T)]} \tag{6}$$

$$f_2 = \frac{8.17 \times 10^{[8 - 1990/(273+T)]}}{1 + 0.0018(S - 35)} \tag{7}$$

3.4 Transmission Loss

In shallow water, sound travels a long way by repeatedly reflecting off the bottom and surface, a process known as multipath propagation [29]. This phenomenon introduces transmission losses. The equation employed in this analysis accounts for the horizontal separation distance (r) between the sound source and receiver, specifically when r is within a range of up to 1 times H (skip distance). In this context, H represents the average water depth of the acoustic study area, which serves as a conservative definition for the purposes of this analysis (skip distance). The skip distance H can be defined by using (8). The transmission losses due to multipath in shallow water is defined using (9) for the case, when r is within the range of H, (10) for the case when $H \leq r \leq 8H$ and (11) when $r > 8H$. Where, d is the mixed layer depth, z is the scenario depth, K_L is the near field anomaly, α_T is the shallow water attenuation coefficient [30].

$$H = \sqrt{\frac{1}{3}(d + z)} \tag{8}$$

$$TL_{Multipath} = 20 \times \log(r) + \alpha \times r + 60 - K_L \tag{9}$$

$$TL_{Multipath} = 15 \times \log(r) + \alpha \times r + \alpha_T(r/H - 1) + 5 \times \log(H) + 60 - K_L \tag{10}$$

$$TL_{Multipath} = 10 \times \log(r) + \alpha \times r + \alpha_T(r/H - 1) + 10 \times \log(H) + 60 - K_L \tag{11}$$

4 Simulation Parameters

To simulate the transmission losses of an UASN, several parameters need to be considered for the simulation model which helps in accurately predicting the transmission losses. Table 1 provides the detailed list of parameters used for simulation along with their ranges.

Table 1. Execution Parameters

Parameter	Range
Depth (meters)	0–100
Temperature (°C)	30–12
Salinity (ppt)	30–35
Frequency(kHz)	0.1–100
pH	7.8
R_t (meters)	100
MLD (meters)	10–95
K_L(dB)	7–20

5 Simulation Results

The transmission of underwater acoustics is significantly influenced by the acoustic velocity in shallow water. The speed at which sound travels through water varies with temperature changes. Figure 1 illustrates this relationship, showing that sound speed increases with greater depth but decreases as the water temperature decreases. For instance, at a specific depth of 10 m and a temperature of 30 °C, the initial sound speed is measured to be 1510 m/s. However, when the temperature is lowered to 14 °C while keeping the depth constant, the sound speed increases to 1540 m/s, as shown in Fig. 1. Interestingly, as the depth increases, the sound speed becomes more closely associated with changes in water salinity rather than changes in temperature.

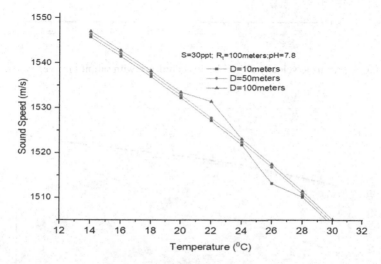

Fig. 1. Acoustic velocity disparities in accordance with temperature in shallow water.

It is observed from Fig. 2, that the salinity variations are also directly proportional to sound speed variations. Higher salinity levels typically result in increased sound speed, while lower salinity levels correspond to decreased sound speed. As the depth increases and salinity levels rise, the sound speed also increases. At a specific salinity ($S = 33$ppt), different sound speed profiles are observed along the depth, ranging from 1534 m/s to 1542 m/s, as depicted in Fig. 2.

The presence of different chemicals in underwater environments can lead to sound absorption. Chemical components such as dissolved gases, salts, and organic matter have distinctive absorption characteristics, influencing the propagation of sound waves in the water medium. These chemicals interact with sound waves, absorbing energy at specific frequencies and resulting in a reduction in sound intensity over distance (see Fig. 3).

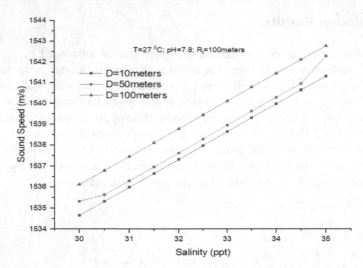

Fig. 2. Acoustic velocity disparities in accordance with salinity in deep water.

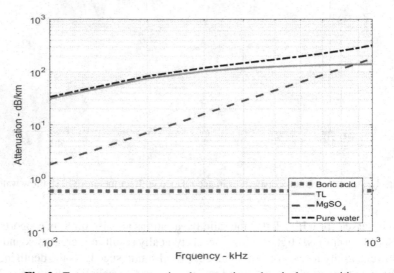

Fig. 3. Frequency *vs* attenuation due to various chemical compositions.

When the mixed layer depth is shallow, it can lead to transmission losses due to high attenuation. Specific impact of mixed layer depth on transmission losses depends on various factors, such as the frequency of the sound waves, the composition of the mixed layer, the characteristics of the seafloor, and the environmental conditions. It is depicted from Fig. 4, that the impact of mixed layer depth on transmission losses under various medium parameters. As the mixed layer depth increase, the transmission losses are also increase for the case of $H \leq r \leq 8H$.

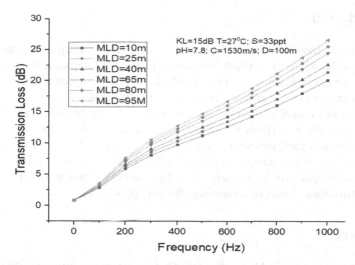

Fig. 4. Transmission Losses in shallow water for multipath model by varying *MLD*.

Near field anomalies occur when sound waves interact with abrupt changes in the water column, such as changes in depth, temperature, salinity, or bottom topography. These anomalies can cause the sound waves to deviate from their expected behavior, resulting in transmission losses. As the value of near field anomaly increase, the transmission losses are reducing gradually (see Fig. 5).

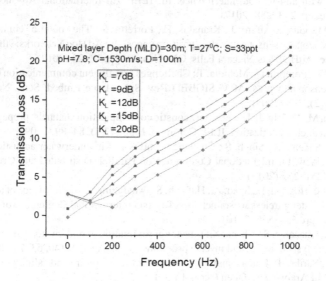

Fig. 5. Transmission Losses in shallow water for multipath model by varying K_L.

6 Conclusion

In this study, we have formulated and examined an acoustic multipath channel model to explore how the mixed layer depth and near field anomaly impact transmission losses in underwater environments. The primary objective is to comprehend the influence of various parameters of the underwater medium, including temperature, salinity, depth, and pH, on transmission losses. Our simulation results reveal intriguing findings. When the temperature decreased from 30 °C to 14 °C, the acoustic velocity increased by 30 m/s. Similarly, a salinity increases from 30 ppt to 35 ppt led to a 7 m/s rise in acoustic velocity. Furthermore, transmission losses surged by 58.8% with an increase in mixed layer depth (MLD) from 10 m to 95 m. Conversely, when the near field anomaly (K_L) increased from 7 dB to 20 dB, these losses decreased significantly by 43.7%.

References

1. Sozer, E.M., Stojanovic, M., Proakis, J.G.: Underwater acoustic networks. IEEE J. Ocean. Eng. **25**(1), 72–83 (2000)
2. Akyildiz, I.F., Pompili, D., Melodia, T.: Underwater acoustic sensor networks: research challenges. Ad Hoc Netw. **3**(3), 257–279 (2005)
3. Venkateswara Rao, Ch., Padmavathy, N.: Effect of link reliability and interference on two-terminal reliability of mobile ad hoc network. In: Advances in Data Computing, Communication and Security. Lecture Notes on Data Engineering and Communications Technologies, vol. 106, pp. 555–565(2022)
4. Rao, C.V., Padmavathy, N., Chaturvedi, S.K.: Reliability evaluation of mobile ad hoc networks: with and without interference. In: IEEE 7th International Advance Computing Conference, pp. 233–238 (2017)
5. Barbeau, M., Garcia-Alfaro, J., Kranakis, E., Porretta, S.: The sound of communication in underwater acoustic sensor networks: (position paper). In: Ad Hoc Networks: 9th International Conference, AdHocNets Niagara Falls, ON, Canada, pp. 13–23 (2017)
6. Akyildiz, I.F., Pompili, D., Melodia, T.: Challenges for efficient communication in underwater acoustic sensor networks. ACM SIGBED Rev. Spec. Issue Embed. Sens. Netw. Wireless Comput. **1**(2), 3–8 (2004)
7. Stojanovic, M., Preisig, J.: Underwater acoustic communication channels: propagation models and statistical characterization. IEEE Commun. Mag. **47**(1), 84–89 (2009)
8. Jindal, H., Saxena, S., Singh, S.: Challenges and issues in underwater acoustics sensor networks: a review. In: International Conference on Parallel, Distributed and Grid Computing Solan, pp. 251–255 (2014)
9. Ismail, N.-S., Hussein, L., Syed, A., Hafizah, S.: Analyzing the performance of acoustic channel in underwater wireless sensor network. In: Asia International Conference on Modelling & Simulation, pp. 550–555 (2010)
10. Wanga, X., Khazaiec, S., Chena, X.: Linear approximation of underwater sound speed profile: precision analysis in direct and inverse problems. Appl. Acoust. **140**, 63–73 (2018)
11. Ali, M.M., Sarika, J., Ramachandran, R.: Effect of temperature and salinity on sound speed in the central Arabian sea. Open Ocean Eng. J. **4**, 71–76 (2011)
12. Kumar, S., Prince, S., Aravind, J.V., Kumar, G.S.: Analysis on the effect of salinity in underwater wireless optical communication. Mar. Georesour. Geotechnol. **38**(3), 291–301 (2020)

13. Preisig, J.: Acoustic propagation considerations for underwater acoustic communications network development. Mobile Comput. Commun. Rev. **11**(4), 2–10 (2006)
14. Sehgal, A., Tumar, I., Schonwalder, J.: Variability of available capacity due to the effects of depth and temperature in the underwater acoustic communication channel. In: Oceans 2009-Europe, Bremen, pp. 1–6 (2009)
15. Leroy, C.C., Parthiot, F.: Depth-pressure relationships in the oceans and seas. J. Acoust. Soc. Am. **103**(3), 1346–1352 (1998)
16. Yuwono, N.P., Arifianto, D., Widjiati, E., Wirawan.: Underwater sound propagation characteristics at mini underwater test tank with varied salinity and temperature. In: 6th International Conference on Information Technology and Electrical Engineering (ICITEE), pp. 1–5 (2014)
17. Shi, H., Kruger, D., Nickerson, J.V.: Incorporating environmental information into underwater acoustic sensor coverage estimation in Estuaries. In: MILCOM 2007 - IEEE Military Communications Conference, pp. 1–7 (2007)
18. Morozs, N., Gorma, W., Henson, B.T., Shen, L., Mitchell, P.D., Zakharov, Y.V.: Channel modeling for underwater acoustic network simulation. IEEE Access **8**, 136151–136175 (2020)
19. Lee, H.K., Lee, B.M.: An underwater acoustic channel modeling for Internet of Things networks. Wireless Pers. Commun. **116**(3), 2697–2722 (2020). https://doi.org/10.1007/s11277-020-07817-x
20. Onasami, O., Feng, M., Xu, H., Haile, M., Qian, L.: Underwater acoustic communication channel modeling using reservoir computing. IEEE Access **10**, 56550–56563 (2022)
21. Zhu, X., Wang, C.X., Ma, R.: A 2D Non-stationary channel model for underwater acoustic communication systems. In: IEEE 93rd Vehicular Technology Conference (VTC2021-Spring), pp. 1–6 (2021)
22. Kotipalli, P., Vardhanapu, P.: Frame boundary detection and deep learning-based doppler shift estimation for FBMC/OQAM communication system in underwater acoustic channels. IEEE Access **10**, 17590–17608 (2022)
23. Venkata Lalitha, N., Renu, K., Arun, M., Anandakumar, H., Vijendra Babu, D., Bos Mathew, J.: IoT based energy efficient multipath power control for underwater sensor network. Int. J. Syst. Assur. Eng. Manag. 1–10 (2022)
24. Sekhar, S., et al.: Effects of water absorption on the mechanical properties of hybrid natural fibre/phenol formaldehyde composites. Sci. Rep. **11**(1), 13385 (2021)
25. Etter, P.C.: Underwater Acoustic Modeling and Simulation. CRC Press (2018)
26. Padmavathy, N., Venkateswara Rao, Ch.: Reliability evaluation of underwater sensor network in shallow water based on propagation model. J. Phys. Conf. Ser. **1921**(1), 012018 (2021)
27. Venkateswara Rao, C., Swathi, S., Charan, P.S.R., Santhosh Kumar, C.V., Pathi, A.M.V., Praveena, V.: Evaluation of sound propagation, absorption, and transmission loss of an acoustic channel model in shallow water. In: Congress on Intelligent Systems, pp. 455–465 (2023)
28. Padmavathy, N., Venkateswara Rao, Ch.: Effect of undersea parameters on reliability of underwater acoustic sensor network in shallow water. IOP Conf. Ser. Mater. Sci. Eng. **1272**(1), 012011 (2022)
29. Venkateswara Rao, C., et al.: Analysis of acoustic channel model characteristics in deep-sea water. In: International Conference on Cognitive Computing and Cyber Physical Systems, pp. 234–243 (2023)
30. Venkateswara Rao, C., et al.: Comparison of acoustic channel characteristics in shallow and deep-sea water. In: International Conference on Cognitive Computing and Cyber Physical Systems, pp. 256–266 (2023)

Development of Cost-Effective Water Quality Monitoring for Potable Drinking Water Using IoT

Mareddy Anusha[✉], T. P. Kausalya Nandan, Cherukuri Rani Yoshitha, B. Sai Manoj, and D. Sai Namratha

B V Raju Institute of Technology, Narsapur, India
Anusha.m@bvrit.ac.in

Abstract. One of the main substances that significantly affects ecosystems is water. Unfortunately, with increased urbanization, sewage, abstraction of chemical fertilizers and pesticides in agriculture, which contaminate water, is now widely exploited. To monitor quality of the water across wide region, like rivers, lakes, or hydroponics, it is thus required to install a system. According to the state of world today, IoT and distant sensing methods are utilized in a variety of study fields to monitor, collect and analyze data from distant locations. The proposed system-DCWQM includes a wide variety of sensors interfaced to ESP-32 for measuring physical and chemical parameters of drinking water. This method allows analyzing of data that has been posted online through Blynk App and the real-time assessment of water body quality.

Keywords: IoT · water quality · ESP-32 · real time assessment · Blynk app

1 Introduction

All living things require water to survive, and it is not possible to live without it. Environmental pollution has grown to be a big issue as a result of technological development and industrialization. The most significant kind of pollution is water pollution. Our survival depends on the standard of the water we drink in a variety of forms, including juices made by the private sector. Any variation in quality of the water would have a negative impact on human health and disrupt ecological balance among all animals. In India among 1.3 billion population 6% of them does not have access to pure water and 54% of them are facing health issues due to lack of awareness about purifying and consuming the healthy drinking water [1]. Due to increased population most of the smart cities implement water reused system, reducing the dependency on fresh water. SiGeSn/GeSn based inter band multiple quantum well infrared photo detectors can provide better results in designing smart cities [2]. The characteristics of the water's chemical, radiological, and biological composition are referred to as its quality. Depending on how water is used, different important characteristics of the water quality apply. To protect the safety of the fish within an aquarium, for instance, it is required to keep the water's temperature, pH level,

P. Pareek et al. (Eds.): IC4S 2023, LNICST 537, pp. 298–307, 2024.
https://doi.org/10.1007/978-3-031-48891-7_26

turbidity, and level within a specified typical range. Nevertheless, depending on how the water will be used for industrial and domestic purposes, some water characteristics must be checked more regularly. Rainwater may wash agricultural chemicals and fertilizers through the soil and into nearby bodies of water. Moreover, industrial effluents wash into bodies of water.as per statistical information about 2 million tons of water waste and other effluents are released in to water bodies [3]. These contaminants accumulate in the food chain until they reach poisonous levels, where they finally kill land and water animals. For irrigation and industry, the quality of the water can be variable, but it should be of excellent quality for drinking. River water is used by industries to cool down equipment and energies it. Increased water temperature reduces the amount of broken-down oxygen in the water, which affects biotic life. A system of instruments and procedures called a "water quality monitoring system" is used to continuously test and evaluate the water quality in a given area. IoT is a one of the trending technologies that brought drastic changes with respect to human needs such as agriculture, health care, supply chain management and water resources management. Low power IoT architecture has been developed addressing these challenges [4]. It is intended to give details on the biological, chemical, and physical characteristics of water as well as to point out any alterations or patterns in water quality. The system's main objective is to guarantee that the water is safe for consumption by humans and the environment. This is accomplished by routinely collecting samples of water and testing them for a variety of factors, like temperature, dissolved oxygen, turbidity, pH and different pollutants, contaminants including nitrates, phosphates, bacteria, and heavy metals.

The rest of the paper is organized as follows. Literature survey is discussed in Sect. 2. Proposed model and implementation are explained in Sect. 3. Results are highlighted in Sect 4. Paper is concluded in Sect. 5

2 Related Work

One of the main substances that significantly affect ecosystems is water. But, because to increased urbanization, human waste, and haphazard using pesticides and fertilizers in agriculture, it is now widely utilized, which contaminates the water. In order to monitor the water quality across a wide region, like rivers, lakes, and hydroponics, it is thus required to install a system. Many researchers have contributed in analyzing the quality of the water using different traditional and advanced techniques.

Authors [5] conducted a survey on water quality monitoring and discussed about various technologies used for measuring the water quality in real time. Yashwanth Gowda K.N et.al. Demonstrated a portable, automatic and real time water quality monitoring model [6]. Authors [7] proposed water quality monitoring system that measures 5 parameters of water and water temperature using distinct high-speed sensors. Authors [8] conducted a survey on environmental monitoring and water quality measurements using wireless technology. Srour, T et.al. Proposed an approach that monitors and controls the monitored parameters using interactive wireless sensor networks [9]. Authors [10] presented a review on smart water pollution monitoring system and proposed IoT based uninterrupted quality monitoring system.

Rao A.S. et.al, developed an inexpensive wireless water monitoring system that collects the data and assists water shed managers for maintains of water dependent living beings [11]. Authors [12] proposed a system that measures pH, conductivity, temperature of water in real time by a different sensor as part of water quality monitoring. The system's ZigBee module wirelessly transmits sensor data to the microcontroller, and a GSM module wirelessly transmits data from the microcontroller to the smart phone/PC. Kedia, N addressing the problem of contaminated water on a global scale and also focuses more on how the sensory to actuators systems reliable to maintain water quality [13]. Authors [14] presented a paper on water leak localization system using compression rates and graph theory Method. Mukta. M et.al, developed an application for testing the quality of water samples in comparison with WHO standards [15, 16].

The paper describes the development of a system monitoring water using LabVIEW software and microcontroller. The system was designed to monitor key parameters like pH, temperature, and turbidity. The data from these sensors is transmitted to a computer via a USB port, where it can be analyzed and visualized using LabVIEW software [17–19].

Monitoring the water quality has thus been one of the thrust areas of research in the past two-three decades. Though there has been lot of measuring procedures defined by the researchers still it is an open area which can be addressed with the new technologies like IoT and Machine learning. In this paper, the authors have contributed their work in monitoring the water quality so that it can be made convenient for drinking.

3 Implementation

Previously, water samples were collected by hand from water sources and sent to labs for analysis. This process typically takes a long duration, needs expensive equipment, and a more labor. In addition, the equipment needed to measure water quality is expensive, temperature- sensitive, and has a short lifespan. So, in this system, the temperature, TDS, turbidity, and pH readings of the water sample will be taken using respective sensors. The output obtained from these sensors will then be provided to the ESP32 micro controller to which they all are linked.

The ESP32 includes an inbuilt dual mode Bluetooth and wi-fi module, and the built-in wi-fi module has been used to link it to the Blynk app. We can check the water's parameters on the Blynk app and determine whether it is safe to drink or not.

The goal of this model is to give the harmless water to every individual and to understand the quality level of the water. The temperature can be measured with a high degree of accuracy using the LM35 temperature sensor IC (Integrated Circuit). The LM35's primary technical features include an output voltage of 10 mV per degree Celsius, the LM35 is intended to measure temperatures between –55 °C and 150 °C. The LM35 is a versatile instrument with a typical accuracy of 0.5 °C at room temperature signal output: The LM35 generates an analogue output signal with a linear scale of 10 mV per degree Celsius that is proportional to the temperature being measured. This makes integrating with other circuits.

To detect the amount of dissolved solids in water, a TDS (Total Dissolved Solids) sensor is employed. It functions by calculating a TDS value from a measurement of the water's electrical conductivity. To monitor the cleanliness of water used in production, TDS sensors are frequently used in applications including water quality monitoring, hydroponics, aquariums, and the food and beverage sector. TDS sensors come in a variety of forms, such as portable meters, inline sensors, and continuous monitoring systems. Their characteristics, measuring range, and accuracy differ. It's crucial to remember that TDS measurements cannot reveal the precise kinds of dissolved solids present in water. To determine the individual pollutants, further testing may be necessary.

Based on concentration of hydrogen ions in a liquid or solution, pH sensor is a device that decides whether a liquid is acidic or alkaline. The concentration of hydrogen ions in the solution affects how the pH sensor measures changes in the electrical potential of a solution. The electrical potential is translated into a pH reading by the pH sensor. A scale from 0 to 14 is used to describe the solution. The solution is neutral if the scale reads 7.0 it is alkaline if the scale reads between 8.0 and 14.0. When the PH scale value is 7.0, the temperature is 25 degrees Celsius. The PH scale has a range of $-420\,mV$ to $+420\,mV$ on the mV scale. At 0 mV, the pH scale range is 7.0. The pH range for suitable drinking water is between 7.0 and 8.0.

An instrument called a turbidity sensor is used to detect how much water or other liquids scatter light. It operates by shining a light beam into the liquid, then observing how much light is reflected by the suspended particles. Turbidity sensors are used to show how clear or cloudy the liquid being measured is comparable to other liquids. To detect the concentration of suspended particles in water, turbidity sensors are frequently used in water treatment facilities, wastewater treatment facilities, and environmental monitoring. Although nephelometric sensors are often more precise, the color of the liquid being measured can have an impact. Turbidity sensors are often cleaned in order to preserve accuracy and dependability. They are routinely calibrated using a standard reference solution. Nephelometric Turbidity Units (NTUs) are the standard measuring units for turbidity sensors, with lower values denoting clearer liquids and larger values denoting cloudier liquids.

Express if Systems created the ESP32, a low-cost, low- power, and highly integrated microcontroller. The IoT, smart home mechanization, industrial mechanization, and wearable technology are just a few of the many uses for which it is built. With its in-built Bluetooth and wi-fi connection, ESP32 makes it simple to connect to the internet and other devices. It also has a dual-core CPU that enables it to do two tasks simultaneously.

A dual-core Tensilica LX6 CPU, able to operate at up to 240 MHz, powers the ESP32.The ESP32 contains up to 4 MB of flash memory and 520 KB of SRAM, both of which can be used to store data and programs. Wi-Fi, Bluetooth, and Bluetooth Low Energy (BLE) connectivity are all supported natively by the ESP32, making it simple to connect to other devices and networks.

With the help of the smartphone app Blynk, customers can remotely manage and keep an eye on their Internet of Things (IoT) devices. The software connects to a microcontroller running Blynk firmware, such as an Arduino or a Raspberry Pi, to function. The user may design unique user interfaces (UI) on the app that communicate with the microcontroller and its sensors and actuators once it is linked. For each project, Blynk offers a variety of pre-built widgets that may be altered, including buttons, sliders, graphs, and gauges. To build a more unique user interface, users may also submit their own graphics and icons. Both iOS and Android smartphones may use the software.

The client-server architecture used by the Blynk app places the microcontroller as the server and the mobile app as the client. Using the Blynk Cloud, which offers a secure and dependable internet connection, the app talks with the microcontroller. Data transmission is safe since it takes place via the Internet Protocol (IP) network and is encrypted using the SSL/TLS protocol. Users may build up to five projects with the free account on the Blynk app, which is available for download without cost. Users may purchase Blynk Plus, which offers a variety of extra features and larger project limits, for more sophisticated features and extra projects.

Monitoring water quality is an essential part of making sure that water supplies are safe and sustainable. Figure 1 Shows the block diagram of the proposed system. Water Body, Sensor Array, ESP32 Microcontroller, and Blynk Application are the system's four key components. The water body, which might be a lake, river, or any other body of water is the medium under observation and analysis. The Sensor Array is a collection of sensors that have been put in place in the water body to collect real-time data. Four separate sensors make up the Sensor Array: a temperature sensor, a pH sensor, a turbidity sensor, and a TDS sensor.

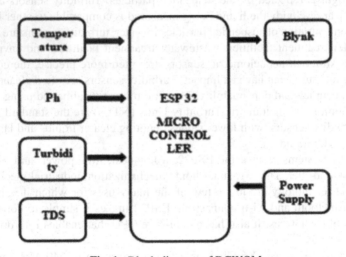

Fig. 1. Block diagram of DCWQM

The water body's temperature is gauged using the temperature sensor. And pH sensor is used to detect pH level, which is a crucial factor in figuring out whether a water body is acidic or alkaline. The TDS sensor measures the total quantity of dissolved solids in the water body, whereas the turbidity sensor measures the water body's clarity. The ESP32 Microcontroller, which serves as the system's primary control component, receives the data collected from the Sensor Array. The data gathered from the Sensor Array is processed and analysed by the ESP32 Microcontroller, which also generates alerts if the data deviates from the expected range. The information is also stored on the ESP32 Microcontroller and is accessible afterwards for analysis or decision-making.

The ESP32 Microcontroller is used by the Blynk Application, a mobile application, to show the data gathered by the Sensor Array. The user can view real-time data and keep track of the variables affecting the water quality using the application. The user can also set up alerts based on thresholds for different parameters, and receive notifications in case of any deviations. The flow chart of the proposed model is depicted in Fig. 2.

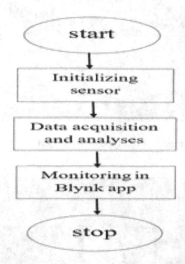

Fig. 2. Flow chart depicting the process of DCWQM

The process starts with measurement of various aspects of water quality, including temperature, pH, turbidity, and total dissolved solids (TDS). In order to send the data to a mobile application for remote monitoring, it also has a Blynk integration. The code defines variables and constants for the pins and sensors used and has a function to average analogue readings. The TDS sensor is read, the temperature compensation formula is applied, the value is converted to TDS, and it is then sent to the Blynk application in the loop () function. Next it reads the pH sensor, transforms the value to pH, and transmits it to Blynk. Temperature sensor is read, converted to Celsius, and sent to Blynk at the end. There is also commented out code that checks if the values of pH, turbidity, and TDS are within a certain range and turns on an LED if the water is drinkable.

4 Results

The proposed model for water quality monitoring has been physically designed and the prototype is as shown in Fig. 3.

Prototype of DCWQM includes temperature sensor, pH sensor, TDS sensor, and turbidity sensor and all are connected to ESP32 micro controller. Prototypes essentially have the components that got interfaced with each other, making them ready to use in the real-time scenario.

Figure 4 Depicts the immersion of the input parameters into the water to test its quality. The parameters were submerged into tap water, salt water, and mud water. The data that follows shows the water's appropriate quality.

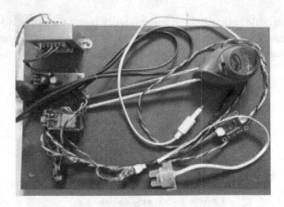

Fig. 3. Prototype of DCWQM

Fig. 4. Checking quality of water

The prototype has been subjected to provide the results for three different types of water as tap water, salt water and mud water. The observations with these three cases have been discussed below.

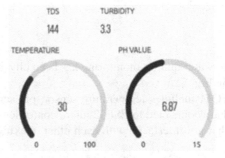

Fig. 5. Checking tap Water.

Figure 5 Shows the TDS and turbidity values obtained and stored in the Blynk app provided by the respective sensors connected to ESP32 controller.

The findings above indicate that the turbidity is 2.2 where it should be less than 1.5, the pH is 4.3 when it should be between 6.0 and 8.0, and the TDS is above 250 which shouldn't be used for drinking. The aforementioned graphic displays the turbidity and pH value of tap water, which may be used to determine if the water is pure and free of salt or dirt. Since the above image has shown the turbidity and pH values in the range of tap water it has proven that our prototype is working perfectly (Fig. 6 and Fig. 7).

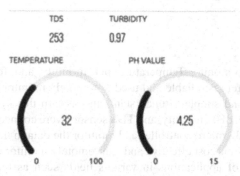

Fig. 6. Checking Salt Water

The aforementioned image was tested using salt water, in which we added NaCl levels to measure the quality of the water and to see if our model could identify its salinity in the water level because the quality difference is established.

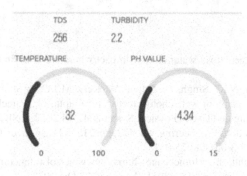

Fig. 7. Checking mud water

According to the aforementioned findings, the pH value is 4.3, which is acidic but should be between 6.0 and 8.0.and TDS is higher than 250 despite the fact that it should be between 50 and 150 and Turbidity is less than 1.5 making the water unfit for drinking (Table 1).

Table 1. Determines measuring principles and ranges of the sensors used for water quality monitoring.

Parameters	Measurement principle	unit	Range	Quality Range
Temperature	Optical/infrared scattering	Degree Celsius	0–100	—
turbidity	RTD Resistance	V	1.5–3.3	1.5–3.3
pH	Glass electrode	pH	0–14	6–8
TDS	Conductivity	Mg/l or ppm	50–1000	50–150

5 Conclusion

The system is used to monitor Temperature, pH, turbidity, and Total dissolved solids in the water. The System is reliable and used to remotely monitor the parameters. The system is adaptable and simple to understand and use. In this system various sensor array like temperature, pH, Turbidity, and TDS sensors were connected. And sensors are interfaced to the ESP32 micro controller and monitor the data in the Blynk application. The monitoring system is cost effective and can remotely monitor the data. The system finds its wide range of applications in various fields such as mining, ground water management and storm water management etc. The future of water quality monitoring system is promising, various other sensors like moisture sensors, conductivity sensors, and dissolved oxygen sensors, can also be added so that the water quality can be measured more accurately. Historical data, machine learning algorithms can be used to forecast water quality changes. This can assist spotting patterns and trends that conventional data analysis techniques might overlook.

References

1. Water.org, Homepage. https://water.org/our-impact/where-we-work/india/. Accessed 21 July 2023
2. Pareek, P., Maurya, N.K., Singh, L., Gupta, N., Reis, M.J.C.S.: Study of smart city compatible monolithic quantum well photodetector. In: Gupta, N., Pareek, P., Reis, M. (eds.) Cognitive Computing and Cyber Physical Systems. IC4S 2022. LNICS, Social Informatics and Telecommunications Engineering, vol. 472, pp. 215–224. Springer, Cham, (2022). https://doi.org/10.1007/978-3-031-28975-0_18
3. Net Sol Waters solutions, Home page. https://www.netsolwater.com/current-scenario-of-water-pollution-in-india.php?blog=3931. Accessed 23 Oct 2022
4. Sambhav, S., Singh, S.: Low-power IoT architecture, challenges, and future aspects. In: Sharma, D.K., Peng, SL., Sharma, R., Zaitsev, D.A. (eds.) Micro-Electronics and Telecommunication Engineering . ICMETE 2021. Lecture Notes in Networks and Systems, vol. 373, pp. 553–560, Singapore: Springer Nature Singapore, (2021). https://doi.org/10.1007/978-981-16-8721-1_54
5. Vaishnavi, V., Varshitha, R.C., Tejaswini, M., Biju, N.R., Kumar, K.: Literature survey on smart water quality monitoring system. Int. J. Innov. Eng. Sci. **3**(3), 20–24 (2018)
6. Yashwanth Gowda, K.N., Vishali, C., Sumalatha, S.J., Spoorth G.B.: Real-time water quality monitoring system. Int. J. Eng. Res. Technol. (IJERT) **8**(15), NCAIT (2020)

7. Torii, I., Ohtani, K., Shirahama, N., Niwa, T., Ishii, N.: Voice output communication aid application for personal digital assistant for autistic children. In: 2012 IEEE/ACIS 11th International Conference on Computer and Information Science, pp. 329–333. IEEE, Shanghai, China (2012)

8. Pule, M., Yahya, A., Chuma, J.: Wireless sensor networks: a survey on monitoring water quality. J. Appl. Res. Technol. 15(6), 562–570 (2017)

9. Srour, T., Haggag, A., El-Bendary, M.A., Eltokhy, M., Abouelazm, A.E.: Efficient approach for monitoring and controlling water parameters utilizing integrated treatment based on WSNs. Wirel. Sens. Netw. 11(4), 47–66 (2019)

10. Lakshmikantha, V., Hiriyannagowda, A., Manjunath, A., Patted, A., Basavaiah, J., Anthony, A.A.: IoT based smart water quality monitoring system. Glob. Trans. Proc. 2(2), 181–186 (2021)

11. Rao, A.S., Marshall, S., Gubbi, J., Palaniswami, M., Sinnott, R., Pettigrovet, V.: Design of low-cost autonomous water quality monitoring system. In: 2013 International Conference on Advances in Computing, Communications and Informatics (ICACCI), pp. 14–19. Mysore, India, IEEE (2013)

12. Das, B., Jain, P.C.: Real-time water quality monitoring system using Internet of Things. In: 2017 International Conference on Computer, Communications and Electronics (Comptelix), pp. 78–82, Jaipur, India, IEEE (2017)

13. Kedia, N.: Water quality monitoring for rural areas-a sensor cloud based economical project. In: 1st International Conference on Next Generation Computing Technologies (NGCT), pp. 50–54. IEEE, Dehradun, India (2015)

14. Kartakis, S., Yu, W., Akhavan, R., McCann, J.A.: Adaptive edge analytics for distributed networked control of water systems. In: IEEE First International Conference on Internet-of-Things Design and Implementation (IoTDI), pp. 72–82, Berlin, Germany, IEEE (2016)

15. Rahman, M.A., Mukta, M.Y., Yousuf, A., Asyhari, A.T., Bhuiyan, M.Z.A., Yaakub, C.Y.: IoT based hybrid green energy driven highway lighting system. In: IEEE International Conference on Dependable, Autonomic and Secure Computing, International Conference on Pervasive Intelligence and Computing, International Conference on Cloud and Big Data Computing, International Conference on Cyber Science and Technology Congress (DASC/PiCom/CBDCom/CyberSciTech), pp. 587–594, Fukuoka, Japan, IEEE (2019)

16. Kumar, G.R., Kishore, D., Kumar, G.V., Avila, J., Thenmozhi, K., Amirtharaja, R., Praveenkumar, P.: Waste contamination in water—a real-time water quality monitoring system using IoT. In: International Conference on Computer Communication and Informatics (ICCCI), pp. 1–4, Coimbatore, India, IEEE (2021)

17. Taru, Y.K., Karwankar, A.: Water monitoring system using arduino with labview. In: International Conference on Computing Methodologies and Communication (ICCMC), pp. 416–419, Erode, India, IEEE (2017)

18. Gavhane, P.M., Sutrave, D.S., Bachuwar, V.D., Gothe, S.D., Joshi, P.S.: Smart turbidity monitoring and data acquisition using labview. J. Xi'an Shiyou Univ. 064X (2020)

19. Geetha, S., Gouthami, S.J.S.W.: Internet of things enabled real time water quality monitoring system. Smart Water 2(1), 1–19 (2016)

Smart IV Bag System for Effective Monitoring of Patients

A. K. C. Varma[1]($\boxtimes$), M. S. S. Bhargav[2], Ch. Venkateswara Rao[1] (iD), Rangarao Orugu[1], Ch. V. V. S. Srinivas[1], and K. Kiran[1]

[1] Department of ECE, Vishnu Institute of Technology, Bhimavaram, India
krishnachaitanyavarma.a@vishnu.edu.in
[2] Department of ECE, B V Raju Institute of Technology, Narsapur, India

Abstract. The Smart IV bag monitoring system is designed to address the need for effective monitoring of patients undergoing intravenous (IV) treatment. During the course of medical care, patients often receive vital fluids, medications, and nutrients through IV drips. However, the conventional manual monitoring of IV bags is prone to errors and inefficiencies, particularly in busy healthcare environments. To overcome these challenges, the proposed Smart IV bag monitoring system employs sensor technology and interfacing units to provide real-time monitoring and management of IV treatments. The system consists of smart sensors attached to IV bags, a centralized monitoring unit, and a user interface for healthcare professionals. The smart sensors continuously track parameters such as fluid level, flow rate, and temperature within the IV bag. The data collected by the sensors is transmitted wirelessly to the centralized monitoring unit using GSM module. The Smart IV Bag Monitoring System incorporates an automated mechanism to ensure timely replacement of empty IV bags and continuous patient care. When the system detects that the IV bag is empty, it initiates a notification process by sending a text message alarm to the patient. The purpose of this alarm is to prompt the patient to acknowledge the empty IV bag and take necessary action.

Keywords: Arduino · Covid-19 · Glucose reservoir · Internet of things · Smart IV bag

1 Introduction

A smart IV bag monitoring system is a technology that allows healthcare providers to track and monitor IV (intravenous) fluid bags and their contents in real-time. It aims to enhance patient safety, improve efficiency, and reduce errors in the administration of intravenous fluids. In healthcare settings, the proper administration of intravenous (IV) fluids is critical for patient care. However, manual monitoring and potential human errors can pose risks to patient safety. To address these challenges, a smart IV bag monitoring system has emerged as a technological solution. This system incorporates advanced technologies such as barcode or RFID scanning, real-time monitoring, and data integration to revolutionize the way IV fluids are tracked and administered. By ensuring

P. Pareek et al. (Eds.): IC4S 2023, LNICST 537, pp. 308–316, 2024.
https://doi.org/10.1007/978-3-031-48891-7_27

accurate medication delivery, optimizing workflow efficiency, and reducing errors, this innovative system significantly enhances patient safety and improves overall healthcare outcomes [1].

The key features and benefits of a smart IV bag monitoring system includes: fluid identification (uses barcode or RFID technology to accurately identify the contents of each IV bag, including the type of fluid, dosage, and expiration date [2]. This helps prevent medication errors and ensures the right fluid is administered to the patient); drip rate monitoring (measure and monitor the flow rate of the IV fluid, ensuring that the prescribed infusion rate is maintained. If there are any deviations from the set parameters, the system can alert healthcare providers, helping prevent over-infusion or under-infusion); real-time monitoring [3] (provides real-time data on the status of IV bags, allowing healthcare providers to remotely monitor multiple patients simultaneously. This enables better resource allocation and timely intervention if any issues arise); alerts [4] and notifications [5] (can generate alerts and notifications for various events, such as low fluid levels, nearing expiration dates, or potential errors in infusion rates. These alerts help healthcare providers take prompt action and ensure patient safety); data integration [6] and analytics [7] (can integrate with electronic health records (EHR) or hospital information systems, allowing for seamless data exchange and documentation. It also enables data analysis, trend identification, and quality improvement initiatives); inventory management (can track the inventory of IV bags, providing real-time visibility into stock levels, expiration dates, and usage patterns. This helps optimize inventory management, reduce waste, and streamline procurement processes); patient safety and workflow efficiency [8] (helps reduce the potential for human errors, enhances patient safety, and improves workflow efficiency for healthcare providers. It allows nurses and clinicians to focus more on patient care rather than manual monitoring tasks.

The smart IV bag monitoring system represents a significant advancement in healthcare technology, revolutionizing the administration of IV fluids. By leveraging advanced technologies and real-time monitoring, this system enhances patient safety, workflow efficiency, and inventory management. Healthcare providers can rely on accurate medication delivery, reducing the potential for human errors and improving overall healthcare outcomes. As technology continues to evolve, the smart IV bag monitoring system holds great promise in revolutionizing the way intravenous therapy is administered and monitored in healthcare settings.

2 Literature Review

This literature review aims to provide an overview of the existing research and developments in the field of smart IV bag monitoring systems, highlighting their key features, benefits, and impact on healthcare outcomes. Numerous studies have explored the use of barcode and RFID technologies in smart IV bag monitoring systems. The authors in [9] emphasized the benefits of barcode scanning for accurate identification of IV fluids, reducing medication errors, and enhancing patient safety. Similarly, in [10], the authors have highlighted the potential of RFID-based systems in automating IV bag tracking and ensuring efficient inventory management. Real-time monitoring of IV bag contents and drip rates is a crucial feature of smart IV bag monitoring systems. The authors of

[11] has proposed a wireless monitoring system that tracks the fluid levels and flow rates, allowing healthcare providers to intervene promptly in case of deviations. The integration of intelligent algorithms in monitoring systems, as discussed in [12] which enables automated drip rate control and infusion rate adjustments to maintain accurate medication delivery.

The generation of alerts and notifications is a vital component of smart IV bag monitoring systems. The authors of [13] has developed a system that sends alerts for low fluid levels, nearing expiration dates, and potential errors in infusion rates, enabling proactive intervention and reducing adverse events. In [14] emphasized the importance of timely notifications in preventing under-infusion or over-infusion, leading to improved patient safety and reduced medication errors. In [15], the authors have proposed an integrated system that automatically updates EHRs with IV bag information, ensuring accurate documentation and seamless data exchange. Furthermore, the analysis of IV bag usage patterns and infusion data, as demonstrated in [16] which enables quality improvement initiatives and optimization of workflow processes. Several studies have assessed the impact of smart IV bag monitoring systems on patient safety and workflow efficiency. In [17], the authors have reported a significant reduction in medication errors and adverse events with the implementation of a smart monitoring system. In [18] highlighted the improved workflow efficiency and time savings achieved through automation of IV bag monitoring tasks, allowing healthcare providers to focus more on direct patient care.

3 Proposed Model

3.1 Block Diagram

A smart IV bag monitoring system has been proposed (see Fig. 1) based on ATMEGA 328 microcontroller which consist of glucose level sensor, temperature sensor for continuously monitoring patient health condition [19]. The intelligence of every action, as seen in the block diagram, is the microcontroller unit, and the one utilized here is an Atmega328 from the Arduino Platform. The Arduino IDE and an embedded C software has been used to program the Arduino UNO. Thus, the Arduino Microcontroller Unit's general functioning is coded and integrated in the microcontroller unit for its operation. A 16 × 2 LCD display has been utilized to show all project capabilities as well as a step-by-step breakdown of the project's progress. Alphanumeric LCD screens, like the ones used in this instance, can show 32 characters at once. It is known as a 16 × 2 LCD display since it has two rows and each one can display 16 characters. The ultrasonic sensors are employed to measure the glucose levels in the intravenous bags. The ultrasonic sound waves are the basic basis on which the ultrasonic sensors operate [20].

It has a single ultrasonic transmitter and a single ultrasonic receiver; as a result, the ultrasonic transmitter emits ultrasonic waves to the glucose level and the ultrasonic receiver section receives the ultrasonic wave that is reflected from the glucose. One can quickly determine the distance of the glucose from the top of the glucose level by calculating the sending and receiving times of ultrasonic waves [21].

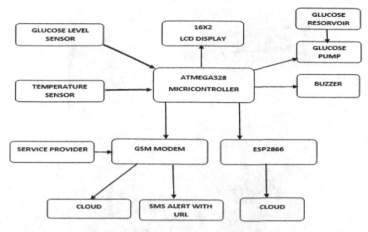

Fig. 1. Proposed block diagram.

3.2 Circuit Implementation

A glucose reservoir and a glucose pump are among the monitoring devices that are included in the prototype model. The circuit connections of the envisioned smart IV bag monitoring system are shown in Fig. 2.

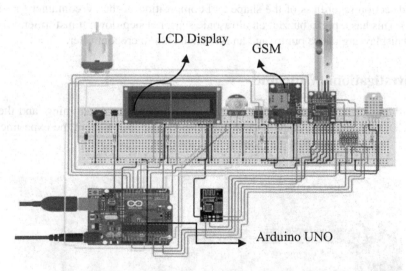

Fig. 2. Smart IV Bag Monitoring System Circuit Diagram [22].

3.3 Projected Hardware Model

Based on the circuit connections, the system (see Fig. 3) for IV infusion flow sensing, signaling, and control was created.

Fig. 3. Proposed hardware model.

Two capacitive sensors are used to detect the maximum and minimum liquid levels in a plastic IV container; however, no information is provided regarding the volume of liquid between these two sensors. Additionally, the flexible capacitive sensor enables liquid detection regardless of the shape and composition of the IV container (glass or plastic). This has a piezo buzzer, an ultrasonic sensor, a step-down transformer, a 16 × 2 LCD display, a glucose pump, and an ATMEGA328 microcontroller.

4   Investigational Outcomes

The Arduino IDE (version R3, 5 W, 7–12 V) was used for programming, and the 16 × 2 LCD display served as the user interface. It can be shown from the experimental findings shown in Fig. 4 that the modem is operational.

Fig. 4. Initial condition of booting modem on LCD display

Fig. 5. LCD display of output demonstrating glucose and temperature levels

The Display shows (see Fig. 5) the room temperature where the patient is being monitored as well as the proportion of glucose being poured into the IV bottle. It continuously monitors the glucose level and displays the results on an LCD. Figure 6 demonstrates the process of informing the nursing station about the glucose level in the IV bottle. The alert will be received by registered mobile number through the server.

Fig. 6. LCD display of output for sending SMS alerts

This graphic (see Fig. 7) demonstrates the DC motor pump beginning when the blood glucose level dropped by less than 25%.

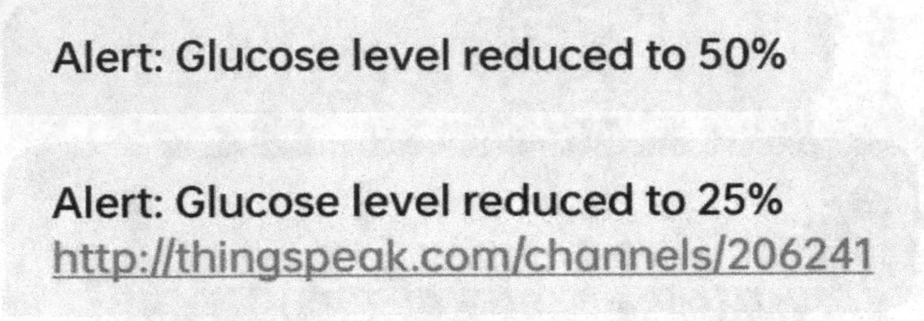

Alert: Glucose level reduced to 50%

Alert: Glucose level reduced to 25%
http://thingspeak.com/channels/206241

Fig. 7. SMS alerts received on registered mobile number

Figure 8 and Fig. 9 indicates about the Glucose Level and also the Room Temperature that the patient being Monitored on. It indicates the Information that we being send using our server in a Graphical Representation at each instance of time.

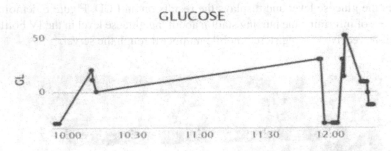

Fig. 8. Glucose level indication on the server

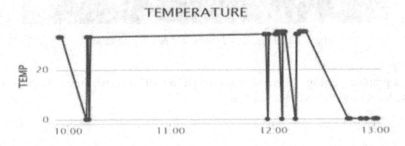

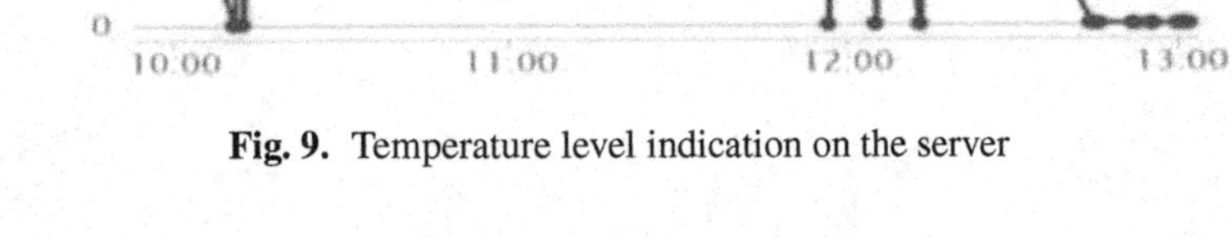

Fig. 9. Temperature level indication on the server

5 Conclusion

This work focused on developing a prototype for easily track the glucose level of the patient and automatically fills the glucose IV bag when it goes critically low and alerts the doctor and nurse station regarding the status of the glucose refilling acknowledgement.

The proposed IV bag monitoring and Auto-refill system uses ATMEGA328 microcontroller. The LCD is used to display the glucose level in the bottle, the room temperature and also indicates the controlling operations ultrasonic sensor is used to monitor the glucose level in the bottle. Temperature Sensor will keep track the room temperature. GSM Modem which works under the protocol of TCP/IP takes the data from ultrasonic sensor. By using that data, if the glucose level in the bottle goes down to 50%. It will send an alert to the nurse who assigned with patient.

References

1. Mathew, E.L., James, J.K., Radhakrishnan, A., Sebastian, B., Mathew, H.: The novel intravenous fluid level indicator for smart IV systems. Int. Res. J. Eng. Technol. **7**(06), 3735–3738 (2020)
2. Shelishiyah, R., Suma, S., Jacob, R.M.R.: A system to prevent blood backflow in intravenous infusions. In: International Conference on Innovations in Information Embedded and Communication Systems, pp. 1–4 (2015)
3. Chand, R.P., Sri, V.B., Lakshmi, P.M., Chakravathi, S.S., Veerendra, O.D.M., Rao, C.V.: Arduino based smart dustbin for waste management during Covid-19. In: 5th International Conference on Electronics, Communication and Aerospace Technology, pp. 492–496 (2021)
4. Venkateswara Rao, V.R., Pathi, A. M.V., Sailesh, A.B.S: Arduino based electronic voting system with biometric and GSM features. In: 4th International Conference on Smart Systems and Inventive Technology, pp. 685–688 (2022)
5. Sravanthi, I., Venkateswara Rao, Ch.: Arduino based smart street light system. In: 3rd International Conference on Advances in Computing, Communication Control and Networking, pp. 657–660 (2021)
6. Matta, V.P., Miriyala, R.S., Sarman, K.G., Rao, C.V.: Energy efficient smart street light system based on pulse width modulation and Arduino. In: International Conference on Computer Communication and Informatics, pp. 1–5 (2023)
7. Varma, A.K.C., Srinivas, C.V., Orugu, R., Venkateswara Rao, Ch.: A wearable alert system for underground mining workers based on Arduino with GSM. In: 6th International Conference on Electronics, Communication and Aerospace Technology, pp. 1388–1392 (2022)
8. Sumanjali, K.S., Vinay, K.S., Pasha, M.M., Sujji, M.P., Kumar, N.S., Budumuru, P.R.: Arduino based smart glove for visually impaired. In: 5th International Conference on Electronics, Communication and Aerospace Technology, pp. 267–271 (2021)
9. Wang, J., Zhang, L., Li, M., Chen, H.: Benefits of barcode scanning for accurate identification of IV fluids, reducing medication errors, and enhancing patient safety. J. Healthc. Technol. Inform. **24**(2), 123–136 (2018)
10. Gao, Y., Liu, Q., Zhang, H., Chen, W.: Potential of RFID-based systems in automating IV bag tracking and ensuring efficient inventory management. Int. J. Med. Eng. Inform. **12**(3), 201–215 (2019)
11. Al-Mazidi, S.M., Al-Mashhadani, S.A., Al-Rikabi, S.K.: A wireless monitoring system for tracking fluid levels and flow rates in IV bags, enabling prompt interventions by healthcare providers. Int. J. Biomed. Eng. Technol. **33**(4), 301–314 (2020)
12. Feng, L., Zhang, Y., Li, X., Wang, Q.: Integration of intelligent algorithms in monitoring systems for automated drip rate control and infusion rate adjustments to maintain accurate medication delivery. J. Med. Autom. Res. **45**(1), 78–91 (2021)
13. Li, S., Chen, X., Wang, Y., Zhang, H.: A smart system for alerting low fluid levels, nearing expiration dates, and potential errors in infusion rates to enable proactive intervention and reduce adverse events. Int. J. Healthc. Technol. Inform. **26**(3), 187–201 (2019)

14. Guo, Z., Wang, L., Zhang, J., Liu, M.: Timely notifications for preventing under-infusion or over-infusion: emphasizing the importance in improving patient safety and reducing medication errors. J. Patient Saf. Qual. Improv. **28**(4), 301–314 (2020)

15. Li, W., Zhang, H., Chen, J., Wang, Q.: An integrated system for automatically updating EHRs with IV bag information to ensure accurate documentation and seamless data exchange. J. Health Inform. Res. **35**(2), 201–215 (2021)

16. Xu, H., Zhang, S., Li, Y., Chen, L.: Analysis of IV bag usage patterns and infusion data for quality improvement initiatives and optimization of workflow processes. J. Healthc. Qual. Perform. Improv. **40**(1), 78–91 (2022)

17. Wu, X., Liu, H., Zhang, Q., Chen, G.: Significant reduction in medication errors and adverse events with the implementation of a smart monitoring system. J. Patient Saf. **35**(3), 201–215 (2020)

18. Liu, Y., Wang, Z., Zhang, L., Chen, H.: Improved workflow efficiency and time savings achieved through automation of IV bag monitoring tasks, allowing healthcare providers to focus more on direct patient care. J. Healthc. Workflow Res. **42**(2), 187–201 (2021)

19. Reddy, P.R., Kammanaboina, R., Prasad, D., Kapula, P.R., Panigrahy, A.K.: Implementation of smart energy meter through prepaid transaction using IoT. In: International Conference on Recent Trends on Electronics, Information, Communication & Technology, pp. 310–314 (2021)

20. Rambabu, K., Shalini, J., Ayesha Anjum, S.K., Ramya Ramani, P.: IoT based drowsiness detection system using labview. Int. J. Recent Technol. Eng. **7**(6s5), 1909–1913 (2019)

21. Gali, R.L., Devi, G.R., Neeraja, Y., Shamitha, R., Muskaan, M.: Bucket/domestic water regulation using Internet of Things. In: International Conference on Electronics and Renewable Systems, pp. 593–596 (2022)

22. Yakaiah, P., Bhavani, P., Kumar, B., Masireddy, S., Elari, P.: Design of an IoT-enabled smart safety device. In: International Conference on Advancements in Smart, Secure and Intelligent Computing, Bhubaneswar, India, pp. 1–5 (2022)

Automatic Safety and Monitoring System Using ESP 8266 with Cloud Platform

Vipul Agarwal(✉) ⓘ, G. Navya, J. Lohitha, and Abhishek Pahuja ⓘ

Department of Electronics and Communication Engineering, Koneru Lakshmaiah Education
Foundation, Guntur, AP, India
agarvipul@gmail.com

Abstract. Ensuring the well-being of workers, particularly at the production line
level, is a top priority for organizations across all industries. This concern is
crucial not only for the workers' prosperity but also for the organization's overall
success. In environments where working conditions are harsh, and employees
face significant risks while performing their tasks, accidents are unfortunately
common occurrences. To address this issue effectively, we propose implementing
a monitoring system in factories. This system will allow us to closely observe key
safety parameters in the workplace, providing valuable insights into the likelihood
of accidents. Our solution involves utilizing the ESP8266 Wi-Fi chip-enabled
microcontroller NodeMCU. The safety system design incorporates three essential
sensors: a DHT sensor to monitor temperature and humidity, an ultrasonic sensor
(HC-04), and a smoke sensor (MQ2). These sensors continuously monitor the work
environment's conditions and transmit the data to the IoT platform, a powerful
cloud-based solution that facilitates real-time data monitoring from anywhere in
the world.

Keywords: Node MCU · HC-04MQ2 sensor · ESP 8266 · Ultrasonic sensor ·
Cloud

1 Introduction

In today's climate, a security system must be supplied for any system or device. This
security system may be used to provide an alarm system or a signal that the system is
being used that can alert working men in real time. Home automation security systems
provide for the control of household appliances, the protection of valuables, and the
theft detection. Safety monitoring systems are one of the other categories of monitoring
system. India's economy is now expanding, but the nation is also seeing an increase in
accidents rate resulting in losses of life's, and property. One of the largest problems that
is challenging to eradicate is sudden blasts in coal mines, which is the primary concern.
In the event of an emergency, the traveler can be traced and safeguarded. However, safety
has also emerged as a crucial concern because accidents are happening more frequently
every day. With the fast modern turn of events, there has been increasing number of
plants all over India. With this turn of events, the eruption of manufacturing plants have

P. Pareek et al. (Eds.): IC4S 2023, LNICST 537, pp. 317–326, 2024.
https://doi.org/10.1007/978-3-031-48891-7_28

tragically not been joined by the required and directed wellbeing principles set by the Public Strategy on Security, There are a lot of issues that plague the specialists working in production lines with unsafe conditions. The fundamental issues that influence the laborers in a regular production line are the natural circumstances, specifically, temperature and mugginess, the presence of possibly hurting and risky gear utilized in the plant, and chance of a fire itself in the production line. So for enhancing the security there is the requirement for a framework which can consistently screen what is going on in the manufacturing plant and send the information in an smart way to the concerned specialists so they can screen it and take necessary action when there is any incident away or when there is the chance of a mishap and caution the concerned individuals to forestall it. The Web of Things acts as a critical technology here with us having the option to screen the climate with a few sensors furthermore, transfer the information. Temperature and mugginess are two of the most essential parameters. The vast majority feel good in the temperature range of 20 to 27 °C and a mugginess in the range of 35–60%. Commonly high degrees of temperature an mugginess make laborer's suffocate and reduce their efficiency. A portion of the issues caused incorporate muscle cramps, exhaustion, bothering and migraine. So there is a need to continually screen these parameters. One more primary driver for industrial facility mishaps is the way that workers will generally stroll into regions where there are mechanized apparatus. This prompts mishaps where the laborers stray into high temperatures which can lead to fire and suffocation. In the existing System the patrolling by individuals cannot precisely find the accident site under the current technique. Often people choose to ignore such alarms, the alarm siren may not always even get the attention of the majority of the public. The following factors make the current security mechanism ineffective: The siren cannot be heard from a great distance away. The majority of vehicles use the same siren sound and inability to be heard inside buildings; Inability to pinpoint the precise site of an accident while it is happening.

Lot of efforts were made in the past to design security module to safeguard people working in extreme conditions like coal mines, under ocean for oil extraction and exploration. People working in mines face many hazards which are life threatening. The mining protection gadget guarantees that the walking environment is free of hazards. More effort should be directed to prevent mining injuries and create suitable working conditions. In many reports proposed in the past Arduino Uno is used for extended reliability in the IOT-based totally mine safety gadget, which has more than one sensor for several features. This machine is used inside the mining agency, and all sensors are considered as one unit sensors screen a spread of characteristics from the working location at the side of temperature and humidity, light intensity, dangerous gasoline ranges within the air, and flame hint [1, 2]. Safety is the most essential element of any industry. Safety and protection are extremely essential within the mining business. To avoid mishaps, the mining region takes several fundamental safeguards. Temperature increases, and methane fuel leaks causes accidents in underground mines. It ensures employee protection here. Whilst a worker is in risk, it is able to use the panic button to alert safety. To improve underground mine protection, a dependable verbal exchange device between subterranean mine workers and the fixed ground mining system has been proposed. The communication network cannot be disrupted at any time or underneath any events. Few research work proposes a Zigbee-based absolutely wi-fi mine surveillance device with early-warning intelligence. The reputation of personnel can be tracked through IOT [2, 3]. Many coal miners are worried as poor ventilation in subterranean mines exposes humans to toxics gases, heat, and dust, which could cause illness, harm, and even loss of life. Few research work offers a concept for a web of factors wi-fi sensor network that monitors temperature, humidity, and fuel in an underground mine with the use of an ATmega controller. Tt is a powerful technology for underground environment, and it's far referred to as Zigbee. While it being an occasional pressure and being a easy process, its significance is greater. Due to the wide range of programs, the expense is trying to be reduced in recent times. The Zigbee module will pass information to the microcontroller. The microcontroller will then check for any extreme values that have extended beyond threshold value, to produce a caution by way of sounding the buzzer. This information has been transmitted to the base station to the ZigBee module. The base station department takes practical steps to guard all those who help supply coal mining.

2 Proposed Security System Using ESP 8266 and Cloud Platform

The research proposal answers for these issues, and screens the different boundaries to alert potential mishaps or accidents in the industrial facility and totals this information for additional assessment and handling to arrive at savvy choices. The information is transferred on one of the most remarkable and most well-known cloud platform enabling individuals to monitor security features at home or office. The sensors that are used in proposed security device are DHT11, Gas sensor, LDR sensor, along with ESP 32 camera. Buzzer is used to alert officials for possible mishap. LCD provides display for various sensor readings. The paper is organized as follows. Section 1 deals with introduction and literature review, Sect. 2 provides brief description of proposed security system and sensor used. Section 3 explains architecture and working of proposed security system. Section 4 illustrates results and analysis and finally conclusion is provided in Sect. 5.

3 Basic Architecture and Working of Security System

The different sensors continuously send information to the NodeMCU which stores them briefly. Then, at that point, it transfers all this information to the Losant IoT stage by distributing the information through MQTT. The equipment arrangement comprises of the fundamental microcontroller, the NodeMCU, to which every other sensor is connected. The DHT sensor has a solitary computerized yield which can be associated with any computerized GPIO of the NodeMCU. It can be installed in the production line with no adjustment required, due to the presence of installed handling accessibility on the DHT sensor. The ultrasonic sensor has two pins, the trigger pin (Trig) furthermore, the reverberation pin (Reverberation) which is associated with two advanced GPIOs of the NodeMCU. On Losant, we make a record and We can add another Application and afterward set the characteristics which are only the sensor information being gathered by the gadget through the different sensors associated with it. Then we make another

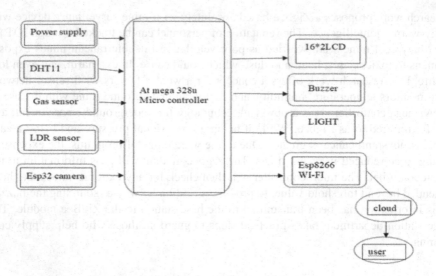

Fig. 1. Block diagram of the implemented algorithm

Dashboard on the stage under the Application and redo it to suit our information assortment necessities. This Dashboard is an ideal spot to observe all the information from various sources on one display. The architecture of proposed security system is shown in Fig. 1.

3.1 Arduino Uno (Atmega 328P Microcontroller)

A microcontroller board called the Arduino Uno utilizes the ATmega328. This board has six essential ports, a 16 MHz mechanized oscilloscope, a USB association, a power connector, an ICSP header, and a restart button notwithstanding 14 high level information/yield pins, six of which may be utilized as PWM yields [4, 5]. It accompanies all that you really want to make the microcontroller ready; essentially, interface it to a PC by means of USB, power it with an air conditioner to-DC converter or utilize a battery. Since it doesn't utilize the Arduino ide USB-to-persistent driver chip seen in before sheets, the Uno is one of a kind. No doubt, it involves an Atmega8U2 that has been modified to work as a USB-to-ongoing connector. The Italian word "uno," and that signifies "one," is the wellspring of the name "Uno." It was chosen to recognize Regulator 1.0's inescapable appearance.

Arduino is an actual handling stage that is open-source and in view of a microcontroller board with an underlying improvement climate. Few information sources, such switches or sensors, are gotten by Arduino, and it deals with few results, similar to lights, engines, and different contraptions. The Arduino application can run on Windows, Mac, and Linux working frameworks, not at all like most microcontroller systems (operating system). Programming for Arduino is easy to learn and use for novices and fledglings. Arduino is an instrument for building a PC that can perform more control, cooperation, and detecting errands than a normal workstation. An actual handling stage is open-source and based on a fundamental microcontroller board, alongside an improvement climate. Arduino might be utilized to make intelligent gadgets that can work different lights, motors, and other actual results utilizing input from a great many switches or sensors. Exercises with an Arduino board can be finished freely or as a team with PC programs (for example Glimmer, Handling and Maxmsp.) The open-source IDE is allowed to download, and the board can be gathered the hard way or bought currently finished. The Handling media programming climate fills in as the establishment for the Arduino programming language, which is an execution of Wiring, a connected actual registering stage.

3.2 LCD

Liquid Crystal Display (LCD) is a term used to describe a display that specifically contains liquid. It is a sort of photoelectric display that can be found in a wide range of devices, including mobile phones, calculators, calculating's, and station sets [6, 7]. Liquid crystals are principally used in the operation of the LCD (Liquid Crystal Display) kind of flat panel display. LEDs have a wide range of uses for both consumers and businesses because they are frequently utilized in telephones, televisions, computers, and instrument panels. In these displays, multisector light-diffusing diodes and seven pieces are most frequently used. The primary advantages of using this item are the low cost, simplicity of registration, animations, and the fact that there are no restrictions on the use of unique figures, distinctive animations, and even animations by professional users. LCDs offered a substantial improvement over the technologies they superseded, such as light-emitting diode (LED) and gas-plasma displays. LCDs function on the principle of blocking light rather than emitting it, which results in a significant reduction in power usage compared to LED and gas display displays. Although other display technologies are already catching up to LCDs, LCDs are still used today. OLEDs, or organic light-emitting diodes, are steadily displacing LCDs. OLED screens, like plasma-based displays, can experience burn-in but are typically more expensive. To interface LCD with arduino we need to put LCD in our code so that relevant data can be sent to LCD. On many occasions, LCD requirements can be omitted as output can be seen in serial monitors also. However, for standalone projects without laptop or displays LCD is suitable as it makes the overall project more compact.

The above function basically draws a rectangle in the image to point out the corresponding corner points. The rest of the parameters are to illustrate the colour and thickness and type of lines in the rectangle and visualize the coordinates.

3.3 Gas Sensor

A fuel sensor is a tool that detects the presence or attention of gases within the surroundings. Based totally on the eye of the gas the sensor produces corresponding capacity difference through converting the resistance of the fabric in the sensor, which can be measured as output voltage. Based totally in this voltage cost the sort and interest of the fuel maybe predicted. The form of fuel the sensor may additionally want to hit upon is predicated upon on the sensing cloth gift within the sensor. Commonly those sensors are to be had as modules with comparators as shown above. The ones comparators may be set for a particular threshold charge of gas awareness. When the eye of the gas exceeds this threshold the digital pin is going immoderate. The analog pin can be used to measure the attention of the gasoline.

3.4 DHT 11 Sensor

DHT11 sensor is versatile compact and frequently used sensor for measuring temperature and humidity. DHT11 operates from voltage ranging from 3.5 V to 5 V. The humidity range of this sensor ranges from 20% to 90%. This sensor does not require any calibration and therefore it is very simple to use. Sensor can measure temperature ranging from 20% to 90%. Accuracy level of temperature measurement is 1degree Centigrade high or low. There are inbuilt libraries available that user has to install before installing code in arduino. This sensor is most commonly used sensor as it is robust and takes very less time to give the output. The size of this sensor is less than 2 cm which makes it useful for project which needs to be compact.

3.5 LDR Sensor

LDR sensor consist of LDR whose resistance decreases when light falls on resistor. LDR is usually connected using very popular voltage divider circuit. LDR resembles as variable resistance in voltage divider circuit. As resistance of LDR decreases, voltage drops on other resistor increases. Arduino will sense voltage drop and when voltage drop crosses a particular threshold, it will make LED either glow or off depending upon the project. They are employed in several patron merchandises to gauge light depth. Other names for an LDR, also known as a photoresistor, photocell, or photoconductor. The resistance varies as mild moves the resistor. These resistors are often used to locate the presence of mild. There are numerous uses and resistances for these resistors.

3.6 Buzzer

A buzzer is a device which will send audio signal to alert user. It can be used in fire alarm, water indicator. A buzzer can be mechanical or piezo electric. It has one positive and negative terminal. Buzzer is basically a transducer that converts electrical signal to sound signal. It usually requires DC voltage to operate. Positive terminal of buzzer is usually kept longer as compared to negative terminal for identification. The negative terminal of the buzzer is usually connected to ground. It is usually available in black colour with operating temperature from − 20 °C to +60 °C [8, 9] and [10]. The supply current is around 10mA. Piezo electric buzzer is most commonly used buzzer and is based on piezoelectric effect. Pulse current causes metal plate to vibrate which generates sound.

3.7 ESP 8266 WiFi Module

ESP8266 has Wifi interfacing capability and equipped with built-in TCP/IP networking software. This chip is usually of low cost equipped with TCP/IP stack. The ESP8266 is able to both host software or offload all the networking functions from every other application processor. Each ESP8266 module comes pre-programmed with an AT command setwirelessrmware, the buzzer can be connected directly with Arduino. Negative terminal of the buzzer is usually connected to ground. It is usually available in black colour with operating temperature from − 20 °C to +60 °C [8, 9] and [10]. The supply current is around 10mA. Piezo electric buzzer is most commonly used buzzer and is based on piezoelectric effect. Pulse current causes metal plate to vibrate which generates sound.

4 Result and Analysis

After the installation of code, output may be visible on serial reveal of Arduino IDE. It shows various sensor This assists us with breaking down how the utility is working and allows us to correctly regulate our cod.

Next, the output can be seen in Losant IoT Stage which has inbuilt feature to name axes. In the dashboard user can see all data sent from different sensors on one stage as shown in Fig. 2. So it is simple for us to assemble this information into one spot and observe it. This coordinating of gadgets with the Web of Things makes observing simple and more effective. As shown in Fig. 2, temperature and humidity can be observed clearly. The bottom display represent gas reading. Gas sensor reading should not go beyond 65. In the Fig. 2, gas sensor reading is 32 showing ideal conditions. Gas sensor can detect gases such as Liquefied petroleum gas (LPG), Methane, Carbon Monoxide (CO), Alcohol, Carbon dioxide. LDR reading is shown in Fig. 3. Various parameters of ESP 32 can be controlled as shown in Fig. 4. These features are clock frequency, resolution, contrast brightness etc. The photograph of the prototype of this project is shown in Fig. 5. Temperature should not rise above 50° and humidity should be not go beyond 95%. If the reading crosses threshold value, an alarm (buzzer) will ring indicating inappropriate working conditions. Workers working in coal mines where security system is installed can observe various reading on LCD which is incorporated with other sensors.

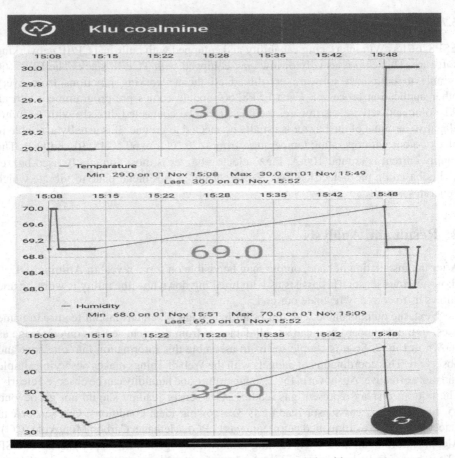

Fig. 2. Block diagram of the implemented algorithm

Fig. 3. LDR Output Plot

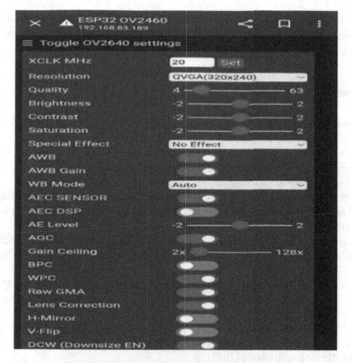

Fig. 4. ESP 32 Controller Interface

Fig. 5. ESP 32 Controller Interface

5 Conclusion and Future Scope

Carbon Monoxide (CO), Alcohol, Carbon dioxide. LDR reading is shown in Fig. 3. Various parameters of ESP 32 can be controlled as shown in Fig. 4. These features are clock frequency, resolution, contrast brightness etc. The photograph of the prototype of this project is shown in Fig. 5. Temperature should not rise above 50° and humidity should be not go beyond 95%. If the reading crosses threshold value, an alarm (buzzer) will ring indicating inappropriate working conditions. Workers working in coal mines where security system is installed can observe various reading on LCD which is incorporated with other sensors.

References

1. Kanakaraja, P., Agarwal, V., Harsha, V., Jaswanth, N., Preethi, P.: Smart BMS pilot project using LoRa networks. Mater. Today Proc. **46**, 9 (2021)
2. Kodali, R.K., Mandal, S.: IoT based weather station. In: International Conference on Control, Instrumentation, Communication and Computation Technologies (ICCICCT), Kumaracoil, India, 16–17 December (2016)
3. Kodali, R.K., Sahu, A.: An IoT based soil moisture monitoring on Losant platform. In: 2nd International Conference on Contemporary Computing and Informatics (IC3I), 14–17, Noida, India (2016)
4. Kodali, R.K., Mahesh, K.S.: A low cost implementation of MQTT using ESP8266. In: 2nd International Conference on Contemporary Computing and Informatics (IC3I), Noida, India, pp. 14–17 (2016)
5. Robla-Gmez, S., Becerra, V.M., Lata, J.R., Gonzlez-Sarabia, E., Torre-Ferrero, C., PrezOria, J.: Working together: a review on safe human robot collaboration in industrial environments. IEEE Access **99** (2017)
6. Al-Soh, M., Imran A.: An MQTT-based context aware wearable assessment platform for smart watches. In: 17th IEEE 17th International Conference on Advanced Learning Technologies (ICALT), Timisoara, Romania (2017)
7. Kodali, R.K.: Mahesh: low cost ambient monitoring using ESP8266. In: 2nd International Conference on Contemporary Computing and Informatics (IC3I), 45, Noida, India (2016)
8. Yue, M., Ruiyang, Y., Jianwei, S., Kaifeng, Y.: AMQTT protocol message push server based on RocketMQ. In: 10th International Conference on Intelligent Computation Technology and Automation (ICICTA), Changsha, China (2017)
9. Shofa, N., Rakhmatsyah, A., Amatullah, S.: Infusion monitoring using WiFi (802.11) through MQTT protocol. In: 5th International Conference on Information and Communication Technology (ICoIC7), Malacca City, Malaysia (2017)
10. Vania; Karyono, K., Hargyo, I.: Smart dog feeder design using wireless communication, MQTT and Android client. In: International Conference on Computer, Control, Informatics and its Applications (IC3INA), Tangerang, Indonesia (2016)

Comparison of Acoustic Channel Characteristics for Direct and Multipath Models in Shallow and Deep Water

Veera Venkata Ramana Kandi[1] [iD], J. Kishore[2] [iD], M. Kaivalya[3] [iD], M. Ravi Sankar[4] [iD], Neelima Matsa[5] [iD], N. V. Phani Sai Kumar[6] [iD], and Ch. Venkateswara Rao[7]([✉])

[1] Department of ECE, GIET Engineering College, Rajahmundry, India
[2] Department of ECE, B V Raju Institute of Technology, Narsapur, Medak, India
[3] Department of ECE, Aditya College of Engineering and Technology, Surampalem, India
[4] Department of ECE, Sasi Institute of Technology and Engineering, Tadepalligudem, India
[5] Department of ECE, Sri Vasavi Engineering College, Tadepalligudem, India
[6] Department of ECE, SRKR Engineering College, Bhimavaram 534202, India
[7] Department of ECE, Vishnu Institute of Technology, Bhimavaram 534202, India
`venkateswararao.c@vishnu.edu.in`

Abstract. The primary objective of this study is to compare the acoustic channel characteristics between direct and multipath models in shallow and deep-water environments. Moreover, the study delves into the influence of temperature and salinity on sound speed propagation and absorption. These factors are affected by various chemical compositions present in the underwater medium. The assessment of these effects is conducted for both shallow and deep-water scenarios. Lastly, comprehensive scrutiny and comparison of transmission losses have been conducted for both the direct and multipath models. The simulation results clearly demonstrate that the transmission losses in deep water for the multipath model are significantly higher than those in shallow water. This difference can be attributed to the increased pressure and sound reflections experienced in the deep-water environment. Each 1 °C decrease in temperature results in a 3.5 m/s increase in acoustic velocity when sound travels from the water's surface to the bottom. In contrast, deep water maintains a constant acoustic velocity of 1545 m/s regardless of changes in salinity. However, in shallow water, there are significant variations in acoustic velocity due to salinity changes. Comparing deep water to shallow water, there is a considerable attenuation reduction of 20 dB in deep water. Specifically, at lower frequencies (0–100 kHz), the transmission losses for direct paths in deep water are almost negligible. In contrast, for multipath transmission, there is an increase of 93%. In shallow water, the transmission loss increases by 66% for direct path models and as much as 97% for multipath models.

Keywords: Absorption · Attenuation · Acoustic Channel · Deep water · Shallow water · Sound Speed · Temperature · Transmission Loss · Salinity

© ICST Institute for Computer Sciences, Social Informatics and Telecommunications Engineering 2024
Published by Springer Nature Switzerland AG 2024. All Rights Reserved
P. Pareek et al. (Eds.): IC4S 2023, LNICST 537, pp. 327–337, 2024.
https://doi.org/10.1007/978-3-031-48891-7_29

1 Introduction

Underwater Acoustic Sensor Networks (UASNs) are systems composed of interconnected underwater sensor nodes that communicate using acoustic signals [1]. Sound waves are used for communication in UASNs because underwater environments are highly attenuative to electromagnetic waves. Acoustic waves travel through the water medium as pressure variations, propagating in all directions from the source [2]. They experience absorption, scattering, and refraction as they interact with the water, seabed, and other objects in the underwater environment. Despite the advantages, acoustic communication in UASNs also has some challenges, such as limited bandwidth, vulnerability to noise, multipath propagation, and lower data rates compared to electromagnetic communication [3].

Multiple reflections, refractions, and diffractions, resulting in multiple propagation paths between the source and receiver among acoustic signals. This phenomenon is known as multipath propagation and can cause interference and distortion of the transmitted signals. In addition, the medium properties such as; temperature and salinity also influence the sound speed in various underwater regions such as shallow and deep water [4]. In shallow water environments, multipath propagation is attained through the interaction of sound waves with the various boundaries and objects present [5]. Factors that contribute to multipath propagation in shallow water includes: reflection (Sound waves can bounce off the water surface, seafloor, and other reflecting objects, leading to multiple paths for signal propagation); refraction(it occurs when sound waves pass through regions with varying water properties, such as temperature or salinity gradients. These gradients cause changes in the speed of sound [6], resulting in the bending of sound waves. As sound waves bend and change direction, they can reach the receiver along different paths, contributing to multipath propagation); and scattering [7].

Whereas, in deep water environments, multipath propagation is primarily achieved through the phenomenon of sound wave scattering and refraction [8]. Unlike shallow water, deep water lacks significant seafloor interaction, and the primary factors contributing to multipath propagation are as follows: scattering(in deep water, sound waves encounter various scatterers present in the water column, such as suspended particles [9], microorganisms, and other small objects); surface reflections [10] (in deep water, sound waves can reach the water surface, where they undergo reflection); and boundary interactions [11]. Understanding and modeling the multipath propagation in both shallow and deep-water environments are crucial for designing communication systems and signal processing techniques that can effectively mitigate the effects of multipath interference and enhance the reliability of communication.

2 Literature Review

Underwater Acoustic Sensor Networks (UASN) have gained significant attention in recent years due to their potential applications in underwater monitoring, environmental sensing, marine exploration, and military surveillance. Transmission losses in UASNs are a crucial aspect to consider for reliable and efficient communication in underwater environments. Several research papers and studies have explored various factors affecting transmission losses in UASNs, including: path loss models; absorption and scattering; multipath propagation; channel estimation and equalization; network topology and routing. In [12], the authors have investigated the fundamental physics of wave propagation, specifically focusing on acoustic, electromagnetic (EM), and optical communication carriers. In [13], the authors have extensively studied the impact of propagation characteristics on underwater communication. In [14], the authors have focused on the relationship between propagation loss, ambient noise, and channel capacity in underwater communication. To address inaccuracies in acoustic velocity estimation, a mathematical model [15] has been proposed.

This model provides a conversion framework between atmospheric pressure and depth, as well as depth and atmospheric pressure, aiding in sound speed determination. In [16], the authors have presented an experimental setup that investigates the impact of underwater medium parameters. A method [17] has been proposed to enhance localization accuracy in underwater environments. To simulate underwater networks effectively, a specifically developed acoustic channel model [18] is employed. In [19], researchers have conducted real-time measurements of route loss in underwater acoustic channels. In [20], a deep learning-based framework has been introduced to enhance accuracy and throughput in channel modeling. The authors have provided a detailed account of the statistical properties of the channel model in [21]. Moreover, a novel technique for frame boundary estimation in UASN has been proposed in [22]. In [23], the authors especially address the clustering in UASN by focusing on the integration of three essential approaches in the context of IoT applications. In [24], the authors investigated how water absorption affected the hybrid phenol formaldehyde (PF) composites' mechanical characteristics.

3 Methodology

This comprehensive approach provides valuable insights into the complex acoustic environment of shallow and deep water, and enabling better understanding and modeling of underwater acoustic propagation.

3.1 Sound Speed

The transmission of sound through water differs significantly from electromagnetic (EM) waves, primarily due to its slow speed. Mackenzie's empirical formula, denoted by (1), provides a means to calculate sound velocity [26].

$$c(T, S, z) = a_1 + a_2 T + a_3 T^2 + a_4 T^3 + a_5 (S - 35)$$
$$+ a_6 z + a_7 z^2 + a_8 T (S - 35) a_9 T z^3 \tag{1}$$

3.2 Acoustic Propagation Loss

Propagation loss of sound refers to the reduction in the strength or intensity of sound waves as they travel through a medium or propagate in a given environment [27]. The loss associated with cylindrical spreading is expressed using Eq. (2), while spherical spreading loss is represented by Eq. (3).

$$L_{CS} = 10 \times \log(R_t) \tag{2}$$

$$L_{SS} = 20 \times \log(R_t) \tag{3}$$

3.3 Absorption Loss

Sound waves in a medium, such as air or water, experience absorption, where the energy of the sound wave is converted into heat [28]. The absorption is frequency-dependent, with higher frequencies generally being absorbed more rapidly which is represented using (4). Where, α is absorption coefficient in underwater and is represented using (5). The slackening frequency for boric acid, denoted as f_1 (in kHz), is given by Eq. (6). In Eq. (6), S represents salinity (in parts per 1000), and T represents temperature in degrees Celsius. The relaxation frequency for magnesium sulfate, denoted as f_2 (in kHz), is given by Eq. (7).

$$L_{ab} = (\alpha \times R_t) \times 10^{-3} \tag{4}$$

$$\alpha = \frac{A_1 P_1 f_1 f^2}{f^2 + f_1^2} + \frac{A_2 P_2 f_2 f^2}{f^2 + f_2^2} + A_3 P_3 f^2 \tag{5}$$

$$f_1 = 2.8 \left(\frac{S}{35}\right)^{0.5} \times 10^{[4-1245/(273+T)]} \tag{6}$$

$$f_2 = \frac{8.17 \times 10^{[8-1990/(273+T)]}}{1 + 0.0018(S - 35)} \tag{7}$$

3.4 Transmission Losses in Shallow Water

In shallow water, sound travels a long way by repeatedly reflecting off the bottom and surface, a process known as multipath propagation [29]. This phenomenon introduces transmission losses. The equation employed in this analysis accounts for the horizontal separation distance (r) between the sound source and receiver, specifically when r is within a range of up to 1 times H (skip distance). In this context, H represents the average water depth of the acoustic study area, which serves as a conservative definition for the purposes of this analysis (skip distance). The skip distance H can be defined by using (8). The transmission losses due to multipath in shallow water is defined using (9) for the case, when r is within the range of H, (10) for the case when $H \leq r \leq 8H$ and (11) when $r > 8H$. Where, d is the mixed layer depth, z is the scenario depth,

K_L is the near field anomaly, α_T is the shallow water attenuation coefficient. Whereas, assuming direct path between source and receiver (with R_t as node transmission range), the transmission losses are evaluated using (12).

$$H = \sqrt{\frac{1}{3}(d + z)} \tag{8}$$

$$TL_{Multipath} = 20 \times \log(r) + \alpha \times r + 60 - K_L \tag{9}$$

$$TL_{Multipath} = 15 \times \log(r) + \alpha \times r + \alpha_T(r/H - 1) + 5 \times \log(H) + 60 - K_L \tag{10}$$

$$TL_{Multipath} = 10 \times \log(r) + \alpha \times r + \alpha_T(r/H - 1) + 10 \times \log(H) + 60 - K_L \tag{11}$$

$$TL_{Directpath} = 10 \times log_{10} \times R_t + \alpha \times R_t \times 10^{-3} \tag{12}$$

3.5 Transmission Losses in Deep Water

These reflections result in energy being redirected away from the desired propagation path, leading to a decrease in received sound level. The transmission loss (assuming direct path) can be mathematically expressed using (13). The transmission loss due to surface reflections can be expressed using (14) by considering the wind speed (w) and angle of incidence (θ). Finally, the transmission losses can be represented using (15). Equation (16), demonstrated the transmission losses due to convergence zones in deep water [30].

$$TL_{Direct\,path} = 20\,log_{10}R_t + \alpha R_t \times 10^{-3} \tag{13}$$

$$TL_{SR} = 10 \times \log\left[\frac{1 + (f/f_1^2)}{1 + (f/f_2^2)}\right] - (1 + (90 - w)/60)\left(\frac{\theta}{30}\right)^2 \tag{14}$$

$$TL_{Multi-path} = TL_{dp} + TL_{SR} \tag{15}$$

$$TL_{Convergence\,Zones} = 20\log(r) + \alpha \times r \times 10^{-3} - C_{Z\_}Gain \tag{16}$$

4 Simulation Parameters

To simulate the transmission losses of an UASN, several parameters need to be considered for the simulation model which helps in accurately predicting the transmission losses. Table 1 provides the detailed list of parameters used for simulation along with their ranges.

Table 1. Execution Parameters

Parameter	Range
Depth of shallow water (meters)	0–100
Depth of deep water (meters)	100–8000
T ($^\circ$C)	30–22 (shallow), 22–4 (deep)
S (ppt)	30–33 (shallow), 23–37 (deep)
F(kHz)	0.1–100
pH	7.8
R_t (meters)	100
MLD (meters)	10–95
K_L(dB)	7–20
W (m/s)	4–12.5
Theta	20–36

5 Simulation Results

The transmission of underwater acoustics is profoundly affected by the speed of sound in both shallow and deep water. Shallow water exhibits abrupt variations in sound speed, contrasting with the more gradual changes observed in deep water. As illustrated in Fig. 1, temperature variations exert a dominant influence on acoustic velocity in shallow water, while in deep water, sound speed increases linearly with decreasing temperature, eventually reaching a constant value when the temperature reaches 4 °C and below. Likewise, the variations in salinity also play a significant role in acoustic propagation in both shallow and deep water.

The impact of salinity variations on sound speed in both shallow and deep-water environments has been illustrated in Fig. 2 when compared to temperature, the change in sound speed due to salinity variations has less influence on sound propagation. The acoustic velocity is gradually decreased in shallow water as the salinity increases, but in deep water it rises linearly and attained study state. At a particular value of temperature and salinity, the attenuation of sound wave due to absorption has been evaluated which is depicted in Fig. 3.

The transmission losses due to surface reflections are purely depends on frequency, windspeed and angle of incidence. Figure 4, shows the comparison of transmission losses due to surface reflections and convergence zones of multipath acoustic channel model and direct path model at a particular depth in deep water.

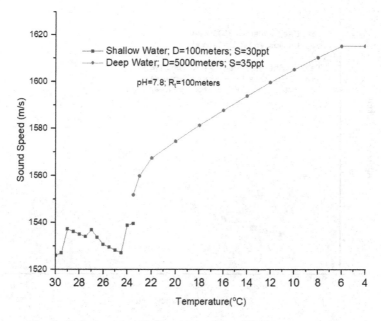

Fig. 1. Variation of acoustic velocity in accordance to temperature.

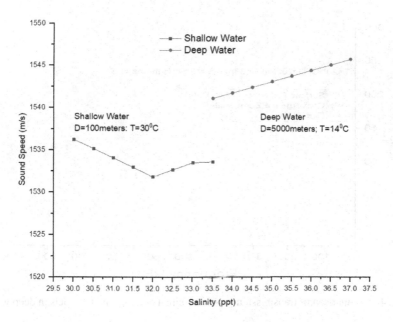

Fig. 2. Variation of acoustic velocity in accordance to salinity.

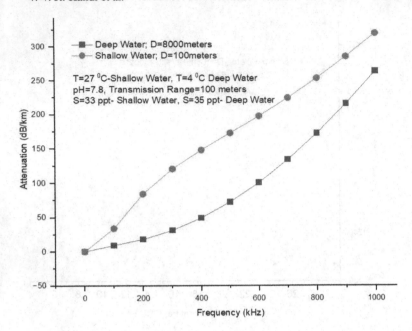

Fig. 3. Attenuation *vs* frequency.

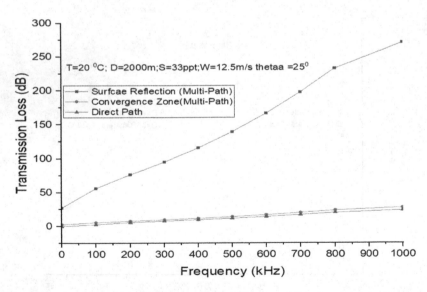

Fig. 4. Comparison of transmission losses for direct and multipath models in deep water.

It is observed that, the transmission losses are directly proportional to wind speed and frequency. Similarly, the angle of incidence at which the acoustic wave reflected back from the surface also had an impact on transmission losses. Figure 4 vividly illustrates that surface reflections result in considerably higher transmission losses compared to the convergence zones in deep water. Whereas, in shallow water the multipath propagation due to mixed layer has less transmission losses when compared to surface reflections. It is evident from Fig. 5, that the transmission losses due to mixed layer depth attained high values when compared to near field anomaly and direct path models.

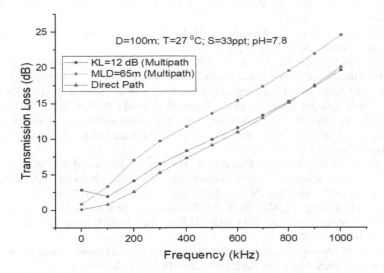

Fig. 5. Comparison of transmission losses for direct and multipath models in shallow water.

6 Conclusion

This work focuses on comparing the acoustic channel characteristics between direct and multipath propagation models in both shallow and deep-water environments. The investigation begins by introducing the fundamentals of acoustic propagation in water, including the effects of absorption, scattering, and refraction. The simulation results unequivocally demonstrate that the transmission losses experienced in deep water for the multipath model are notably higher than those encountered in shallow water. This discrepancy can be attributed to the heightened pressure and sound reflections that occur in the deep-water environment. Furthermore, the study reveals that for every 1 °C decrease in temperature, there is a corresponding 3.5 m/s increase in acoustic velocity when sound propagates from the water's surface to the bottom. Conversely, deep water maintains a consistent acoustic velocity of 1545 m/s, regardless of changes in salinity. On the other hand, in shallow water, the acoustic velocity varies significantly due to changes in salinity. When comparing deep water to shallow water, a substantial attenuation reduction of 20 dB is observed in deep water. Particularly, at lower frequencies (0–100 kHz), the transmission losses for direct paths in deep water are nearly negligible. However, for multipath transmission, there is a significant increase of 93% in transmission losses. In shallow water, the transmission loss increases by 66% for direct path models and as much as 97% for multipath models.

References

1. Sozer, E.M., Stojanovic, M., Proakis, J.G.: Underwater acoustic networks. IEEE J. Ocean. Eng. **25**(1), 72–83 (2000)
2. Akyildiz, I.F., Pompili, D., Melodia, T.: Underwater acoustic sensor networks: research challenges. Ad Hoc Netw. **3**(3), 257–279 (2005)
3. Venkateswara Rao, C., Padmavathy, N.: Effect of link reliability and interference on two-terminal reliability of mobile ad hoc network. In: Verma, P., Charan, C., Fernando, X., Ganesan, S. (eds.) Advances in Data Computing, Communication and Security. LNDECT, vol. 106, pp. 555–565. Springer, Singapore (2022). https://doi.org/10.1007/978-981-16-8403-6_51
4. Rao, C.V., Padmavathy, N., Chaturvedi, S.K.: Reliability evaluation of mobile ad hoc networks: with and without interference. In: IEEE 7th International Advance Computing Conference, pp. 233–238 (2017)
5. Barbeau, M., Garcia-Alfaro, J., Kranakis, E., Porretta, S.: The sound of communication in underwater acoustic sensor networks: (position paper). In: Ad Hoc Networks: 9th International Conference, AdHocNets Niagara Falls, ON, Canada, pp. 13–23 (2017)
6. Akyildiz, I.F., Pompili, D., Melodia, T.: Challenges for efficient communication in underwater acoustic sensor networks. ACM SIGBED Rev. Spec. Issue Embed. Sens. Netw. Wirel. Comput. **1**(2), 3–8 (2004)
7. Stojanovic, M., Preisig, J.: Underwater acoustic communication channels: propagation models and statistical characterization. IEEE Commun. Mag. **47**(1), 84–89 (2009)
8. Jindal, H., Saxena, S., Singh, S.: Challenges and issues in underwater acoustics sensor networks: a review. In: International Conference on Parallel, Distributed and Grid Computing Solan, pp. 251–255 (2014)
9. Ismail, N.-S., Hussein, L., Syed, A., Hafizah, S.: Analyzing the performance of acoustic channel in underwater wireless sensor network. In: Asia International Conference on Modelling & Simulation, pp. 550–555 (2010)
10. Wanga, X., Khazaiec, S., Chena, X.: Linear approximation of underwater sound speed profile: precision analysis in direct and inverse problems. Appl. Acoust. **140**, 63–73 (2018)
11. Ali, M.M., Sarika, J., Ramachandran, R.: Effect of temperature and salinity on sound speed in the central Arabian sea. Open Ocean Eng. J. **4**, 71–76 (2011)
12. Kumar, S., Prince, S., Aravind, J.V., Kumar, G.S.: Analysis on the effect of salinity in underwater wireless optical communication. Mar. Georesour. Geotechnol. **38**(3), 291–301 (2020)
13. Preisig, J.: Acoustic propagation considerations for underwater acoustic communications network development. Mob. Comput. Commun. Rev. **11**(4), 2–10 (2006)
14. Sehgal, A., Tumar, I., Schonwalder, J.: Variability of available capacity due to the effects of depth and temperature in the underwater acoustic communication channel. In: Oceans 2009-Europe, Bremen, pp. 1–6 (2009)
15. Leroy, C.C., Parthiot, F.: Depth-pressure relationships in the oceans and seas. J. Acoust. Soc. Am. **103**(3), 1346–1352 (1998)
16. Yuwono, N.P., Arifianto, D., Widjiati, E., Wirawan: Underwater sound propagation characteristics at mini underwater test tank with varied salinity and temperature. In: 6th International Conference on Information Technology and Electrical Engineering (ICITEE), pp. 1–5 (2014)

17. Shi, H., Kruger, D., Nickerson, J.V.: Incorporating environmental information into underwater acoustic sensor coverage estimation in estuaries. In: MILCOM 2007 - IEEE Military Communications Conference, pp. 1–7 (2007)
18. Morozs, N., Gorma, W., Henson, B.T., Shen, L., Mitchell, P.D., Zakharov, Y.V.: Channel modeling for underwater acoustic network simulation. IEEE Access **8**, 136151–136175 (2020)
19. Lee, H.K., Lee, B.M.: An Underwater acoustic channel modeling for Internet of Things networks. Wirel. Pers. Commun. **116**(3), 2697–2722 (2020). https://doi.org/10.1007/s11277-020-07817-x
20. Onasami, O., Feng, M., Xu, H., Haile, M., Qian, L.: Underwater acoustic communication channel modeling using reservoir computing. IEEE Access **10**, 56550–56563 (2022)
21. Zhu, X., Wang, C.-X., Ma, R.: A 2D non-stationary channel model for underwater acoustic communication systems. In: IEEE 93rd Vehicular Technology Conference (VTC2021-Spring), pp. 1–6 (2021)
22. Kotipalli, P., Vardhanapu, P.: Frame boundary detection and deep learning-based doppler shift estimation for FBMC/OQAM communication system in underwater acoustic channels. IEEE Access **10**, 17590–17608 (2022)
23. Venkata Lalitha, N., et al.: IoT based energy efficient multipath power control for underwater sensor network. Int. J. Syst. Assur. Eng. Manag., 1–10 (2022)
24. Sekhar, S., et al.: Effects of water absorption on the mechanical properties of hybrid natural fibre/phenol formaldehyde composites. Sci. Rep. **11**(1), 13385 (2021)
25. Etter, P.C.: Underwater Acoustic Modeling and Simulation. CRC Press, Boca Raton (2018)
26. Padmavathy, N., Venkateswara Rao, Ch.: Reliability evaluation of underwater sensor network in shallow water based on propagation model. J. Phys. Conf. Ser. **1921**(1), 012018 (2021)
27. Venkateswara Rao, C., et al.: Evaluation of sound propagation, absorption, and transmission loss of an acoustic channel model in shallow water. In: Congress on Intelligent Systems, pp. 455–465 (2023)
28. Padmavathy, N., Venkateswara Rao, Ch.: Effect of undersea parameters on reliability of underwater acoustic sensor network in shallow water. IOP Conf. Ser. Mater. Sci. Eng. **1272**(1), 012011 (2022)
29. Venkateswara Rao, C., et al.: Analysis of acoustic channel model characteristics in deep-sea water. In: International Conference on Cognitive Computing and Cyber Physical Systems, pp. 234–243 (2023)
30. Venkateswara Rao, C., et al.: Comparison of acoustic channel characteristics in shallow and deep-sea water. In: International Conference on Cognitive Computing and Cyber Physical Systems, pp. 256–266 (2023)

Clustering Based Hybrid Optimized Model for Effective Data Transmission

Nadimpalli Durga(✉) , T. Gayathri , K. Ratna Kumari, and T. Madhavi

Shri Vishnu Engineering College for Women, Bhimavaram, India
{ndurgacse,gayathritcse,kratnacse,madhavi.v}@svecw.edu.in

Abstract. The Internet of Things (IoT) is a system of unified gadgets that can conversation data and operate in tandem thanks to the web. When it comes to the longevity of a network, smooth data production is crucial, and wireless sensor networks (WSN) play a key character in the IoT in this regard. Despite the IoT's usefulness in many areas, it still faces obstacles in the form of security, energy, load balancing, and storage. Clustering and multi-hop routing are two methods used in the architecture of an IoT-assisted WSN to reduce energy consumption. This research therefore provides a novel effective hybrid optimization strategy for choosing cluster heads. In to adjust the white shark optimizer's (WSO) stochastic behaviour while it seeks out food, the suggested method makes use of the whale optimization approach (WOA). The new HWSO was also tested against a group of contemporary meta-heuristic methods, such as the artificial optimizer (GTO), the coyote optimization algorithm (COA), and the original WSO. Finally, the proposed network is put through its paces by making use of NS-3.26's extensive simulation features. Improvements in packet delivery ratio (PDR), latency, energy consumption, number of dead nodes, and longevity of the network may be shown in the simulation results.

Keywords: Wireless networks · White shark optimizer · Internet of Things · Clustering · Whale optimization approach · Coyote optimization algorithm · Data Transmission

1 Introduction

These days, IoT and mobile edge computing (MEC) are commonplace tools for anticipating and meeting future technical requirements. Innovations in technology have mostly targeted information transmission needs, lightening the burden on the network, and increasing throughput [1]. Reliable platforms, smart cities, and transportation all rely on networks that can handle minimal delays, store large amounts of data, and operate in mission-critical areas of operation [2]. Scientists are paying more attention to the development of autonomous networks for Internet of Things devices as their prevalence grows [3]. The primary objective of the network executive idea is to design a multi-hop network that makes efficient use of electricity by linking nodes at the source and the destination using mobile phone relay nodes (RNs) [4]. Increased interest in (WSNs) can

© ICST Institute for Computer Sciences, Social Informatics and Telecommunications Engineering 2024
Published by Springer Nature Switzerland AG 2024. All Rights Reserved
P. Pareek et al. (Eds.): IC4S 2023, LNICST 537, pp. 338–351, 2024.
https://doi.org/10.1007/978-3-031-48891-7_30

be attributed to the quick expansion of MEMS technologies and wireless networks in recent years [5]. The IoT would not be complete without WSN. As a result, WSN applications have connected with the human, digital, and material spheres. Typically, WSNs are made up of a large sum of minor sensors spread out crossways a big region, and base stations (BS) that collect data from these sensor nodes.. Clustering is the process of dividing a network into smaller, more manageable pieces [7]. CH is the initial node in the routing process, and it is responsible for delivering the message to its eventual destination [8].

Clustering consumes some of the network's radio resources [9]. During network clustering, nodes can do computations to arrange its nearby nodes clusters. The cluster structure's robustness relies heavily on the care taken during cluster formation and CH selection [10]. Rearranging cluster nodes (CMs) is one way to make structural changes to an existing cluster. The cluster maintenance signaling also makes use of the network's radio resources. Costs associated with forming and keeping a cluster are measured in terms of the time spent translating control messages. The effectiveness of a clustering model is measured in terms of numerous factors, such as the size of the cluster and how long it takes to construct [11]. Using AI techniques, we can classify networks into different groups [12]. The primary problem with these approaches is that they are computationally intensive and so cannot guarantee optimal results. They move too slowly toward optimal results.

Energy consumption and transmission power are inseparable. If we select either a high or low transmission power, SNs will expend more energy. To reduce energy waste, the transmission power should be kept between its extremes. The concept of biological evolution forms the basis for evolutionary algorithms (EA) [13]. Evolutionary approaches, learning classifier systems, and genetic algorithms are all part of EA. Evolutionary algorithms are a good option to try if previous methods have failed. In the face of seemingly intractable [14] problems, evolutionary techniques are often embraced. A near optimum solution to an unresolved situation is acceptable even if EA is computationally expensive. An atmosphere conducive to the natural selection of effective solutions to the issue at hand will be established [15]. Using these biological algorithms, the optimal solution to problems associated with constructed habitats may be found. The scalability issue is addressed by grouping nodes together and allowing them to share their geographical coordinates.

The following are some clarifications that help highlight the limitations of the described methods:

Premature and sluggish convergence rates are a problem for several of the described meta-heuristic optimization techniques, leading to entrapment in local optima.

In addition to needing a lot of data to train the neural network-based method, several other published systems have the drawback of taking a long period and requiring a lot of work to implement.

To overcome these restrictions, this study introduces a hybrid strategy that combines the white (WSO) with the (WOA) in order to alter the behaviours of the unique WSO while it forages for food. These are some of the ways in which this work contributes:

- To improve the misuse phase of the standard (WSO), we present a hybrid version of the algorithm that uses the whale optimization algorithm's spiral updating position strategy.
- The suggested HWSO is tested against the generic white shark optimizer (GTO), the clustered optimal algorithm (COA), and the white shark optimizer itself for accuracy.

2 Related Works

Work objects were presented by Arunachalam et al. [16] to provide a unique energy-efficient process by combining the features of sophisticated aggregation approaches. To pick the CH optimally according to energy, distance, and weight value, a spider monkey optimization (SMO) based protocol is designed. The suggested technique has been evaluated using a number of performance indicators, and the findings have been compared to those of other, more current state-of-the-art models to demonstrate its superiority.

Using a (E-RARP) and a clustering method (GEC), Gunigari and Chitra [17] have created a hybrid Energy Efficient and Reliable (ACO). E-RARP is a novel ACO-based routing protocol for WSNs that is both energy efficient and reliable. The proposed protocol ensures consistent connectivity and high-quality lines of communication to boost power.

In order to balance network security and energy efficiency, Nagaraj et al. [18] propose the secure encryption (SERPPA). When it comes to backing up and keeping tabs on the network's nodes, SERPPA has a central entity known as a cluster head.

Nirmaladevi and Prabha [19] have been concentrating on developing a routing system that can function even when selfish nodes are present. The hierarchical clustering method is used by (SN-TOCRP). The Fuzzy-based Crowd Search Algorithm is inspired by nature and is used to choose cluster Heads (CH). To identify nodes that are acting in their own self-interest, we introduce an authentication approach that verifies the authority of the cluster leaders.

Oppositional Cuckoo Search Optimisation based Clustering with Classification Model for Big Data Analytics in Healthcare Environment has been proposed by Gayathri [20], who also creates a Map Reduce (MR)-based, enhanced metaheuristic algorithm-based clustering and classification model.

3 Proposed System

3.1 Environment-Adaptive Hierarchical Clustering

In this scenario, it is assumed that M nodes have been appropriately placed in the placement region and that their coordinates are known in advance. These M nodes would be considered the first clusters in the aggregate hierarchical clustering algorithm's theoretical framework:

$$C_i = \{X_i\}, i \in M \tag{1}$$

where C_i denotes the cluster's i^{th} iteration. Each round of clustering would use the biggest Euclidean distance between any two groups as the clustering cost. The two nearest clusters would then merge into a novel cluster until the termination illness is met or the requisite sum of clusters is attained.

Assume that C(M + a) includes X_i, X_j after many clustering processes, furthest distant from one another among the nodes they enclose. Using the HC procedure's idea of the biggest distance among clusters, we may express the largest reserve among these two clusters as $D(C\_(M + a), C\_(M + b))$.

$$D(C_{M+a}, C_{M+b}) = D(X_i, X_k) \tag{2}$$

where $M + a$ and $M + b$ characterize the label of the clusters.

In this paper, the chief distance among the organized nodes D_{max} is designated and attuned to serve as the clustering finish threshold T.

$$= \sigma max \left\{ \frac{T = \sigma D_{max}}{\sqrt{(X_i - X_j)^2 + (Y_i - Y_j)^2}} \right\}, \ i, j \in M, i \neq j \tag{3}$$

where $X_i = (X_i, Y_i)$, $X_j = (X_j, Y_j)$ signifies the nodes, and s signifies the applied factor, which is defined as the ratio within the confidence distance.

3.2 HCEH-UC Routing Procedure

This work proposes a distributed data transmission mode modification approach for WSN nodes to use in order to achieve optimal data transmission. The fatigued node might be recharged in time to make for the next cycle, and the cluster head may be switched between nodes based on the amount of energy left. To ensure the proper functioning of WSNs with high-energy efficiency, each cluster may adaptively complete the data collecting cycle. Therefore, the proposed HCEH-UC routing algorithm allows for continuous coverage of the desired region by EH-WSNs. Table 1 displays the radio energy consumption model used to determine the WSNs' data transmission energy requirements.

Table 1. The energy ingesting in radio broadcast and reception manner.

Mode	Energy-Consumption
transmission/reception manner	$50(\text{nJ} \cdot \text{bit}^{-1})$
free-space evidence amplification (ε_{fs})	$10(\text{pJ} \cdot \text{bit}^{-1} \cdot \text{m}^{-2})$
multipath-fading info amplification (ε_{mp})	$0.0013(\text{pJ} \cdot \text{bit}^{-1} \cdot \text{m}^{-4})$

A wireless propagation model is recognized using the multipath attenuation models, and the energy required to communicate k bits of data is given by the formula ETx:

$$E_{Tx}(k, d) = E_{Tx-elec}(k) + E_{Tx-amp}(k, d) = \{E_{elec} * k + \varepsilon_{fs} * k * d^2,$$

$$d < d_0 E_{elec} * k + \varepsilon_{mp} * k * d^4, d \geq d_0 \tag{4}$$

where d is the distance to be sent, d_0 is the distance is the broadcast energy, and ETxamp is the intensification energy needed to transmit data to distance d:

$$d_0 = \sqrt{\frac{\varepsilon_{fs}}{\varepsilon_{mp}}} \tag{5}$$

At the same period, the energy obligatory to receive k − bit data can be portrayed as (6):

$$E_{Rx}(k) = E_{Rx-elec}(k) = E_{elec} * k \tag{6}$$

The data is then compressed by the CHs before being sent to the base station or the end user end. The total amount of data that a cluster of Q nodes needs to send to the base station in one cycle is kBs bit, the base station is dBs, the amount of data that the qth node in the cluster needs to send to the cluster cycle is kqs bit, and the distance among q and s is dqs.

The energy used by the cluster head node s during a single cycle of the network can be broken down into the following parts (ERx = energy used to receive data from other nodes, E_Df = energy used for data fusion, and ETx = energy used to transmit data package to base post B):

$$E_{Bs} = E_{Rx}(k_{Bs}) + E_{Df}(k_{Bs}) + E_{Tx}(k_{Bs}, d_{Bs}) = E_{elec} * k_{Bs} + E_{DA} * k_{Bs}$$
$$+ E_{Tx-elec}(k_{Bs}) + E_{Tx_{amp}}(k_{Bs}, d) \tag{7}$$

where E_{DA} signifies the energy ingesting continuous for data fusion.

Assume the node q conveys kqs bit data to the cluster s in one data broadcast cycle:

$$\sum_{q=1}^{Q} k_{qs} = k_{Bs}. \tag{8}$$

Thus, the energy ingesting E_{qs} of the qth node to transmission these data can be labelled as (9):

$$E_{qs} = E_{Tx-elec}(k_{qs}) + E_{Tx-amp}(k_{qs}, d), q \neq s. \tag{9}$$

The cluster head node should go to sleep mode to conserve energy when its battery life is too low to sustain routing operations. The data transmission job is so taxing that the sleep node can hardly carry it out. Therefore, the novel cluster head node would be chosen based on the location data and status of the other nodes in the cluster in an effort to achieve UC-EH, or continuous target coverage. Let's pretend that the Estimation is the power needed to transmit data between nodes.:

$$E_{estimation}(s) = E_{Bs} + \sum_{q=1}^{Q} E_{qs}, q \neq s \tag{10}$$

where the $E_{estimation}(s)$ includes the energy obligatory E_{qs} for nodes energy can be characterized by E_{rest}, and the r is adopted to signify the likelihood of being designated as a cluster head for the qth.

$$\rho(q) = \{1 - \frac{E_{estimation}(q)}{E_{rest}(q)}, q \in G0, \qquad q \notin G \tag{11}$$

where G is the group of nodes that haven't been chosen to send data during the current round of transmission. As the distance between nodes and the grew, so did Eestimation. Since the successor CH's data delivery energy needs to be low and the successor CH's leftover energy needs to be high, the node with the highest probability r develops the successor CH.

The suggested clustering technique would produce node clusters that are uniformly dispersed over the target detection region, with nodes in each cluster being closer together than they are to the base station. As a result, the energy needs of the cluster's leader node are significantly higher than nodes. The energy amassed by the cluster head must sustain communication with the base station, communication with the cluster nodes, and interaction with the cluster's successor cluster head.

3.3 Proposed Clustering Algorithm: Hybrid Model

3.3.1 White Shark Optimizer (WSO)

Braik et al. [23] presented the (WSO), one of the newest meta-heuristic methods. The authors were inspired to create this algorithm by the complex behaviors of great white sharks, which include the use of highly developed senses of hearing and smell. White sharks are gorgeous and well adapted hunters; they use their powerful jaws and fins to capture prey such as dolphins, small whales, crabs, seabirds, and seals. Great white sharks employ a technique of surprise and rapid movement to ambush their victim, after which they strike with devastating force.

There are three different actions that must be taken in order to devour the prey (food source): moving towards the prey utilizing the hesitations generated by the prey's movement in the waves, randomly searching for depths, and identifying nearby prey. The great white sharks can then use these processes to adjust their postures and find the optimal solutions. The WSO may be modeled by creating an initial population of solutions in a matrix. Where N is the population size and d is the issue dimension, the size of the initial population matrix is N:

$$w = \left[w_1^1 w_2^1 \cdots w_d^1 w_1^2 w_2^2 \cdots w_d^2 \vdots w_1^n \vdots w_2^n \vdots \cdots \vdots w_d^n \right] \tag{12}$$

where w_j^i characterizes the ith white shark site in the jth measurement. It can be intended based on the lower (lb_j) and upper (ub_j) limits of the search as:

$$w_j^i = lb_j + rand \times (ub_j - lb_j) \tag{13}$$

where rand is an integer chosen at random between zero and one. Using Eq. (12), an initial solution's fitness is computed, and an updating procedure is set into motion if the new location is superior to the old one. The great white shark detects the position of its food by the hesitance of its waves and then swims in undulating motions at a pace:

$$v_{k+1}^i = \mu\left(v_k^i + \rho_1\left[w_{gbest_k} - w_k^i\right] \times c_1 + \rho_2\left[w_{best}^{v_k^i} - w_k^i\right] \times c_2\right) \tag{14}$$

where v_{k+1}^i and v_k^i are w_ki is the site of the ith white shark in repetition k; c_1 and c_2 are random statistics in the range [0, 1]; w_best(v_ki) site to the swarm during iteration k; and v_ki is the index vector sum i for sharks obtaining the best location, and it can b:

$$v = [n \times rand(1, n)] + 1 \tag{15}$$

The parameters ρ_1 and ρ_2 that control the w_{gbest_k} and $w_{best}^{v_k^i}$ best effects on w_k^i; they can be subtracted as shadows:

$$\rho_1 = \rho_{max} + (\rho_{max} - \rho_{min}) \times e^{-\left(\frac{4k}{K}\right)^2} \tag{16}$$

$$\rho_2 = \rho_{min} + (\rho_{max} - \rho_{min}) \times e^{-\left(\frac{4k}{K}\right)^2} \tag{17}$$

where ρ_{min} and ρ_{max} are K is the maximum number of iterations, _min = 0.5, _max = 1.5, and = the initial and inferior velocities to improve the great white sharks' movements. The convergence rate of the WSO may be analyzed with the use of the correction factor, denoted by the term in Eq. (14):

$$\mu = \frac{2}{\left|2 - t - \sqrt{t^2 - 4t}\right|} \tag{18}$$

where t is the acceleration factor of the procedure.

Great white sharks, as was previously said, devote the vast majority of their time to hunting for high-value prey. As a result, their locations shift as they approach their prey, which they do by listening to the waves caused by the prey's movements or by detecting the prey's scent. Great white sharks in this scenario wander to seemingly random locations while they hunt for food; this behavior may be modeled as follows [21]:

$$w_{k+1}^i = \{w_k^i \times \neg \oplus w_0 + ub \times a + lb \times b \, if \, rand < mvw_k^i$$

$$+ \frac{v_k^i}{f} \quad if \, rand \geq mv \tag{19}$$

where $\neg$ is the negation operator, Eqs. (20) and (21) define binary vectors a and b, Eq. (22) computes a logical vector w_0, and Eq. (23) determines the frequency of the great white shark's wavy movements.

$$a = sgn\left(w_k^i - ub\right) > 0 \tag{20}$$

$$b = sgn\left(w_k^i - lb\right) < 0 \tag{21}$$

$$w_0 = \oplus(a, b) \tag{22}$$

$$f = f_{min} + \frac{f_{max} - f_{min}}{f_{max} + f_{min}} \tag{23}$$

where f_{max} and f_{min} are the highest and lowest frequencies at which the great white shark's movements undulate. The great white shark's propulsion power, denoted by the parameter mv, is raised iteratively as shown in [23]:

$$mv = \frac{1}{a_0 + e^{\left(\frac{0.5K-5}{a_1}\right)}} \tag{24}$$

where a_0 and a_1 are two parameters for controlling exploration and exploitation. The use of mv speeds up the search process and fortifies the WSO's exploratory and exploitative tendencies. Because of this benefit, the author decided to use this sort of method to address the addressed issue. Here's how you may model how a great white shark might move to get closer to its prey:

$$w_{k+1}^i = w_{gbest_k} + r_1 \vec{D}_w \times sgn(r_2 - 0.5) \quad if \quad r_3 < S_s, \tag{25}$$

where w_{k+1}^i characterizes the ith countless white shark new site with regard to its prey:

$$\vec{D}_w = \left| rand \times \left(w_{gbest_k} - w_k^i\right)\right| \tag{26}$$

The parameter S_s in Eq. (27) is active to label the olfactory senses when subsequent its prey thoroughly; it can be intended as follows [23]:

$$S_s = \left|1 - e^{\frac{-a_2 k}{K}}\right| \tag{27}$$

where a_2 is a parameter used to control the examination/misuse behaviors.

3.3.2 The Proposed Hybrid WSO-Based Methodology

In the simplest procedure of the WSO, the sharks move toward their prey spot using a single approach, which may cause the algorithm to miss additional favorable nodes in the vicinity. Therefore, in this study, the WSO has been combined with a different method based on a spiral shaped path in order to improve the exploitation behavior of the original WSO. According to the whale optimization algorithm (WOA), the application of the spiral-shaped route was motivated by whale prey spots. The following correlation represents the great white shark's spiraling journey to its meal:

$$W_{t+1}^i = \vec{D}.e^{hl}.\cos \cos(2\pi l) + W_t^* \tag{28}$$

$$\vec{D} = \left| \vec{W^*} - \vec{W} \right| \tag{29}$$

where $\vec{D}$ is the great white shark's prey's distance from the shark, the constant h used to define the logarithmic spiral's form, and a random value l in the interval [1]. The great white shark's approach to its prey, as in Eq. (25), may be altered using the spiral equation in the following way:

$$w_{t+1}^i = \{w_{gbest_k} + r_1 \vec{D}_w \times sgn(r_2 - 0.5) \ if \ r_3 < S_s \vec{D} .e^{bl} .cos \ cos(2\pi l) + w_t^* \ if \ r_3 < S_s \tag{30}$$

The primary framework of the proposed HWSO is summarized in pseudo code (see Algorithm 1) below when applied to the parameter estimation and optimization issue of the battery model. The first random set of solutions is generated by assigning bounds to the model parameters. Then, the relevant values for the goal function's starting point are determined using Eq. (11).

4 Results and Discussion

We evaluate the effectiveness of our suggested protocols by simulating them with varying numbers of nodes, varying simulation times, varying the fraction of faulty/failure nodes, and varying the speed at which the nodes may move. Our simulation lasted for 100 s, during which time 100 mobile nodes were spread out throughout a 500 m by 500 m region of the network as shown in Table 1. We analyze nodes with a given 100 energy to start, and CBR traffic is assumed as the source type (Table 2).

Table 2. Default simulation parameters.

Parameter	Value
Topology dimension	500 m × 500 m
Mobility perfect	Random way point perfect
Propagation Model	Free space propagation perfect
Sum of Nodes	100
Simulation Period	100 s
Early Energy	100 J
Mobility Speed	10 m/s
Sum of Fault nodes	0
Pocket loss rate	0
Antenna perfect	Omni directional
MAC type	802.11
Traffic Kind	CBR

4.1 Performance Metrics

In the simulation studies, we employ the following performance measures [22, 23].

4.1.1 Packet Delivery Ratio (PDR)

The packet delivery rate is the fraction of total data packets transmitted that arrived at their target node. PDR demonstrates a protocol's efficiency in delivering data over a network. Here is how the PDR is calculated:

$$PDR = \frac{\sum P_d}{\sum P_s} \times 100 \tag{31}$$

In this equation, P_d characterizes the sum of packets delivered and P_s characterizes the sum of packets sent.

4.1.2 Throughput

How many megabits per second (Mbps) of data have been sent and received across the network in a certain time period. It measures both quality and efficiency. A low percentage of lost data packets during transmission indicates a high throughput. Here is how we quantify it:

$$G = \frac{\sum B_r \times 8}{T} \times 10^6 (Mbps) \tag{32}$$

In this equation, G is the throughput, B_r is the entire sum of bytes conventional, and T is the simulation time.

4.1.3 End-to-End Delay (E2E)

The time it takes for a packet of data to arrive at its final destination after being sent from the sending node. It's sometimes referred to as "One-Way Delivery" (OWD). Delays of any kind fall under this category. The sending node checks its routing database to see if there is a path to the destination before sending any data packets. The source node initiates route discovery by broadcasting RREQ signals to its nearby nodes if no route is known to exist. This will keep happening until one of the nodes sends back the sender in accordance with the AOMDV protocol. Based on the fitness function in Eq. (11), the source node will determine the optimal route and only use this destination node. The whole end-to-end (E2E) lag time is computed as:

$$E2E = \frac{\sum_{i=0}^{n} R_i - S_i}{n} \tag{33}$$

Here, n is the total sum of packets positively received, R_i is the time at which the ith packet arrived at its destination node, and Si is the time at which the ith packet was transmitted from its source node [23]. The timestamps S_i and R_i are gathered from the application layers.

4.1.4 Energy Consumption

During the simulation period, this is the cumulative total of the energy used by the network's nodes. The formula is as follows:

$$E = \sum_{i=0}^{m} I_i - E_i \qquad (34)$$

The energy consumption, denoted by E, the initial energy, denoted by Ii, and the final energy, denoted by E_i, of node i at the conclusion of the simulation period for all m nodes are all denoted by these symbols.

4.2 Validation Analysis of Proposed Model

In this research work, the existing models are implemented and their results are averaged for every parameter that is shown in Table 3, 4, 5 and 6.

Table 3. Throughput analysis of different models

Simulation Period (Seconds)	20	40	60	80	100
GTO	69	65	61	59	58
COA	70	68	65	63	60
WSO	72	71	69	68	67
HWSO	83	80	77	74	71

In above Table 3 represent that the Throughput analysis of different nodes and different models. In this analysis the proposed model reaches the 20th node throughput as 83 and the 40th node as 80 and 60th node throughput as 77 and also the 80th node throughput as 74 and finally the 100th node throughput as 71 repressively.

Table 4. Energy consumption with sum of nodes by different models

Number of nodes	20	40	60	80	100
GTO	35	40	45	50	60
COA	32	38	40	48	58
WSO	30	36	38	46	55
HWSO	22	26	30	38	42

In above Table 4 represent that the Energy consumption with sum of nodes analysis we used different nodes and different models. The proposed model in range of 20th node the value reached 22 and 40th node value as 26 and the 60th node value as 30 and the 80th node value as 38 and finally the 100th node the energy consumption value as 42 respectively. In Fig. 3 our proposed model takes less delay if the no of nodes increased.

Table 5. End-to-end delay with sum of nodes.

Sum of nodes	20	40	60	80	100
GTO	40	43	48	50	55
COA	35	38	43	48	50
WSO	30	35	38	43	53
Proposed	25	29	34	38	42

Table 6. PDR with sum of nodes.

Sum of nodes	20	40	60	80	100
GTO	76	55	50	45	40
COA	75	50	48	43	44
WSO	74	53	46	40	38
Proposed	83	65	60	58	55

In above Table 5 represents End-to-end delay with sum of nodes and Table 6 represent that the PDR with number of nodes. In this analysis we used different nodes and different models. The proposed model in range of 20th node the value reached 83 and 40th node value as 60 and the 60th node value as 60 and the 80th node value as 58 and finally the 100th node the PDR value as 55 respectively.

5 Conclusion

The determination of this research was to develop a novel hybrid optimization approach for improving WSNs' energy efficiency and durability. The Internet of Things (IoT) network's main level consists of randomly distributed nodes that exchange data using various gathering mechanisms. The described method then largely uses a hybrid model to choose CHs and arrange clusters. To do this, we proposed a novel hybrid meta-heuristic strategy based on a WSO and the WOA. The WOA improved the WSO's stochastic behavior in its hunt for food. The suggested method's significance lay in the fact that it solved many of the issues seen in earlier approaches, the most significant of which was becoming stuck in local optima. Once the best paths have been determined, the CHs will use them to send data to the BS. Instead of using a standard search approach, the suggested protocol employs an energy-saving strategy in which the best CHs are selected using an enhanced an efficient function. These steps improve the efficiency of the procedure. We compared the effectiveness of the proposed protocol to that of other well-known cluster-based conventions to show that it holds up across a variety of presentation criteria. The artificial (GTO), the coyote optimization algorithm (COA), and the rudimentary WSO were all used to compare and contrast the method to other optimizers. Future improvements to the suggested method's energy efficiency might come from data aggregation and sleep scheduling systems.

References

1. Srivastava, A., Singh, A., Joseph, S.G., Rajkumar, M., Borole, Y.D., Singh, H.: WSN-IoT clustering for secure data transmission in e-health sector using green computing strategy. In: 2021 9th International Conference on Cyber and IT Service Management (CITSM), pp. 1–8. IEEE (2021)
2. Gulati, K., Boddu, R.S.K., Kapila, D., Bangare, S.L., Chandnani, N., Saravanan, G.: A review paper on wireless sensor network techniques in Internet of Things (IoT). In: Proceedings, vol. 51, pp. 161–165 (2022)
3. Ullah, A., Azeem, M., Ashraf, H., Jhanjhi, N.Z., Nkenyereye, L., Humayun, M.: Secure critical data reclamation scheme for isolated clusters in IoT-enabled WSN. IEEE Internet Things J. **9**(4), 2669–2677 (2021)
4. Arya, G., Bagwari, A., Chauhan, D.S.: Performance analysis of deep learning-based routing protocol for an efficient data transmission in 5G WSN communication. IEEE Access **10**, 9340–9356 (2022)
5. Agarwal, V., Tapaswi, S., Chanak, P.: Intelligent fault-tolerance data routing scheme for IoT-enabled WSNs. IEEE Internet Things J. **9**(17), 16332–16342 (2022)
6. Wang, X., Chen, H.: A survey of compressive data gathering in WSNs for IoTs. Wirel. Commun. Mob. Comput. (2022)
7. Kuthadi, V.M., Selvaraj, R., Baskar, S., Shakeel, P.M., Ranjan, A.: Optimized energy management model on data distributing framework of wireless sensor network in IoT system. Wirel. Pers. Commun. **127**(2), 1377–1403 (2022)
8. Nandan, A.S., Singh, S., Malik, A., Kumar, R.: A green data collection & transmission method for IoT-based WSN in disaster management. IEEE Sens. J. **21**(22), 25912–25921 (2021)
9. Mahajan, H.B., Badarla, A.: Cross-layer protocol for WSN-assisted IoT smart farming applications using nature inspired algorithm. Wirel. Pers. Commun. **121**(4), 3125–3149 (2021)
10. Chandnani, N., Khairnar, C.N.: An analysis of architecture, framework, security and challenging aspects for data aggregation and routing techniques in IoT WSNs. Theor. Comput. Sci. **929**, 95–113 (2022)
11. Shafiq, M., et al.: Robust cluster-based routing protocol for IoT-assisted smart devices in WSN. Comput. Mater. Continua **67**(3), 3505–3521 (2021)
12. Khalaf, O.I., Abdulsahib, G.M.: Optimized dynamic storage of data (ODSD) in IoT based on blockchain for wireless sensor networks. Peer Peer Netw. Appl. **14**, 2858–2873 (2021)
13. Verma, S., Kaur, S., Rawat, D.B., Xi, C., Alex, L.T., Jhanjhi, N.Z.: Intelligent framework using IoT-based WSNs for wildfire detection. IEEE Access **9**, 48185–48196 (2021)
14. Lenka, R.K., Kolhar, M., Mohapatra, H., Al-Turjman, F., Altrjman, C.: Cluster-based routing protocol with static hub (CRPSH) for WSN-assisted IoT networks. Sustainability **14**(12), 7304 (2022)
15. Jothikumar, C., Ramana, K., Chakravarthy, V.D., Singh, S., Ra, I.H.: An efficient routing approach to maximize the lifetime of IoT-based wireless sensor networks in 5G and beyond. Mob. Inf. Syst., 1–11 (2021)
16. Arunachalam, G.S., Ramalingam, G., Nanjappan, R.: A classy energy efficient spider monkey optimization based clustering and data aggregation models for wireless sensor network. Concurr. Comput. Pract. Exp., e7492 (2023)
17. Gunigari, H., Chitra, S.: Energy efficient networks using ant colony optimization with game theory clustering. Intell. Autom. Soft Comput. **35**(3) (2023)
18. Nagaraj, S., et al.: Improved secure encryption with energy optimization using random permutation pseudo algorithm based on Internet of Thing in wireless sensor networks. Energies **16**(1), 8 (2023)

19. Nirmaladevi, K., Prabha, K.: A selfish node trust aware with optimized clustering for reliable routing protocol in Manet. Meas. Sens., 100680 (2023)
20. Gayathri, T., Bhaskari, D.L.: Oppositional cuckoo search with optimized based clustering and classification model for big data analytics in health care environment. J. Comput. Sci. Inf. Eng. **25**(4) (2022)
21. Braik, M., Hammouri, A., Atwan, J., Al-Betar, M.A., Awadallah, M.A.: White shark optimizer: a novel bio-inspired meta-heuristic algorithm for global optimization problems. Knowl. Based Syst. **243**, 108457 (2022)
22. Hamza, F., Vigila, S.M.C.: Cluster head selection algorithm for MANETs using hybrid particle swarm optimization-genetic algorithm. Int. J. Comput. Netw. Appl. **8**(2), 119–129 (2021)
23. Arulprakash, P., Kumar, A.S., Prakash, S.P.: Optimal route and cluster head selection using energy efficient-modified African vulture and modified mayfly in manet. Peer Peer Netw. Appl. **16**(2), 1310–1326 (2023)

A Web-Based Vaccine Distribution System for Covid-19 Using Vaxallot

B. Valarmathi[1] (ID), N. Srinivasa Gupta[2]([✉]) (ID), G. Prakash[3] (ID), A. BarathyKolappan[4], and N. Padmavathy[5]([✉]) (ID)

[1] Department of Software and Systems Engineering, School of Computer Science Engineering and Information Systems, Vellore Institute of Technology, Vellore, Tamil Nadu, India
valarmathi.b@vit.ac.in

[2] Department of Manufacturing Engineering, School of Mechanical Engineering, Vellore Institute of Technology, Vellore, Tamil Nadu, India
srinivasagupta.n@vit.ac.in

[3] Department of Database Systems, School of Computer Science and Engineering, Vellore Institute of Technology, Vellore, Tamil Nadu, India
g.prakash@vit.ac.in

[4] Systems Engineer, Tata Consultancy Services, Chennai, Tamil Nadu, India

[5] Department of Electronics and Communication Engineering, Vishnu Institute of Technology, Bhimavaram, Andhra Pradesh, India
padmavathy.n@vishnu.edu.in

Abstract. Vaxallot seeks to implement a system to distribute vaccines across high-risk groups accounting for various parameters and prove to be superior to what conventional systems are capable of today. It is a Python flask-based tool backed by infrastructure and data resources from the Covid India central repository; all it needs is a single channel input and a single parameter of the value produced, and the algorithm will take care of the rest. Since it's Python-based and has an active integration with google sheets, live value updating could be possible for the real-time output of the distribution. The novelty of the proposed mechanism is the unique priority index, a score that accounts for an array of factors associated with the pandemic and is computed for regions in question here; this makes way for better distribution of vaccines. The application has an exclusive segment centered on handling excess units, if any. Moreover, since the application is developed to suit the needs of dynamic demographics, any region can roll out this application for purposes they desire to serve the masses. Since it isn't bound by a coronavirus, it can be used by the healthcare industry as they deem fit.

Keywords: Vaccine · Distribution · Coronavirus · Pandemic · Priority Index · Excess Handling

1 Introduction

The ongoing pandemic has been devastating destroying millions of innocent lives; what's going to be even more devastating is the vaccine distribution system once a potential vaccine comes to the rescue. Many factors will influence the distribution; the supply

P. Pareek et al. (Eds.): IC4S 2023, LNICST 537, pp. 352–373, 2024.
https://doi.org/10.1007/978-3-031-48891-7_31

chain mechanism would most likely be disrupted due to the volume of vaccines to be distributed to people.

Planning for vaccine distribution and ensuring its effectiveness are essential for stopping the spread of infectious illnesses. Vaccines are vital for people to receive proper and routine immunization, and excellent planning for vaccine distribution is crucial for delivering successful vaccination and healthcare care.

Limited quantities may occasionally be purchased in poor nations where vaccine supplies must occasionally be imported. To reduce the spread of the virus and the number of fatalities brought on by the disease, it is necessary to distribute vaccines in a methodical manner. Through effective allocation, this strategy can improve the number of human lives saved while reducing the negative environmental effects of vaccine production and delivery. Although mathematical techniques have been established for vaccine allocation, it will be difficult for a non-expert to use them. In this study, a web application, often known as a web app, is used to build an optimization model for COVID-19 vaccine allocation.

Thereby, what we really need now is a vaccine management system that can guide us in the distribution mechanism accounting for all variables. This paper shall refer to the proposed system as Vaxallot. Vaxallot is an application on the web backed by user-friendly Sheets by Google; the objective is to simplify the distribution process while amplifying the overall efficacy; how we choose to tackle the problem at hand is described in the subsequent sections.

2 Background

A supply chain management tool for mass administration of the possible COVID-19 vaccine is what's at stake here, clustering high-risk groups, prioritizing population with respect to demographic parameters and accounting for parameters like costing vaccine hesitancy, supply chain management is what the idea is in essence. Expanding on the base idea, it is a web-based tool that is backed with regional metrics on the contraction of coronavirus.

The algorithm will prioritize regions with high intensity and relatively significant volumes of people and will offer the administration of the region insights into which region has to receive the vaccine first and which can wait longer for the vaccine.

The application in entirety consists of four phases, the pre-processing phase, the input phase, the computation phase and the output phase. In the preprocessing phase, missing values are present in the CSV are filled and data are normalized using the Min-Max normalization technique, wherein the minimum and maximum values of a field is fetched and used to normalize the relation in entirety. In the input phase, the application is open for input from the user, in specific terms, total available quantity of vaccines in any given unit at any given time would be a fit for an input parameter. The computation phase is where the numbers are put together to arrive at the unique aforementioned priority index.

Computation of priority index is a unique attribute of this project work, the following is how the priority index is computed for, considerations: Confirmed Cases be 'C',

Recovered Cases be 'R', Deaths be 'D', Population be 'T'. It is shown in the Eq. (1).

$$Priority\ Index\ =\ (C\ -\ (R+D))/T \tag{1}$$

The above formula is what's used in the program, it is then translated to a readable figure and vaccine volume distribution is based upon the computed priority index. In a sample scenario, wherein a region's confirmed total cases stand at 100, recovered cases stand at 75, deaths stand at 5 and population stand at 500, the priority index would be computed as follows:

$$Priority\ Index\ =\ (100\ -\ (75+5))/500 = 0.04$$

It's noteworthy that the priority index isn't backed by a benchmark or yardstick so to speak of, hence the results would vary greatly between regions but the efficacy would remain the same. The equation has been derived via trial and error and continuous experimentation. In the output phase, Once the computation is successful, vaccine distribution by sub region displays as a tabular column and illustrated graphically for the perusal of the end-user.

3 Literature Survey

Here, all the reviews of other works have been grouped according to the model and they have used in respective works. The groups are given below like

a) Vaccine distribution,
b) Vaccine allocation,
c) Decision making,
d) Vaccine development
e) Others

a) Vaccine distribution
This paper [11] identified the type and quantity of vaccines that were purchased, distributed, and administered at public pharmacies. Telephone interviews were conducted with 1704 public pharmacies in 17 provinces. A 17-person hypothesis revealed details about the vaccine used in pharmacies, while all other vaccines were given to other distributors.

This paper [12] analyzed the challenges and efforts in vaccine development and distribution when an emergency situation is met in a region and suggests appropriate measures to tackle them and possibly mitigate the situation effectively.

This paper [13] provided an overview of vaccine distribution chains in low income and middle-income countries and describes the challenges posed by such distribution chains, as well as relevant research documentation and task management activities organized according to seven classification criteria: decision level, methodology, component structured, uncertainty and integrated symptoms, operational measures, real use of health, and countries and integrated vaccines.

This paper [14] was considered in the absence of a global flu vaccine monitoring program that made it difficult to monitor progress towards the '03 Health Assembly

vaccine, analyzes the Influenza Vaccine Supply International Task Force which developed a system to test the worldwide distribution of flu vaccines. The most recent dose distribution data for '14 & '15 were used to modify last analysis.

This paper [15] explored the feasibility of human immunization vaccines worldwide, sophisticated manufacturing methods, extremely careful quality control mechanisms, and reliable distribution channels that might be needed to ensure the products are effective when used. It also researches the technologies used to produce different types of vaccines that significantly affect vaccine costs, industrial level of stability and global availability.

This paper [17] investigated the inefficiency of the World Health Organization's Expanded Program on Immunization Program (EPI) which is a major concern in many developed countries, resulting in a majority not being fully vaccinated and creating a major risk. There is also increased interest in these nations in the development of test kits and efficiency.

This paper [18] proposed an algorithm for DE to address the problem of vaccine distribution. Different age groups in people have different disease problems and different levels of touch. For best results, it is important to increase the distribution of vaccines in general to large clusters. The old model of infectious diseases has examined the effectiveness of the proposed algorithm and developed a number of simulations.

This paper [19] proposed a hybrid compartmental model that looks at different age groups in people with a variety of ailments and different levels of touch. This model was proposed in 2016 and is relatively new. The paper investigates the plausibility of such a model in a practical world.

b) Vaccine allocation
This paper [16] developed a model of the COVID-19, designed for the distribution of the plausible vaccine. This model, known as DELPHI-V-OPT, incorporates a predictable model into a predetermined model to facilitate the delivery of vaccines to all parts of the world and risk groups.

This paper [20] proposed an age-appropriate SEIR model to explore the various approaches to age distribution of Indian vaccines. They used regional metrics and coefficients for transmission of diseases between 28 January 2020 and 31 August 2020 in COVID-19 India cases. Comparative estimates were used to analyze parameters associated with morbidity based on prioritizing age groups in vaccine distribution strategies.

c) Decision making
This paper [5] considered a series of supply chain mechanisms for vaccines, including distributors and retailers. The distributor decides to use a cold chain to deliver the drugs. The retailer on the other hand performs a test when he receives the vaccine. Firstly a basic model is developed to study the conditions in which the distributor will move the terms through a cold chain. The objective is to extend knowledge of the vaccine supply chain mechanism.

This paper [8] investigated responses from countries participating in the training that indicated that the revised training materials and flow of work with the technical support followed enabled them to align their charts with their own procedures and WHO recommendations. This was overseen by National Vaccination Advisory Team.

d) Vaccine development

This paper [3] suggested that in order for the vaccine to be tested before treatment, one of the most challenging aspects of the development of a vaccine that constantly evaluates the effectiveness of previous vaccines against Varicella-Zoster Virus (VZV): a high-dose vaccine and a vaccine that is intended to strengthen the immune system.

This paper [4] investigated the production of vaccines due to the limited commercial market and the approval of regulatory authorities that took years to achieve. It also expresses that the concerns about the ability to accelerate productivity in response to the growing Ebola epidemic in the Democratic Republic of the Congo have been a little encouraging.

e) Others

This paper [1] ensured that the cost of the goods is guaranteed to be suitable for the intended use prior to vaccine trials. Post safety test of a vaccine and determination of the extent of immune response, Phase 2 of the trial is carried out. A vaccine trial is to be considered for safety, this also includes a series of discussions about possible interventions or strategies to address skepticism surrounding the subject matter.

This paper [2] considered a time when there has been an increase in the number of articles on vaccine formulation. Specifically concerning the Herpes zoster vaccine for herpes zoster, where two immunization strategies have been shown to promote immune defense against technological production challenges, and increased vaccine production. In the case of an Ebola vaccine, it was decided to reduce the dose of the vaccine or to use the drugs in severe cases, the effectiveness of the vaccine was limited.

This paper [6] reviewed one hundred and eighty-eight research questionnaires. In all, ten articles included comprehensive investigative tools for questioning, confidence or self-doubt. Self-confidence - Health Care and Other Areas of Health Assurance, Indices, Safety, Attitudes, Training and communication. It introduced a list of key questions like Do you believe that vaccinated vaccines can be dangerous? Are you worried about the terms?

This paper [7] considered that promising vaccine should ultimately be evaluated on a large scale. For clusters at risk, the process of testing the goals before the major phase II trials can be greatly accelerated by the existing studies. It is difficult, if not impossible, to give a definite estimate of time it can save in the process of development through challenging research. The risks of a SARS COV-2 challenge study can be easily surpassed significantly improving COVID-19 understanding and accelerating the development of a vaccine.

This paper [9] examined study subjects that have a participatory selection process with the exception of two on-line courses in Mexico and France, and three American courses offered to vaccine administrators and public health administrators who do not limit the number of study directors' participants. The ability to organize courses especially in Africa includes the residence of the participants and the distance and the struggle to revitalize the content of the courses provided by other intellectual and facilitator responsibilities was evaluated.

This paper [10] analyzed public meetings in October 2012 and February 2013 in addition to monthly conferences, it helped to formulate a definition of vaccine deficit, an effective model of drug-induced factors and indications of drug skepticism, namely a pilot in the WHO region of the United States by April 2013. A systematic review of vaccine submissions had been completed, and based on the available evidence, a systematic review of intervention to address drug skepticism was initiated.

The following table illustrates the various techniques that are at play currently. The existing work in distribution system literature survey is shown in the Table 1.

Table 1. Literature Survey of the existing work in distribution system

Serial No.	Name, Publisher, Month & Year	Objective	Methods & Techniques	Paper Link
1	The complexity and cost of vaccine manufacturing – An overview Elsevier, July 2017	Make certain vaccine efficacy stays constant throughout manufacturing and distribution	Licensure and prequalification criteria	https://doi.org/10.1016/j.vaccine.2017.06.003
2	Strategies for addressing vaccine hesitancy – A systematic review, Elsevier, August 2015	Minimize dosage while address key hesitancy concerns	Hesitancy intervention mechanism	https://doi.org/10.1016/j.vaccine.2015.04.040
3	Vaccine development: From concept to early clinical testing, Elsevier. December 2016	Delivery of vaccines via continuous monitoring of efficacy during trials	Consideration for adaptive immunity	https://doi.org/10.1016/j.vaccine.2016.10.016
4	Enabling emergency mass vaccination: Innovations in manufacturing and administration during a pandemic, Elsevier, May 2020	Handle production capacity limitations during mass administration	Mass vaccination mechanism and accounting for adjuncts	https://doi.org/10.1016/j.vaccine.2020.04.037
5	Cold chain transportation decision in the vaccine supply chain, Elsevier, May 2020	Study conditions in which the vaccine would be transported	Retailer inspection within supply chain management infrastructure	https://doi.org/10.1016/j.ejor.2019.11.005
6	Measuring vaccine hesitancy: The development of a survey tool, Elsevier, August 2015	Develop a survey tool for vaccines in general	Devising a survey tool for hesitancy measure	https://doi.org/10.1016/j.vaccine.2015.04.037
7	COVID-19 vaccine development: Time to consider SARS-CoV-2 challenge studies? Elsevier, July 2020	Handling obstacles during large-scale testing and administration	Ethical concerns and Human challenge studies within the development	https://doi.org/10.1016/j.vaccine.2020.06.007

(*continued*)

Table 1. (*continued*)

Serial No.	Name, Publisher, Month & Year	Objective	Methods & Techniques	Paper Link
8	Building immunization decision-making capacity within the World Health Organization European Region. Elsevier, July 2020	Implement a sound global immunization policy	Capacity building and Evidence-based decision mechanism	https://doi.org/10.1016/j.vaccine.2020.05.077
9	Advanced vaccinology education: Landscaping its growth and global footprint, Elsevier, June 2020	Analyze worldwide footprint and develop methods accordingly	General training and purposeful landscaping models	https://doi.org/10.1016/j.vaccine.2020.05.038
10	Review of vaccine hesitancy: Rationale, remit, and methods, Elsevier, August 2015	Address vaccine hesitancy via systemic intervention	Systemic immunization studies	https://doi.org/10.1016/j.vaccine.2015.04.035
11	Community pharmacy involvement in vaccine distribution and administration, Elsevier, May 2009	Continuously monitor pharmacy inventory	Applicability of non-traditional setting in the vaccine landscape	https://doi.org/10.1016/j.vaccine.2009.02.086
12	Challenges and efforts in vaccine development and distribution, Elsevier, September 2017	Implement an emergency handling mechanism	Vaccine hesitancy, excess production, and logistical inconsistencies	https://doi.org/10.1016/j.vaccine.2017.07.091
13	Vaccine distribution chains in low- and middle-income countries, Elsevier, December 2019	Tackle distribution challenges during a lack of financial support	Direct access to immunization mechanism	https://doi.org/10.1016/j.omega.2019.08.004
14	Survey of distribution of seasonal influenza vaccine doses in 201 countries (2004–2015): The 2003 World Health Assembly resolution on seasonal influenza vaccination coverage and the 2009 influenza pandemic have had very little impact on improving influenza control and pandemic preparedness, Elsevier, August 2017	Implement a possible global monitoring system	Continuous monitoring and vaccine recommendation systems	https://doi.org/10.1016/j.vaccine.2017.07.053
15	Vaccine production, distribution, access, and uptake, The Lancet, August 2011	Ensure availability of vaccines on a global level	Quality control and reliable distribution channels	https://doi.org/10.1016/S0140-6736(11)60478-9

(*continued*)

Table 1. (*continued*)

Serial No.	Name, Publisher, Month & Year	Objective	Methods & Techniques	Paper Link
16	Optimizing Vaccine Allocation to Combat the COVID-19 Pandemic Medrxiv, November 2020	Effectively allocate vaccine units to handle the virus	Response capture hybrid model	https://doi.org/10.1101/2020.11.17.20233213
17	A planning model for the WHO-EPI vaccine distribution network in developing countries, Taylor and Francis, May 2013	Mitigate inefficiencies of the EPI network	Linear programming and capacity expansion	https://doi.org/10.1080/0740817X.2013.813094
18	Optimal Vaccine Distribution Strategy for Different Age Groups of Population: A Differential Evolution Algorithm Approach, Hindawi, August 2014	Develop an age-structured model of distribution	Differential Evolution Algorithm	https://doi.org/10.1155/2014/702973
19	Efficient Vaccine Distribution Based on a Hybrid Compartmental Model. PLOS ONE, May 2016	Develop a distribution mechanism using a compartmental model	Epidemic Compartmental Model	https://doi.org/10.1371/journal.pone.0155416
20	Comparing COVID-19 vaccine allocation strategies in India: a mathematical modeling study, Medrxiv, November 2020	The contrast between various strategies involved in the distribution	Expanded SEIR Model	https://doi.org/10.1101/2020.11.22.20236091

4 Dataset Description and Sample Data

The dataset being used for this project is available on covid19india.org, since it is real-time and is trusted by medical and governmental entities, we chose to use that it represents some rows in the dataset. The dataset consists of five attributes and they are district, population, confirmed cases, recovered cases and deaths cases. The dataset is normalized by using min-max and mean normalization technique. In the original dataset, some values have been intentionally removed to illustrate the missing values handling mechanism. Delete the rows or columns with null values to manage missing values. Columns can be completely removed if they contain more than half of the rows as null values. Rows that have one or more columns with null values can also be removed. The sample dataset used for the proposed work is shown in the Table 2.

Table 2. Sample Dataset

District	Population	Confirmed	Recovered	Deaths
Ariyalur	860579	3148	2694	36
Chennai	5297274	143602	129677	2896
Coimbatore	3942171	19948	15584	3320
Cuddalore	2970742	14865	11052	156
Dharmapuri	1717801	1560	1205	160
Dindigul	2462144	7501	6367	143
Erode	2566988	4082	2957	52
Kanyakumari	2132226	10404	9367	1970
Karur	1213522	1948	1517	30
Krishnagiri	2142982	2715	2036	36
Madurai	3463607	14988	13555	368

5 Proposed Algorithm with Flowchart

The proposed algorithm consists of four phases and is shown in the Fig. 1. The four phases, along with a detailed process flow explanation is presented below. They are

- Phase 1: Preprocessing
- Phase 2: Input
- Phase 3: Computation
- Phase 4: Output

5.1 Phase 1: Preprocessing

In preprocessing phase, the system reads the values from the csv file. The csv file will be operated upon if it consists of missing values. These missing values are handled using Min-Max scaling. In this approach, we try to scale our data between zero and one. The benefit of having such a range, is that we will end up with significantly slighter standard deviations, which can suppress outliers.

Min-Max and Mean Normalization:
Normalization is a technique used to scale data between 0 and 1. Min-Max is a normalization strategy which linearly transforms x to y, which could be substituted by the absolute of *(x-minimum)/(maximum-minimum),* these are values in the set of observed values of x. This means, the min value is equated to zero and the max value is equated to one. Therefore, the entire range of values from minimum to maximum is mapped to the range 0 to 1. It becomes simple to compute a mean value in accordance to the already computed min-max value. Example: For a region X with an unknown total population, minimum of 1000 and maximum of 2000 in population, Min-Max would return 1and therefore mean to be computed will be closer to the maximum, hence the mean of the

mean of min and max and max, which is (mean (mean(min, max), max)) will be substituted to region X. Min-Max and mean normalisation technique is beneficial when the feature distribution is unclear when compared to weighted average normalisation.

5.2 Phase 2: Input

Soon after preprocessing the data, the system is open to read input vaccine production volume at a given facility at any given point in time. It's noteworthy that the system, owing to the purposes of minimalism doesn't require any more input than this. Example: Input total production of 10,000 units.

5.3 Phase 3: Computation

Like the name dictates, in this phase, several computational processes are implemented, starting with computing priority indices for all given sub regions beneath the main region followed by sorting the relation by highest to lowest priority, followed by first round of assigning volume of vaccines to be allotted to a certain sub region followed by second round of assigning volume of vaccines for the purpose of handling excess.

Example: With the given values, a priority index of 0.003175 is computed for Region X, in accordance to the priority index a volume of 1202 is allocated to Region X after handling for excess.

5.4 Phase 4: Output

Once the computation is successful, vaccine distribution by sub region displays as a tabular column and illustrated graphically for the perusal of the end-user. Example: The data from phase 3 is tabulated and plotted and is displayed for the end-user.

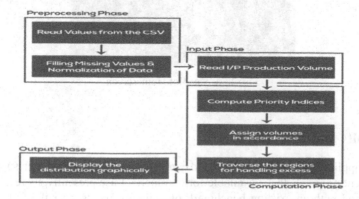

Fig. 1. Flowchart for the proposed Algorithm

6   Pseudocode

The proposed system's pseudo code is given below.

Step 1: Compute the priority index using for all regions Eq. (1) in Section (II)
Step 2: Input the available volume of vaccines
Step 3: Allocate the units in accordance to the computed priority index
Step 4: Reroute to handle excess volume of vaccines
Step 5: Tabulate the computed values and
Step 6: Plot the graph to illustrate distribution schema
Step 7: Display the plot and table in the output screen

7   Application Screenshots

Figure 2(a) represents the home screen of the application that appears soon after pre-processing phase is completed, once an appropriate value is given and distribution is initiated by clicking the "Distribute Now" button, it triggers the computation phase and finally with the data in order, a graphical illustration as given in Fig. 2(b) is drawn along with a tabular column which is presented in the next section.

Fig. 2. a. Vaxallot Homescreen

8   Results and Discussion

Since the application is first of its kind and can't directly be compared with supply chain management programs that's available for commercial use presently, its results cannot be compared with an existing benchmark of any sort, whatsoever the following is a representation of how its results vary with global standards. In the Fig. 3, given 20,000 units, the system computes priority index (PI) and lists it under 'PI' and computes volume to be distributed accordingly. The first column indicates index number as it was before the data was processed for demographical purposes.

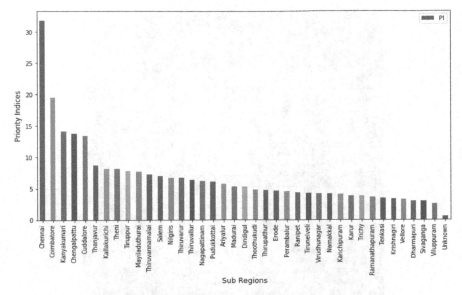

Fig. 2. b. Vaxallot Graphical Representation

The OpenPro software link is given as Instance: https://openpro.com/distribution/.

OpenPro is a popular Supply Chain Distribution Management Software, it's referenced here to illustrate an existing tool that can implement linear distribution. It's noteworthy that Table 3 is a representative simulation sourced out of a popular spreadsheet tool when the algorithm is implemented. From the Fig. 3 and Table 3, we can observe differential assignment in Vaxallot and proportionate assignment in a standard supply chain software. Moreover, for a total of 39 districts and 20,000 units of vaccines available, Vaxallot distributed it amongst all districts and was left with no excess whilst the conventional program post distribution was left with 3616 units (512 * 32). Hence, Vaxallot is clearly better than a conventional application that intends to distribute vaccines amongst consumers.

The comparison of volume of vaccine distribution using conventional program and Vaxallot with PI value is shown in the Table 4. For example, in XYZ district, the number of confirmed cases is 12, but the number of vaccines given to XYZ district is 512. Out of 512, only 12 vaccines are used and remaining 500 vaccines are not used. In ABC district, the number of confirmed cases is 712, but the number of vaccines given to ABC district is also 512. But, in ABC district, all the 512 vaccines are used, still ABC district need 12 vaccines. Here, the vaccine distribution is not effectively used. For more or less confirmed cases, the vaccine distribution is only 512 by using the conventional program.

The volume of vaccine distribution using conventional program is 512 for all the districts (i.e.) the number of vaccines are equally distributed to all the districts. The vaccine distribution using Vaxallot is the proposed method. The proposed method will solve this problem. If more number of confirmed cases is present means more vaccine will be districted to that district by using priority index. The priority index value will be high for high number of confirmed cases present in that district by using Vaxallot.

	District	Population	Confirmed	Recovered	Deaths	PI	Volume
2	Chennai	8297274	143602	129577	2994	31.78	2404
3	Coimbatore	3942171	19949	18984	3320	19.49	1476
10	Kanyakumari	2132226	10404	9357	1970	14.10	1068
1	Changalpattu	2870237	29994	28916	463	13.79	1044
4	Cuddalore	2970742	14995	11062	166	13.36	1012
26	Thanjavur	2742716	7691	9984	1230	9.63	963
8	Kallakurichi	2870237	7292	6060	960	9.14	617
24	Theni	1420326	13339	12342	163	9.09	612
33	Tiruppur	2926119	3792	2428	920	7.74	836
14	Mayiladuthurai	2870237	12498	10966	433	7.66	830
29	Thiruvannamalai	2909958	12211	10966	182	7.21	646
22	Salem	3949644	13006	10463	194	6.92	623
17	Nilgiris	939349	1999	1867	140	6.70	609
30	Thiruvarur	1441276	4660	3669	60	6.67	604
28	Thiruvallur	4260069	26941	24692	446	6.34	479
15	Nagapattinam	1942763	3413	2347	67	6.09	461
19	Pudukkottai	1844913	6996	6399	116	6.03	456
0	Ariyalur	930679	3148	2694	36	6.69	431
13	Madurai	3463607	14999	13666	359	6.20	394
6	Dindigul	2462144	7801	6367	143	6.19	393
31	Thoothukudi	1996201	11894	11066	116	4.74	369
27	Thirupathur	2870237	3339	2924	690	4.66	361
7	Erode	2666989	4092	2967	62	4.87	349
18	Perambalur	644964	1446	1396	190	4.62	343
21	Ranipet	2870237	11667	10699	139	4.36	329
32	Tirunelveli	3608046	10623	9243	189	4.19	317
37	Virudhunagar	2214208	13469	12741	200	4.14	313
16	Namakkal	1939926	2796	2029	46	4.13	313
9	Kanchipuram	4699007	19529	17066	276	4.06	307
11	Karur	1213622	1949	1617	30	3.90	299
34	Trichy	3103411	8336	7301	126	3.74	293
20	Ramanathapuram	1642927	6039	4896	110	3.69	271
25	Tenkasi	2870237	8969	8196	111	3.44	260
12	Krishnagiri	2142982	2716	2006	36	3.34	253
35	Vellore	4497417	11949	10663	183	3.27	247
5	Dharmapuri	1717901	1660	1206	160	3.00	229
23	Sivaganga	1626676	4346	4006	113	2.96	224
36	Viluppuram	3943316	9660	7769	82	2.47	185
38	Unknown	2870237	2229	2146	40	0.48	36

Fig. 3. Output from Vaxallot for 20,000 Vaccine Units

The priority index value will be low for low number of confirmed cases present in that district by using Vaxallot.

The volume of vaccine distribution using conventional program (i.e.) the output from Linear Distribution Supply Chain Software (Simulated) for 20,000 Vaccine Units is 512. For all the districts, the value is 512. In particular district, the confirmed cases are more means the number of vaccines is distributed to that district is 512. If the confirmed cases are less means the number of vaccines are distributed to that district is 512. For example, in particular district, the number of confirmed cases is 200, but the vaccine given is 512. In this case, vaccine distribution not proper manner. That is the drawback of the convention program. In order to avoid this, the proposed method will distribute

Table 3. Output from Linear Distribution Supply Chain Software (Simulated) for 20,000 Vaccine Units

District	Population	Confirmed	Recovered	Deaths	Volume of vaccine distribution using conventional program
Ariyalur	860579	3148	2694	36	512
Chengalpattu	2570237	28994	25916	463	512
Chennai	5297274	143602	129677	2896	512
Coimbatore	3942171	19948	15584	3320	512
Cuddalore	2970742	14865	11052	156	512
Dharmapuri	1717801	1560	1205	160	512
Dindigul	2462144	7501	6367	143	512
Erode	2566988	4082	2957	52	512
Kallakurichi	2570237	7292	6050	850	512
Kanchipuram	4558007	18628	17056	275	512
Kanyakumari	2132226	10404	9367	1970	512
Karur	1213522	1948	1517	30	512
Krishnagiri	2142982	2715	2036	36	512
Madurai	3463607	14988	13555	368	512
Mayiladuthurai	164985	2446	2328	18	512
Nagapattinam	1842753	3413	2347	57	512
Namakkal	1968325	2795	2028	46	512
Nilgiris	838349	1989	1567	140	512
Perambalur	644354	1446	1335	180	512
Pudukkottai	1844913	6886	5889	116	512
Ramanathapuram	1542927	5039	4595	110	512
Ranipet	2570237	11567	10588	138	512
Salem	3969544	13005	10453	194	512
Sivaganga	1526575	4345	4006	113	512
Tenkasi	2570237	5959	5185	111	512
Thanjavur	2742715	7691	6554	1230	512
Theni	1420325	13338	12342	153	512
Thirupathur	2570237	3339	2824	680	512
Thiruvallur	4250039	26841	24592	446	512
Thiruvannamalai	2809958	12211	10366	182	512
Thiruvarur	1441276	4560	3659	60	512
Thoothukudi	1995201	11894	11065	116	512
Tirunelveli	3508046	10523	9243	189	512
Tiruppur	2826119	3792	2425	820	512
Trichy	3103411	8336	7301	125	512
Vellore	4487417	11949	10663	183	512
Villuppuram	3943115	8660	7769	82	512

(continued)

Table 3. (*continued*)

District	Population	Confirmed	Recovered	Deaths	Volume of vaccine distribution using conventional program
Virudhunagar	2214208	13458	12741	200	512
Unknown	2570237	2229	2145	40	512

Table 4. Comparison of the volume of vaccine distribution using the conventional program and Vaxallot with PI value

District	Population	Confirmed	Recovered	Deaths	PI	Volume of vaccine distribution using Vaxallot	Volume of vaccine distribution using conventional program
Chennai	5297274	143602	129677	2896	31.75	2404	512
Coimbatore	3942171	19948	15584	3320	19.49	1476	512
Kanyakumari	2132226	10404	9367	1970	14.1	1068	512
Chengalpattu	2570237	28994	25916	463	13.78	1044	512
Cuddalore	2970742	14865	11052	156	13.36	1012	512
Thanjavur	2742715	7691	6554	1230	8.63	653	512
Kallakurichi	2570237	7292	6050	850	8.14	617	512
Theni	1420325	13338	12342	153	8.09	612	512
Tiruppur	2826119	3792	2425	820	7.74	585	512
Mayiladuthurai	2570237	12498	10966	433	7.65	580	512
Thiruvannamalai	2809958	12211	10366	182	7.21	545	512
Salem	3969544	13005	10453	194	6.92	523	512
Nilgiris	838349	1989	1567	140	6.7	508	512
Thiruvarur	1441276	4560	3659	60	6.67	504	512
Thiruvallur	4250039	26841	24592	446	6.34	479	512
Nagapattinam	1842753	3413	2347	57	6.09	461	512
Pudukkottai	1844913	6886	5889	116	6.03	456	512
Ariyalur	860579	3148	2694	36	5.69	431	512
Madurai	3463607	14988	13555	368	5.2	394	512
Dindigul	2462144	7501	6367	143	5.19	393	512
Thoothukudi	1995201	11894	11065	116	4.74	358	512
Thirupathur	2570237	3339	2824	680	4.65	351	512
Erode	2566988	4082	2957	52	4.59	348	512
Perambalur	644354	1446	1335	180	4.52	343	512
Ranipet	2570237	11567	10588	138	4.35	329	512
Tirunelveli	3508046	10523	9243	189	4.19	317	512
Virudhunagar	2214208	13458	12741	200	4.14	313	512
Namakkal	1968325	2795	2028	46	4.13	313	512

(*continued*)

Table 4. (*continued*)

District	Population	Confirmed	Recovered	Deaths	PI	Volume of vaccine distribution using Vaxallot	Volume of vaccine distribution using conventional program
Kanchipuram	4558007	18628	17056	275	4.05	307	512
Karur	1213522	1948	1517	30	3.8	288	512
Trichy	3103411	8336	7301	125	3.74	283	512
Ramanathapuram	1542927	5039	4595	110	3.59	271	512
Tenkasi	2570237	5959	5185	111	3.44	260	512
Krishnagiri	2142982	2715	2036	36	3.34	253	512
Vellore	4487417	11949	10663	183	3.27	247	512
Dharmapuri	1717801	1560	1205	160	3	228	512
Sivaganga	1526575	4345	4006	113	2.96	224	512
Viluppuram	3943115	8660	7769	82	2.47	186	512
Unknown	2570237	2229	2145	40	0.48	36	512

the vaccine to each district in an effective manner. The comparison of volume of vaccine distribution using conventional program is shown in the Fig. 4.

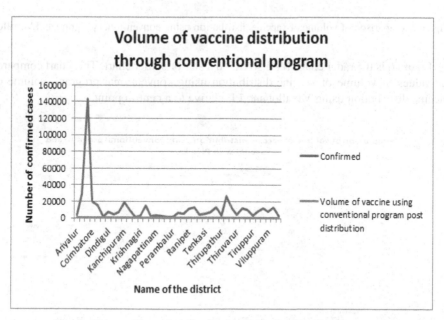

Fig. 4. Volume of vaccine distribution through conventional program

The volume of vaccine distribution using conventional program (i.e.) the output from Linear Distribution Supply Chain Software (Simulated) for 20,000 Vaccine Units is 512. For all the districts, the value is 512. In the proposed method (Vaxallot with PI), in particular district, the confirmed cases are more means more number of vaccines are distributed to that district. If the confirmed cases are less means less number of vaccines are distributed to that district. The comparison of volume of vaccine distribution using conventional program and Vaxallot is shown in the Fig. 5.

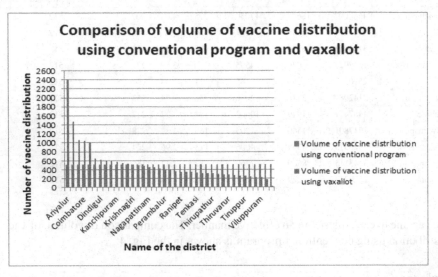

Fig. 5. Comparison of volume of vaccine distribution using conventional program and Vaxallot

Figure 6 is the radar chart. It's also called a star or spider chart. This chart compares the values of volume of vaccine distribution using conventional program, volume of vaccine distribution using Vaxallot and PI relative to a central point.

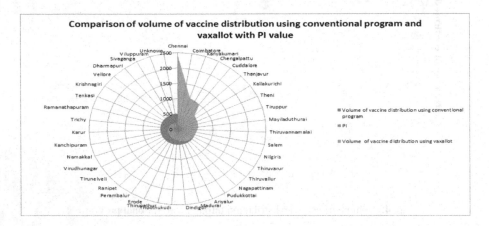

Fig. 6. Comparison of volume of vaccine distribution using conventional program and Vaxallot with PI value

9 Comparative Study

The comparative study of Vaxallot and Conventional system is shown in the Table 5. The parameters for the comparative study of Vaxallot and Conventional system are given below.

A) Scalability
B) Custom Index
C) Preprocessing
D) Excess Handling
E) Usability

A) Scalability
Input information solicited by Vaxallot is limited to available quantity of vaccines and nothing more as opposed to popular distribution ecosystems that require many parameters to proceed with processing, in environments that can't provide all parameters as requested by the software, the algorithm is prone to fail. Since Vaxallot is developed to be accommodated for any condition, it will run successfully. **Example:** When Vaxallot requests for total volume in Fig. 2(a), other applications requests for parameters like age, available beds, number of medical facilities and etc.

B) Custom Index
Unlike conventional tools and utilities, Vaxallot doesn't handle multiple parameters on individual channels. It fuses available values into what's called as the 'Priority Index', an indexing system that sorts all regions according to the computed sensitivity threshold. **Example:** As seen in Table 4, the column 'PI' represents Priority Index, the regions are sorted for better readability and vaccine volume is computed accordingly.

C) Preprocessing
Vaxallot also comes with an inbuilt preprocessing module that seeks to prevent wastage and distribute with careful consideration. This is something almost all supply chain systems lack. **Example:** As seen in the Table 6, with a conventional algorithm applied over the source data without normalization, it returns nothing for fields with missing values, thereby contributing to wastage, Vaxallot overcomes this problem by instituting an explicit minmax algorithm and bridging the inconsistencies.

D) Excess Handling
Administration of vaccines usually take place in several rounds but rounds don't have multiple distribution cycles, thereby introducing surplus units that get carried over to next round and in entirety this inefficiency could have adverse effects, that's why Vaxallot comes with an excess handling mechanism that ensures zero wastage. **Example:** Comparing Fig. 3 and Table 3, we could infer that Table 3 has a linear distribution mechanism and that it doesn't cater regions with missing values, thereby suffering from an excess of close to 3616 units. Figure 3 overcomes this problem by handling excess units.

E) Usability

Most supply chain applications out there are operated locally, meaning a user will have to download and install the application on their system. Vaxallot operates on the cloud, from the information it handles and to its algorithm. **Example:** Total size of Vaxallot v1.0 on cloud is 428 kB as opposed to popular distribution applications whose size is in gigabytes.

Table 5. Comparative study of Vaxallot and Conventional system

Vaxallot	Conventional Distribution System
Vaxallot is scalable, meaning the system can handle growing population in any given region and any given time	Conventional systems will need additional resources to be ported to bigger regions
Vaxallot distributes units in accordance to its exclusive priority index parameter that incorporates crucial factors to consider	In the traditional space it's either linear distribution where factors aren't considered or distribution with some of the many factors considered
Vaxallot also comes with a preprocessing mechanism that effectively deals with missing values in the data source	Conventional systems operate on data as available and wouldn't cater anomalies
Vaxallot can handle excess produce by effectively rerouting them into the system	Most systems out there don't have a standing mechanism to handle surplus
Vaxallot is a lightweight and an on-the-web realtime solution that's backed by data stored on the cloud, hence the system would be up to date, any given time	Popular Supply Chain Distribution tools are bulky softwares that needs to be downloaded onto a local system backed by local data

Table 6. Conventional algorithm applied over the original source data without normalization

Region of Interest	Total Population	Confirmed	Recovered	Deaths	Volume
Ariyalur	860579	3148	2694	36	512
Chengalpattu	(value retracted intentionally)	28994	25916	463	NA
Chennai	5297274	143602	129677	2896	512
Coimbatore	3942171	19948	15584	3320	512
Cuddalore	2970742	14865	11052	156	512
Dharmapuri	1717801	1560	1205	160	512
Dindigul	2462144	7501	6367	143	512
Erode	2566988	4082	2957	52	512
Kallakurichi	(value retracted intentionally)	7292	6050	850	NA

(continued)

Table 6. (*continued*)

Region of Interest	Total Population	Confirmed	Recovered	Deaths	Volume
Kanchipuram	4558007	18628	17056	275	512
Kanyakumari	2132226	10404	9367	1970	512
Karur	1213522	1948	1517	30	512
Krishnagiri	2142982	2715	2036	36	512
Madurai	3463607	14988	13555	368	512
Mayiladuthurai	2570237	12498	(value retracted intentionally)	433	NA
Nagapattinam	1842753	3413	2347	57	512
Namakkal	1968325	2795	2028	46	512
Nilgiris	838349	1989	1567	140	512
Perambalur	644354	1446	1335	180	512
Pudukkottai	1844913	6886	5889	116	512
Ramanathapuram	1542927	5039	4595	110	512
Ranipet	(value retracted intentionally)	11567	10588	138	NA
Salem	3969544	13005	10453	194	512
Sivaganga	1526575	4345	4006	113	512
Tenkasi	(value retracted intentionally)	5959	5185	111	NA
Thanjavur	2742715	7691	6554	1230	512
Theni	1420325	13338	12342	153	512
Thirupathur	(value retracted intentionally)	3339	2824	680	NA
Thiruvallur	4250039	26841	24592	446	512
Thiruvannamalai	2809958	12211	10366	182	512
Thiruvarur	1441276	4560	3659	60	512
Thoothukudi	1995201	11894	11065	116	512
Tirunelveli	3508046	10523	9243	189	512
Tiruppur	2826119	3792	2425	820	512
Trichy	3103411	8336	7301	125	512
Vellore	4487417	11949	10663	183	512
Villuppuram	3943115	8660	7769	82	512
Virudhunagar	2214208	13458	12741	200	512
Others	(value retracted intentionally)	2229	2145	40	NA

Conventional algorithm applied over the original source data without normalization is shown in the Table 6.

10   Conclusion and Future Work

With Vaxallot, if not all challenges, crucial challenges concerning vaccine distribution were addressed. The proposed algorithm seeks to implement differential distribution and the basis of distribution was determined using a specialized parameter called the 'Priority Index'. With that in effect, total volume of vaccines available was effectively distributed amongst the consumers within the listed Sub Regions.

Once a mechanism with a firm understanding of the requisite of the population spread across the world is established, a web-based system could be rolled out. This shall hence pave the way to effectively deal with the ongoing pandemic situation.

Despite is exposed to various attacks in the past, we've only looked at a short-term solution that wears out over time. The application was developed with adequate consideration for future usage in critical conditions and circumstances, therefore, the solution can be adapted & used for possible future incidents too. Possible future work would include specific utility tools for warehouses and logistical ventures, information pertaining to cold storage of the units and incorporation of other purposeful features.

References

1. Plotkin, S., Robinson, J.M., Cunningham, G., Iqbal, R., Larsen, S.: The complexity and cost of vaccine manufacturing - an overview. Vaccine **35**(33), 4064–4071 (2017)
2. Jarrett, C., Wilson, R., O'Leary, M., Eckersberger, E., Larson, H.J.: Strategies for addressing vaccine hesitancy – a systematic review. Vaccine **33**(34), 4180–4190 (2015)
3. Cunningham, A.L., et al.: Vaccine development from concept to early clinical testing. Vaccine **34**(52), 6655–6664 (2016)
4. Hosangadi, D., et al.: Enabling emergency mass vaccination: innovations in manufacturing and administration during a pandemic. Vaccine **38**(26), 4167–4169 (2020)
5. Lin, Q., Zhao, Q., Lev, B.: Cold chain transportation decision in the vaccine supply chain. Eur. J. Oper. Res. **283**(1), 182–195 (2020)
6. Larson, H.J., et al.: Measuring vaccine hesitancy: the development of a survey tool. Vaccine **33**(34), 4165–4175 (2015)
7. Schaefer, G.O., Tam, C.C., Savulescu, J., Voo, T.C.: COVID-19 vaccine development time to consider SARS-CoV-2 challenge studies? Vaccine **38**(33), 5065–5088 (2020)
8. Mosina, L., et al.: Building immunization decision-making capacity within the world health organization European region. Vaccine **38**(33), 5109–5113 (2020)
9. Asturias, E.J., Duclos, P., MacDonald, N.E., Nohynek, H., Lambert, P.-H., The Global Vaccinology Training Collaborative: Advanced vaccinology education landscaping its growth and global footprint. Vaccine **38**(30), 4664–4670 (2020)
10. Schuster, M., Eskola, J., Duclos, P.: Review of vaccine hesitancy: rationale, remit and methods. Vaccine **33**(34), 4157–4160 (2015)
11. Westrick, S.C., Watcharadamrongkun, S., Mount, J.K., Breland, M.L.: Community pharmacy involvement in vaccine distribution and administration. Vaccine **27**(21), 2858–2863 (2009)
12. Käser, T.: Challenges and efforts in vaccine development and distribution. Vaccine **35**(40), 5396 (2017)
13. De Boeck, K.., Decouttere, C., Vandaele, N.: Vaccine distribution chains in low- and middle-income countries: a literature review. Omega **97**, 102097 (2020)

14. Palache, A., et al.: Survey of distribution of seasonal influenza vaccine doses in 201 countries (2004–2015): the 2003 world health assembly resolution on seasonal influenza vaccination coverage and the 2009 influenza pandemic have had very little impact on improving influenza control and pandemic preparedness. Vaccine **35**(36), 4681–4686 (2017)
15. Smith, J., Lipsitch, M., Almond, J.W.: Vaccine production, distribution, access, and uptake. Lancet **378**(9789), 428–438 (2011)
16. Bertsimas, D., et al.: Optimizing vaccine allocation to combat the covid-19 pandemic. Healthc. Manag. Sci., 1–27 (2020)
17. Chen, S.-I., Norman, B.A., Rajgopal, J., Assi, T.M., Lee, B.Y., Brown, S.T.: A planning model for the WHO-EPI vaccine distribution network in developing countries. IIE Trans. **46**(8), 853–865 (2014)
18. Hu, X., Zhang, J., Chen, H.: Optimal vaccine distribution strategy for different age groups of population: a differential evolution algorithm approach. Math. Probl. Eng., 1–7 (2014)
19. Yu, Z., Liu, J., Wang, X., Zhu, X., Wang, D., Han, G.: Efficient vaccine distribution based on a hybrid compartmental model. PLoS ONE **11**(5) (2016)
20. Foy, B.H., et al.: Comparing COVID-19 vaccine allocation strategies in India: a mathematical modelling study. Int. J. Infect. Dis. **103**, 431–438 (2021)

Author Index

© ICST Institute for Computer Sciences, Social Informatics and Telecommunications Engineering 2024
Published by Springer Nature Switzerland AG 2024. All Rights Reserved
P. Pareek et al. (Eds.): IC4S 2023, LNICST 537, pp. 375–378, 2024.
https://doi.org/10.1007/978-3-031-48891-7

Printed in the United States
by Baker & Taylor Publisher Services